Sheet Metal Press Tools
Design and Making

A PRACTICAL APPROACH

Other CBS books by the same author

1. Production Processes (PB)
2. Plastics Technology for Diploma Level Students and Technicians

Sheet Metal Press Tools
Design and Making

A PRACTICAL APPROACH

Midhat Luqman

Life Fellow
The Institution of Engineers (India)
Allahabad

CBS

CBS Publishers & Distributors Pvt Ltd

New Delhi • Bengaluru • Chennai • Kochi • Kolkata • Mumbai
Bhopal • Bhubaneswar • Hyderabad • Jharkhand • Nagpur • Patna • Pune
• Uttarakhand • Dhaka (Bangladesh)

Sheet Metal Press Tools
Design and Making
A PRACTICAL APPROACH

ISBN: 978-81-239-2513-4

Copyright © Author and Publisher

First Edition: 2015

Reprint: 2020

Published by Satish Kumar Jain and Produced by Varun Jain for

CBS Publishers & Distributors Pvt Ltd
4819/XI Prahlad Street, 24 Ansari Road, Daryaganj, New Delhi 110 002, India.
Ph: 011-23289259, 23266861, 23266867 Fax: 011-23243014 Website: www.cbspd.com
e-mail: delhi@cbspd.com; cbspubs@airtelmail.in
Corporate Office: 204 FIE, Industrial Area, Patparganj, Delhi 110 092
Ph: 011-4934 4934 Fax: 011-4934 4935 e-mail: publishing@cbspd.com; publicity@cbspd.com

Branches

- **Bengaluru:** Seema House 2975, 17th Cross, K.R. Road,
  Banasankari 2nd Stage, Bengaluru 560 070, Karnataka
  Ph: +91-80-26771678/79   Fax: +91-80-26771680   e-mail: bangalore@cbspd.com
- **Chennai:** 7, Subbaraya Street, Shenoy Nagar, Chennai 600 030, Tamil Nadu
  Ph: +91-44-26680620, 26681266   Fax: +91-44-42032115   e-mail: chennai@cbspd.com
- **Kochi:** 42/1325, 1326, Power House Road, Opp. KSEB Power House
  Ernakulam 682 018, Kochi, Kerala
  Ph: +91-484-4059061-65   Fax: +91-484-4059065   e-mail: kochi@cbspd.com
- **Kolkata:** 6/B, Ground Floor, Rameswar Shaw Road, Kolkata-700 014, West Bengal
  Ph: +91-33-22891126, 22891127, 22891128   e-mail: kolkata@cbspd.com
- **Mumbai:** 83-C, Dr E Moses Road, Worli, Mumbai-400018, Maharashtra
  Ph: +91-22-24902340/41   Fax: +91-22-24902342   e-mail: mumbai@cbspd.com

Representatives

• Bhopal	0-8319310552	• Bhubaneswar	0-9911037372	• Hyderabad	0-9885175004
• Jharkhand	0-9811541605	• Nagpur	0-9421945513	• Patna	0-9334159340
• Pune	0-9623451994	• Uttarakhand	0-9716462459	• Dhaka (Bangladesh)	01912-003485

Printed at: Rashtriya Printers, Dilshad Garden, Delhi, India

to

my beloved wife

Nuzhat Luqman

who left me to cherish beautiful

memories with her and

the untiring efforts to make

my book successful

Preface

Technological progress plays a significant and imperative role, it influences to impart an edge to the development and prosperity of the country and its people. Prosperity of the people thus goes hand-in-hand with technological development and its implementation in various fields of activities such as agriculture, education, health, power generation and production of a variety of industrial and non-industrial products. All modes of transportation and household gadgets, electrical and electronic items, defence hardware and all engineering-related fields are incorporated by implying several production processes to produce components required for the mentioned fields and others.

There are enormous number of components which are produced by various processes such as forging, moulding, machining and sheet metal press working. As progress is taking place, the change and enhancement is quite prominent and continuous in production processes and services too. The wheel of progress has to move on, thus requirement of the day is to equip and instill in our young generation the engineering knowledge and help them cultivate expertise through education and practical training. Imparting training and achievement can be done through varied science and technology institutes, engineering colleges, or shop floor training centers in industry. Books thus play an extensively important medium of transfer of knowledge and know-how. This publication is meant to smoothly connect theoretical and practical elements into one natural flow. Bearing all this in mind, I want to share my knowledge which is practically oriented and easy for my readers to understand.

This book *Sheet Metal Press Tools Design and Making: A PRACTICAL APPROACH* is planned to provide as much practical insight as possible. Theoretical aspects are also touched upon. Quality consciousness and cost awareness has appreciably increased many-folds, if compared to what it was about half a century back. Improvement in quality standards and cost reduction efforts is almost a continuous process. In the course of this process, new products came up with challenges for designers and producers.

Quality of a sheet metal component very much depends on the design and workmanship of dies and tools by which components are to be produced. Reliability, life and ease of maintenance of a sheet metal working tool are important requisites. It calls for use of most suitable types of steel, other materials, design and making techniques. In this book, emphasis is laid on explaining various materials of construction of tools, tools design, machining methods, assemblies and selection of power presses, automation, safety and quality aspects.

Hope, this book would go a long way in providing useful knowledge, know-how and information which are spread over twelve chapters comprising text, figures, diagrams, photographs and references.

Midhat Luqman

Acknowledgements

Writing a book for publication needs good planning and fixing a target time frame for submission of work to the publisher. At planning stage, it was realized that to prepare this book *Sheet Metal Press Tools Design and Making: A PRACTICAL APPROACH*, I would need support from my well wishers.

It would be discourteous on my part if I do not mention organizations and my friends who have wholeheartedly supported me during the preparation of the manuscript.

My thanks are due to Dr SH Mohsin, Professor of Mechanical Engineering, and Director General, SS Institute of Technology and Management, Aligarh. He helped me in finalizing approach to sheet metal press tools design.

My ex-colleague, Mr Suhail Ansari, has quite a long experience in production of sheet metal components and is an expert of working on AutoCad. He converted all hand drawn figures into AutoCad generated figures.

On a few occasions, information is picked up from Wikipedia, the online free encyclopedia. References are provided accordingly.

Further, my thanks are due for Mr Najmul Azam, USA based computer software expert who has guided me in solving computer related issues from time to time. Else, it would have been impossible to keep on working on the manuscript smoothly.

I would like to thank Mrs Amna Hina Wilayat, UAE-based logo designer and English language advisor, who has helped me in the preparation of the manuscript.

It is my pleasure to mention a student, Mr Azwar Wilayat, who has a flair for designing presentations. On his own initiative, he has designed title cover for this book. It would give an idea of content display for the publisher to develop a suitable title cover as deemed fit.

Last, but not the least, I would like to thank my wife, Nuzhat Luqman, who supported and encouraged me during this ordeal, especially in making the monograph effective and impressive. My children remained a symbol of strength for me during the entire course.

Midhat Luqman

Contents

Introduction

This chapter consists of the following aspects:
1.1 Brief history
1.2 Old practical techniques
1.3 Development of analytical approach
1.4 Development of machine tools

1.1 BRIEF HISTORY

Primitive man passed through stone age, agricultural age, iron age and then industrialization age. Man knew about presence of pure gold particles in nature since around 3400 BC. Other metals such as we know today were also there at that time, but in the form of variety of ores. Such ores were generally spread over all the places on the earth. Concentration of a particular ore was at particular place. Before iron age, primitive man hardly knew about pure metals other than gold. Nature gave clues to man by way of natural conversion of iron ores into iron metals which could be found in traces and small bits in soil around volcanoes. Heat, pressure, other chemical elements, gases and water vapor from volcanoes processed iron ores to form iron in crude form and in the shape of bits and traces. In most of the historic narrations and in museum, generally there are references of artifacts and utensils which were used by primitive man.

Gradually with passing time, man realized that iron can be extracted from **hematite**, an iron ore, by applying heat. Iron so obtained was not of much use as it was not strong and hard.

Next concern of man was how to give various shapes and improve upon the malleability and strength of iron for making useful items for variety of use such as weapons, tools and useful articles. By the time, a number of processes were known by working on gold. By hammering small particles of gold, a lump could be formed. Heating of gold particles was done to melt and make a homogeneous mass. This know-how was then applied to iron also. With passing time, the concept of metallurgy started growing. First of all, there were few metalsmiths, foundrymen who did various trials by adding various chemical elements such as carbon, tin, copper, zinc, etc. Such trials were not in a scientific manner as we know today. Gradually, better educated persons started taking interest in the metallurgy. With the passage of time, qualities of iron and steel were defined in more systematic and scientific manner. By the beginning of 19th century, a number of companies around the world were developing and producing steels in a big way for those times. **Steel maker** Sheffield of England is a very well-known name.

By mid 19th century, scientific test methods were developed for which machines and equipments were needed. Consequently, testing machines came into existence for testing mechanical, thermal, magnetic and chemical properties. National and international standard specifications were developed. These are still in use with refinements. Swedish and Tata India steels are famous for their consistent quality, documentation and pricing.

The knowledge of various grades of steel and mechanical properties is highly desirable for a tool designer and maker. Various steel producers in India, namely Assabsripad and producers of steels in other countries provide detailed description of their steels. These may also be studied by die and tool designer and maker.

1.2 OLD PRACTICAL TECHNIQUES

Much before iron, copper was preferable metal with craftsmen as it is easy to work on it. In olden days, there were no power rolling machines to produce sheets and wires. Only hand driven innovative wire and sheets rolling machines were used. Big sheets could not be rolled. Copper sheets were also made by hammering operation. This method of copper sheets production gave sheets of non-consistent quality and surface finish. Brass is an alloy of copper, zinc and other chemical elements. Craftsmen used brass sheets to make utility and decorative items. They had become such masters of their craft art that they could produce beautiful pieces of artistic items and utilities. Need of producing comparatively large containers necessitated development of manufacturing process. A cooking utensil (bhagona), for example, could be produced by manual process known as spinning. In spinning process, a round disc of copper, brass or aluminum is held between a high speed (about 1200 rpm) rotating jig and 'tool rod 'or 'spinning rod' which is used by craftsmen to press rotating sheets in a peculiar manner so that grains of material flow gradually and in a particular fashion to take shape of supporting or forming mandrel or jig. Spinning is really an art because spinners have to apply his body weight through long rod having a particularly shaped and polished spinning end 'nose'. Spinning tool is supported near the rotating disc. This arrangement magnifies the pressure of tool on the rotating disc. It is not only the amount of weight which is given by spinner, he manipulates the tool in such a way that sheet gradually takes the shape of mandrel.

It is worth mentioning that often metal gets **hard while working** on it, so the spinner gets a feeling that further flow of grains of material is not possible. In case grains are forced to move then stress cracking in metal may result. In such a case, if further spinning is required to be done then it becomes necessary to relieve the stress of the component by heating it at an appropriate temperature and time. After **stress relieving**, further spinning can be done to complete the spinning to give final shape to the component. Another old method of giving shape to copper sheet metal was to spot hammer on the surface on an iron or stone anvil. It needed a large number of strokes in a peculiar fashion so that sheet starts getting shape of desired object, say a cooking utensil. It was also considered an art as it depended upon the skill and feel of craftsmen. Another example of old method of giving intricate shapes to already formed component was to support display surface on a hard wax which may get pressed but does not crack. Craftsmen then used small shaping bits to delicately hammer down the contours and shapes. Examples of such types of works are trophies and metal buttons with logos and shapes like eagle, swastika or other desired shapes. This technique is still used where a few pieces are to be produced. Nowadays, dies and presses are used to produce large number of such parts or articles.

1.3 DEVELOPMENT OF ANALYTICAL APPROACH

Analytical approach is a way to analyze in a systematic manner about the presence of ores of various metals all over the world. Many countries started analyzing various ores and metals almost at the same time. Formal metallurgy laboratories were developed mostly in educational institutions. In many countries, institutes of metallurgy were developed where dedicated teachers and scientists were teaching as well as carrying out research work. Research consisted of study of molecular structure, grain dispersion, grain structure and various properties.

The process of analytical approach is a continuous process in the sense that it needed theoretical and mathematical knowledge and availability of infrastructure and equipments to carry out analysis of ore to metal conversion technology and then to analyze metal for their properties. Over a period of time, testing equipments came into existence and kept on getting developed through joint effort of scientists and technicians. Organizations and governments of that time provided almost all type of help. This included development of standards for tests, standards for various properties and for testing equipments.

Oxidation was probably the foremost concern for researchers and scientists of olden days. They had known by experience and experiments that chemical elements (metals) such as iron, tin, lead, copper, mercury, silver and gold were susceptible to oxidation in varying rate. Iron oxidizes readily, whereas gold is highly resistive to oxidation. These comments are attributes which are of insignificant value nowadays. Today's analytical chemistry and testing technologies has made it possible to carry out quantitative analysis to produce results in definite terms of measurement.

Most of the analytical work is done on iron ores and iron as it is the most sought metal especially for industrial purposes. Industrial purposes may range from ship building, railway tracks, bridges to intricate parts of machineries. With the advancements of science and technology, alloying of iron developed to produce steels and stainless steels of various compositions. For example, a high chromium high carbon steel may have percentage of chromium as high as 14%, carbon 1.1 –2.5% and small percentages of other metals such as tungsten, nickel, vanadium, etc. **Mechanical properties** of steels of various compositions are determined quite accurately. Such properties are systematically documented.

Effect of various chemicals on metals are analytically determined and documented.

Physical, thermal, magnetic and electrical properties are also determined by analytical testing and documentation of results. Data so generated are utilized in the design of various **machine parts**, tools and dies and other machine elements such as shafts, gears, cams, valves, fasteners, etc.

In dies and tools making, metals other than steels are also used such as bronze, sintered tungsten carbides. Since steels are the main metals for construction of dies and tools, therefore, more emphasis is laid on briefly describing properties and test.

Mechanical properties include tensile strength, shear strength, flexural strength, ductility, hardness, compressive strength, etc.

Thermal properties include specific heat, thermal conductivity, coefficient of expansion, etc. Magnetic properties include its ability to retain magnetism or does not get permanently magnetized. Magnetic properties are defined as below:

Soft iron having very low carbon percentage have very high susceptibility and low retaining ability. Steels having higher percentage of carbon more than 1.2 % and other elements such as chromium, vanadium, nickel, etc. have low susceptibility and high retaining ability.

Electrical property would include ability to conduct electric current. It is also described as per the reference below.

Electrical steel, also called lamination steel, silicon electrical steel, silicon steel, relay steel or transformer steel, is speciality *steel* tailored to produce certain *magnetic* properties such as a small *hysteresis* area (small energy dissipation per cycle or low *core loss*) and high *permeability*.

The material is usually manufactured in the form of *cold-rolled* strips less than 2 mm thick. These strips are called *laminations* when stacked together to form a core. Once assembled, they form the *laminated cores* of *transformers* or the *stator* and *rotor* parts of *electric motors*. Laminations may be cut to their finished shape by a punch and die or in smaller quantities may be cut by a laser or by *wire erosion*.

Chemical property may include effects of acids, alkalis, organic and petroleum products. Rusting of steel is also one of the chemical reactions of atmospheric oxygen. Extent of rusting of steel is very much influenced by atmospheric temperature, humidity and presence of other chemicals such as present in sea water. Contact of iron and steel with some gases such as chlorine also promotes rusting.

1.4 DEVELOPMENT OF MACHINE TOOLS

Machine tools is the name given to machines which are used to remove material from the stocks. Removal of materials may be done in many ways such as turning, shaping, milling and grinding. Lathe came into existence around the beginning of 18th century, when a piece of wood was made circular by rotating it manually and using a sharp stone chip as tool to remove wood from rotating piece. With the discovery of iron, various types of small tools such as chisel and hammer came into existence and were extensively used for various purposes.

The development of machine tools was not confined to one or two countries. It took place almost simultaneously in many countries. During late year 1700, development of machine tools took place in England when the need for developing and producing textile machinery was a priority. The first commercial machine tools were manufactured around the year 1800 in England.

Ancient Indians knew how to make wooden axils, spokes and other articles of weapons. In those times, turning of materials was the main concern, hence, a machine came into existence which is now known as lathe. During 19th century, a great development took place in metallurgy, especially iron and steel. Lathes of olden days were constructed with materials which lost accuracies and shapes rapidly. With production of superior qualities of steel, lathe parts with durable performance were made. Lathe parts made of carbon steels were not retaining accuracies for long time. This deficiency in lathe was overcome by the use of better performing steels containing elements such as chromium, nickel, calcium in small percentage.

This type of composition of steels was having much better tensile strength, toughness and hardness. Nitrating steels were developed which gave a hard skin and tough core. **Nitride steels** are still used to make lathe parts such as journal, tool post, gears, screws and other parts.

Working on early lathes was mostly depended on skill of 'workman', a turner. Machine was hardly having a roll in repeated performance and sequence of operation. For example, turning a length of stock needed human effort and judgment while job was rotated by physical effort of animal and man muscle power.

With the invention of steam engines, machine tools were powered by this. Control was still in the hands of operator/turner.

With passing time, source of powering machine tools became electric motors. At the same time, repetitive operation of machine tools was performed with the help of punch cards.

Around mid 20th century, computer came into existence. Technology got developed to interface computers and machine tools. This gave rise to **numerical control machines (NC)**. Further development in technology saw **computerized numerical control** machines (CNC). With the advent of CNC machine tools, accurate and repetitive machining became possible. This was and now is great advantage in mass production of parts and components in the fields such as automobile, aircrafts, two wheelers, weaponry, production machines, etc.

The development of machine tools did not remain confined to lathe, which is considered as 'Mother of machine tools'. Other machine tools such as saws, drill presses, shaper, milling, gear shaper, planner, hobbing machine, grinders, honing, lapping machine came into existence.

In India, a big leap towards development of machine tools was taken around 1949 when government of India established a machine tools making company, HMT (Hindustan Machine Tools). Gradually, many small private manufacturers of machine tools came into existence to share production in the country. Now many types of general and special purpose sophisticated machines are produced by indigenous manufacturers and companies of other countries also got license to share investment. In this way, highly sophisticated know-how came in our country.

Machine tools need **cutting or forming tools** to perform cutting or forming operation. In the beginning of machine tools era, cutting tools of carbon steels were used. These tools performed cutting actions inefficiently as they could not stand temperature more than 250 degree Celsius and became blunt. Tools could perform much better with the development of tool steels such as High Speed Tools (17% tungsten, 3.5% chromium, 0.95% vanadium, 0.6–0.85% carbon). Cast alloys which contained certain percentage of cobalt, chromium, tungsten also give satisfactory performance. Next generation material for making cutting tool is cemented tungsten carbide bits fixed to tool supports. Cemented tungsten carbide bits are made by combining micro particles of tungsten carbide and binder cobalt. Special processes involving heat, pressure and time is carried out to produce bits. Although bits are quite expensive as compared to cobalt/vanadium tools and high speed steels but withstand much higher machining speed and temperatures, hence making its use viable.

With the development of Electronics in India, other types of machine tools came into existence. Electric Discharge Machine (EDM) or spark erosion machine and wire cut machine tools play an important role in machining of sheet metal dies and tools.

2

Materials of Construction

This chapter consists of the following aspects:

2.1 METALS

Metals may be described as those chemical elements which are in solid state at standard temperature and pressure. Electrical current passes through metals with varying degrees, depending on its conductive properties. All metals and non-metals are discovered at various periods of time in countries and years. Modern development in metallurgy and technology enabled scientists and researchers to organize chemical elements (natural and man made) in accordance with their **atomic numbers** which is the number of protons in a nucleus of an atom of an element. Chemicals known so far (natural) are 112 in numbers. All these elements are arranged to display in a particular manner which is known as Periodic Table, as we know today, was first prepared by Dmitri Mendeleev in the year 1869. At that time, only 66 elements were included in periodic table. From that time till now, a number of refinements are done in periodic table such as inclusion of elements which came to earth with celestial bodies and man made through nuclear fusion.

Table 2.1 provides data for some selected elements. Only those data are picked up which are more important from sheet metal press tool design and making.

Presently, the main concern is towards metals which are usually used for sheet metal components (Table 2.1) and in making of sheet metal press tools. From tool designing point of view, it is imperative to know about composition of steel alloys and their properties.

Table 2.1: Some metals for components

Properties	Copper Cu	Iron Fe	Zinc Zn	Titanium Ti
Physical	Solid	Solid	Solid	Solid
Density (near rt)	8.96 g.cm^3	7.847 g.cm^3	7.14 g.cm^3	4.506 g.cm^3
Melting point	1084.62°C 1357.77 K	1538°C 1811 K	419.35°C 692.68 K	1668°C 1941 K
Boiling point	2562°C 2835 K	2862°C 3134 K	907°C 1180 K	3287°C 3560 K
Heat of fusion	13.26 KJ.mol^{-2}	13.81 KJ.mol^{-2}	7.32 KJ.mol^{-2}	14.15 KJ.mol^{-2}
Heat of vaporization	300.4 KJ.mol^{-2}	340 KJ.mol^{-2}	123.6 KJ.mol^{-2}	425 KJ.mol^{-2}

Miscellaneous

Crystalline	Face centered cubic	Body centered cubic	Hexagonal close pack	Hexagonal close pack
Magnetic	Diamagnetic	Ferromagnetic	Diamagnetic	Paramagnetic
Electrical resistivity	(20°C) 16.78 nΩ.m	(20°C) 96.1 nΩ.m	(20°C) 59.0 nΩ.m	(20°C) 420 nΩ.m
Thermal conductivity	401 W.m^{-1}.K^{-1}	80.4 W.m^{-1}.K^{-1}	116 W.m^{-1}.K^{-1}	21.9 W.m^{-1}.K^{-1}
Thermal expansion	(25°C) 165 μ.m^{-1}.K^{-1}	(25°C) 11.4 μ.m^{-1}.K^{-1}	(25°C) 30.2 μ.m^{-1}.K^{-1}	(25°C) 8.6 μ.m^{-1}.K^{-1}
Young's modulus	110.128 GPa	211 GPa	108 GPa	116 GPa
Shear modulus	48 GPa	82 GPa	43 GPa	44 GPa
Bulk modulus	140 GPa	170 GPa	70 GPa	110 GPa
Poisson ratio	0.34	4	2.5	6
Mohr's hardness	3.0	4	2.5	6
Vicker's hardness	369 MPa	6085 MPa	–	970 MPa
Brinell hardness	35 HB=874 MPa	490 MPa	412 MPa	716 MPa

2.2 PROPERTIES OF METALS

An alloy consists of two or more elements in solid condition. The major portion in an alloy is metal iron in its pure form which is soft and ductile. Due to this condition, iron in its pure form has very limited practical use, so to make iron a practical metal, its properties are modified by alloying the element such as carbon, tungsten, chromium, vanadium, cobalt, tin, cadmium, etc. But the main role is of carbon for converting iron into steel, having improved properties. The significant aim of producing alloys is to reduce brittleness, improve hardness, increase wear resistance, resistance to tempering back to softness, resistance to corrosion, better tensile and compressive strength, better retention of sharpness. In general terms, few metal alloys are mentioned below:

Steel——Iron and Carbon

Brass——Copper and Zinc

Bronze——Copper and Tin

Typically, there are **alloys** having a dozen elements for high electrical applications such as space launch vehicles and atomic plants for generation of electricity.

Steel makers throughout the world produce a variety of steels having **composition** of alloying elements, especially metals to meet specific needs of end users. Steels are used in a number of end uses such as ship building, railway rails, high rise buildings, bridges, auto parts, aircrafts engines, a variety of machinery such as hand and power presses, plastic processing machines, plastic moulds and **sheet metal press tools**.

In this book, emphasis is laid on steels which are normally used in making sheet metal press tools.

Steels normally used for making tool parts such as **cutting** and shearing dies and punches may have the following **characteristics:**

- High compressive strength
- High surface hardness with tough core
- Good dimensional stability in hardening
- High wear resistance
- Good resistance to tempering back
- Attains hardness up to 60–61 HRC

Steels that have above characteristics may typically have the following composition:

- Carbon, C ————— 2.05%
- Chromium, Cr ——— 12.5%

- Iron, Fe ——————— 83.05%
- Silicon, Si ——————— 0.3%
- Manganese, Mg—— 0.8%
- Tungsten, W———— 1.3%

General purpose oil hardening non-distorting **tool steel** has manganese, chromium, tungsten as alloying elements.

Its characteristics are:

- Good machining ability
- Good dimensional stability
- Better tensile strength
- Attains hardness around 58–59 HRC

Table 2.2: EN series steels	
Steel	*Notes*
BS 970 070M20 (EN 3A)	A mild steel used for general purpose. Suitable for lightly stressed fasteners, shafts, etc. which can be easily machined and welded. Available hot rolled, normalized, cold drawn or square, flat and hexagon.
BS 970 080M40 (EN8)	A medium strength steel suitable for stressed pins, shafts, studs, keys, etc. Available as rolled or normalized. Supplied as square bar or round bar or flat.
BS 970 080M15 (EN 32)	A case hardening mild steel suitable for general engineering applications. When case hardened, results in a hard surface with tough core. Used for making gears, cams and rollers, etc. Supplied as black round bar and sections.
BS 970 709M40 (EN 19)	A 1% typical chromium molybdenum steel with higher molybdenum. Can be induction hardened. Used for gears and high strength shafts. Suitable for higher strength applications when resistance to shock is required. Available annealed. Supplied as black round or square bar and bright round or square and hexagon.
BS 970 722924 (EN 40B)	A 3% chromium molybdenum nitrating steel. Provide good tough core strength with a hard nitride surface for wear resistance. Supplied as rolled, annealed and hardened and tempered condition in black square and round bar.

Typically, grades, composition and characteristics of steels are internationally standardized and there is large number of varieties and grades of steels. EN series (Table 2.2) of steels provide steel nomenclature and specification to meet specific as well as general requirements of end users. EN stands for European standard for steels. For the purpose of making sheet metal press tools and plastic injections and compression moulds, there are a number of grades of steels. Few typical steels are mentioned below in Table 2.3.

Important Note

Description in Table 2.2 is of general nature. Selection of steel should be done based on latest technical data to be obtained from steel manufacturer.

Tata Iron and Steel company of India produces a large number of variety of steels for various end users. Tata also produces products such as sheets, rails, bars, girders, sections, etc. Steels suitable for dies and tools are produced by M/s Assabsripad Steels in India. It is a company producing high quality of steel in collaboration with a Swedish steel making company, ASSAB. Basically steels of different properties are produced by varying percentage of **alloying elements** and heat treatment. It is beneficial to know the effect of alloying of various elements on steel properties.

Table 2.3: Principal effects of major alloying elements for steel

Elements	Percentage	Primary functions
Aluminum	0.95–1.30	Alloying element in nitrating steels
Bismuth	—	Improves machining ability
Boron	0.001–0.003	A powerful hardening ability agent
Chromium	0.5–2 4–18	Increases hardening ability Increases corrosion resistance
Copper	0.1–0.4	Corrosion resistance
Lead	—	Improves machining ability
Manganese	0.25–0.40	Combines with sulfur and phosphorus to reduce the brittleness. Also helps to remove excess oxygen from molten steel. Increases hardening ability by lowering transformation points and causing transformation to be sluggish.
Molybdenum	0.2–5	Stables carbides, inhabits grain growth. Increases the toughness of steel thus

Contd.

Table 2.3: Principal effects of major alloying elements for steel (Contd.)

Elements	Percentage	Primary functions
		making molybdenum a very valuable alloy metal for making the cutting parts of machine tools and also the turbine blades of turbojet engines. Also used in rocket motors.
Nickel	2–5	Toughened
	12–20	Increases corrosion resistance
Silicon	0.2–0.7	Improves toughness
	2.0	Spring steel
	Higher percentage	Improves magnetic properties
Sulfur	0.08—0.15	Free machining properties
Titanium	—	Fixes carbon in insert particles. Reduces martensitic hardness in chromium steels
Tungsten	—	Also increase the melting point
Vanadium	0.15	Stables carbides, increases strength while retaining ductility, promotes fine grain structure, increases toughness at high temperatures.

Table 2.3 has provided an idea about function of various alloying elements in a particular range of percentage in total alloy. Table 2.2 describes few EN series steels. For a tool designer and specially tool maker, it is highly desirable to know the composition of steels which he may select to make tool parts. If tool maker continues with the habit of knowing the percentage of various elements in an alloy then by making tools and having knowledge of its performance, he may be in a better position to select steel by its composition rather than going by manufacturer name and grade.

No doubt that tool designers generally specify particular steel for various tool parts but practically, it is always not possible to have specified steel at hand. In such a case, a tool maker may select another steel of any other producer and grade. It becomes possible for tool maker because of his knowledge and know-how about composition of various steels.

To give an example, author usually designed and made sheet metal press tools and plastic injection moulds, using carbon steels, oil hardening and non-shrinking and high carbon high chromium steels (HCHCr). Normally, ASSAB XW5 is used for making cutting punches and die rings. At times, ASSAB XW5 was not available. In that case, steels having composition near to ASSAB XW5 is used. Poldi 2002 is alternative steel because its composition resembled ASSAB XW5. The author used ASSAB DF2 as a regular choice. When it was not available, looked for other steels of EN series and selected one which had a composition of ASSAB DF2. Table 2.4 provides composition and properties of few steels. The study of contents of this table may further enhance ability of readers to have better selection of tool steel as tool maker.

Table 2.4: Composition and properties of few steels

Properties	Carbon steel	Alloy steel	Stainless steel	Tool steel
Density (1000 kg/m³)	7.85	7.85	7.75–8.1	7.72–8.0
Elastic module (GPa)	190–210	190–210	190–210	190–210
Poisson's ratio	0.27–0.3	0.27–0.3	0.27–0.3	0.27–0.3
Thermal expansion (10^{-6}/k)	11–16.6	9.0-15	9.0–20.7	9.4–15.1
Melting point (°C)	–	–	1371–1454	–
Thermal conductivity (w/m–k)	24.2–65.2	24–48.6	11.2–36.7	19.9–48.3
Specific heat (J/kg–k)	450–2081	452–1499	420–500	–
Percentage elongation (%)	10–32	4–31	12–40	5–25

Many terms used in the table to indicate properties may not be familiar to some readers. It is therefore desirable to briefly review the following terms:

- Density
- Tensile strength
- Compressive strength
- **Shear strength**
- Elastic modules
- Yield strength
- Hardness
- Cutting edge sharpness

Density

Density of a material is the weight of its mass per unit volume. It may be specified in Metric or British units, which are as follows:

Metric unit—Gram force/unit volume, gmf/cm^3
British unit—lb force/unit volume, $lbf/inch^3$

Elastic Modules

Elastic module is the tendency to deform in a non-permanent condition when a force is applied. For example, if a rubber band is stretched too little, increase in length would attain its original length. What would happen if the rubber band is stretched too much? On releasing pulling force, it will not return to its original length. Similar may be the case with thin copper wire but the elongation in this case would be so small that it is difficult to see it. Same logic may be applied to the other metals such as iron, steel, zinc, brass, etc.

Modules of elasticity is generally expressed in the following mathematical expression:

Elastic Module = Stress/Strain

where

Stress = Force per unit area
Strain = Ratio of change, say in length

Tensile Strength

Tensile strength may be explained in the following manner:

Suppose an iron rod of small diameter (say, 5 mm) is held at one end and the other end is held by a pulling device, both the ends are firmly held. System is such that pulling force may be gradually increased or decreased. Diameter of rod may be accurately measured at any stage of pulling. On gradually increasing the tension (pulling force) and measuring the diameter, it would be revealed that up to a certain value of pulling force, diameter of rod gets minutely reduced and it attains back original dimension of rod. This is called elastic deformation. If the pulling force is increased beyond this elastic limit than the deformation, in this example, diameter of rod would not attain its original value on release of force, so the force applied just before permanent deformation is tensile strength of the material of the rod. Tensile strength is expressed in force per unit area of material under test. It may be expressed in:

SI unit as Pa or

Multiple MPa (Mega pascal)

British system as

Newton per square meter = N/m^2

Or lbf/in^2 or psi or kpsi (Kilo pound per square inch)

Note: Unit conversion table is provided in the Appendix.

Yield Strength

It is applied stress (force per unit area) and due to application of this, deformation becomes non-reversible.

Compressive Strength

Compressive strength of steel is its ability to withstand compressive force without deformation. If a load is axially applied to a tool specimen, it will contract in length and expand in diameter; on removing the load, specimen would attain its original shape and dimension. In case, if the specimen does not attain back its original shape and dimensions then it may be concluded that applied force has exceeded the compressive strength of specimen of metal, in this case, steel.

Hardness

Let there be small plates of copper, aluminum, iron and steel. All the four are scratched by a sharp and hard needle, applying almost the same force. It would be observed that scratch on aluminum is deeper than copper. Scratch on steel is least deep. Variation in the depth of scratch is due to extent of intergranular bond which is quite a complicated characteristic of metal in pure or in alloy condition. It is therefore found that different metals require different type of hardness testing method.

For checking hardness of aluminum, a ball shaped indenter may be useful. Ball dimension and loads are specified and standardized. Hardness testing machine is provided with the arrangement of applying selected load on indenter and also to indicate the travel of indenter in metals, in this case, aluminum. First of all, a small pre-load is applied so that ball sits on surface properly. At this stage, zero reading is set either on a micrometer dial indicator or digital display. Now suppose the same arrangement is applied on iron and steel, there will be hardly any reading worth translating it into hardness. This is because of the fact that intergranular bond is very much stronger in steel as compared to aluminum. Therefore, to measure hardness of steel, a conical or pyramid shaped indenter is used with higher loads on indenter.

Hence, hardness of a metal, say steel, may be defined as the ability of metal to resist deformation.

Various types of hardness scales are as follows:

Rockwell A, B, C, Vicker's, Brinell, etc.

Shore scale is meant for soft materials such as plastic, rubber, etc. Hardness of steel is generally measured on Rockwell C scale where depth of indentation may vary from 45 to 64 HRC (Rockwell C scale hardness number). This method of hardness measurement is easy and direct reading may be taken. There is no need of any calculation as in case of Vicker's pyramid number.

Shear Strength

Shear strength of a metal is its ability to resist deformation due to sliding action of grains in the direction of applied force. Scissors is a classical example of material deformation by shearing action; shear strength is generally measured in force per unit area of shearing surface. In SI system, Pa per meter square is the unit and in BS system, lbf per square inch is the unit of shearing strength.

Cutting Sharpness

Sharpness is generated w.r.t a cross-section through the cutting edge. Materials employed for cutting are specially made steels.

2.3 HEAT TREATMENT

Heat is a very important application in the production of steel. Iron is heated in a furnace together with various alloying elements to shape steel of desired end properties. While steel is being heated to molten state, it undergoes transformation of its granular structure at different temperatures ingredients. Once steel is completely made in a furnace, it is mechanically processed to give various shapes such as sheets, bars, sections, etc.

It is highly desirable to understand transformation process in steel of a part when heat treated.

Steels used for machining parts of a die or mould are generally of such hardness that conventional machining such as turning, shaping, milling, etc. is possible. Steels with 170–210 BHN hardness are normally used. Parts made of such steels are not generally used as they are for the purpose of cutting edges, radii and other dimensions, so to achieve long performance life, machined parts are subjected to heat treatment. Heat treatment is a general term which refers to various treatments such as stress relieving, annealing, normalization, quenching, hardening, tempering, flame hardening, case hardening, etc. A clear concept of all the above operations may be attained by understanding as to what happens to the **microstructures of steel** when heated. Microstructures of steel may be seen by preparing a sample piece properly lapped and seeing it under high power microscope.

Figures 2.1, 2.2 and 2.3 are provided to give an idea as to how **microstructures** of different types of steels look like.

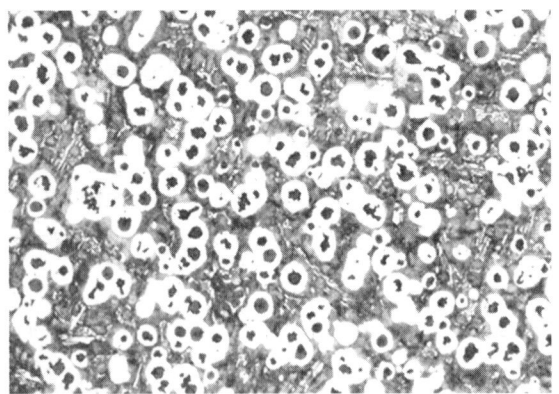

Fig. 2.1: Nodular cast iron etched
Source: www.metallographic.com

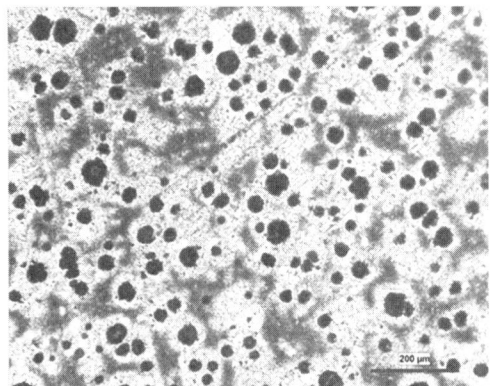

Fig. 2.2: Ductile iron
Source: Wikimedia commons

For having much better concept and knowledge, it is highly recommended to visit the following websites on Internet:

* en.wikipedia.org
* *www.metallographic.com*
* Google search
* Youtube

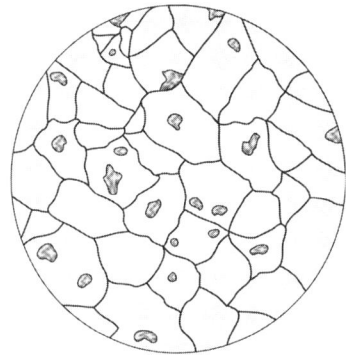 

(a) Transverse section magnification
not stated (Longmuir)

(b) Transverse section
magnification not stated (Guillet)

Fig. 2.3: Wrought iron
Source: commons.wikipedia.com

Photographs of various microstructures may be seen on the following website on Internet. Write on Google search bar 'Steel microstructures images'. A lot of images are there. Also write on Google Search bar, 'Internet microscope'.

For schools: 'Micrograph' and pwatlas.mt.umist.ac.uk/.../ microstructures/low-carbon-steel.html.

Let us take an example of carbon steel as to what and how changes in microstructures take place when steel is heated. Variable factors which may be considered are percentage of carbon in alloy, temperature which steel part achieves, heat soaking time and the manner in which heated part is quenched.

There are a number of microstructures which are formed at various percentages of carbon and temperature at which steel is heated. Microstructures at various stages give different mechanical properties to steel such as toughness, hardness, tensile strength, etc. Figure 2.4 shows formation of microstructures at various percentages of carbon and temperature.

Various regions shown in Fig. 2.4 are different microstructures and are given names as follows:

- Pearlite
- Ferrite
- Cementite
- Austenite

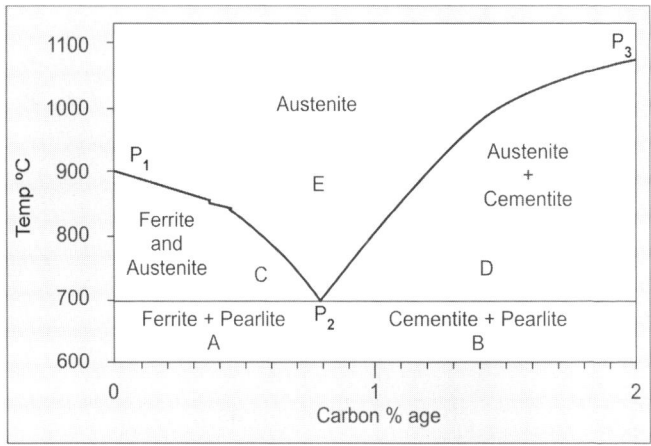

Fig. 2.4: Formation of steel microstructure

Regions are given codes arbitrarily for the ease of explaining. Region A is formed when steel with carbon content 0.8% is heated up to 700°C. This region consists of a mixture of Pearlite and Ferrite microstructures of grains. When percentage of carbon lies between 0.8 and 2.0 and heated up to 700°C then this region consists of microstructures of Pearlite and Cementite.

Region C has a mixture of microstructures of Austenite and Ferrite. It is formed when steel containing carbon up to 0.8% and its temperature ranges from 700 to 900°C. Note that ratio of mix of Austenite and Ferrite change with actual percentage of carbon in steel and the temperature up to which steel is heated.

Microstructures of region D consist of a mix of Cementite and Austenite.

Microstructures of region E consist of only Austenite which occurs above lines P_1P_2 and P_3. Lines P_1P_2 and P_2P_3 may be called as transition line which is formed on account of various combinations of carbon percentage and heating temperatures.

Typical and brief description of various microstructures are given below:

a. Pearlite

Ferrite and Cementite are combined to form a microstructure called Pearlite. When seen under high power microscope, grain structure is visible which somewhat looks like thumb impression of human being.

b. Ferrite

Solid steel at room temperature has a condition when it has body centered cubic crystal. Ferrite steels contain minimum amount of carbon and hence, such steels are soft.

c. Cementite

It is a very hard and brittle alloy of iron and carbon. During hardening heat treatment, cementite causes hardness.

d. Austenite

It is the phase (condition) of steel which re-crystallize and has a face centered cubic crystal structure.

Microstructure phases (condition) as described above are also tabulated in Table 2.5.

Table 2.5: Microstructure phases

S.No.	Name given to micro grain structure of steels	Type of structure	Crystal structure	Description
1.	Austenite phase	Face centered cubic crystal		Greater amount of desolved carbon. Increased formability.
2.	Cementite			Compound of iron and carbon. Very hard and brittle.
3.	Ferrite	Body centered cubic crystal		Phase at which solid steel has a BCC crystal. Contains minimum carbon, relatively soft.
4.	Pearlite	Combination of Ferrite and Cementite		Steel with exact 0.77%. Consists of

Contd.

Table 2.5: Microstructure phases (*Contd.*)

S.No.	Name given to micro grain structure of steels	Type of structure	Crystal structure	Description
			Resemble human finger print	uniform Pearlite at room temperature.
5.	Martensite	Distorted body centered tetragonal crystal structure		It is very hard and brittle.
			Atoms are in center and one each at corners of dis-torted cubic form	

Heat Treatment Shop

Almost all tool rooms, making sheet metal press tools or plastic injection moulds are equipped with heat treatment facilities. Generally, there are two furnaces—one for heating steel parts, say up to 1100°C and the other furnace is meant for tempering hardened parts. Tempering furnaces may attain temperature up to, say 500°C. Furnaces are generally operated by electric power. Size of heating chamber is chosen at the time of selection and purchase of furnace. Choice of chamber dimensions depends on possible largest dimension of part to be hardened and tempered. In practice, larger length, width and depth are selected so that a big size part could be heat treated if required. Similarly, tempering furnace of appropriate heating chamber dimensions is selected, purchased and installed.

At least two quenching tanks must be there. One containing quenching oil and the other containing *brine water*. In India, quenching oils are produced by Indian Oil Corporation. Various grades of quenching oils may be searched. One of it is given further (Product from Indian Oil Corporation of India).

Industrial Speciality Oils

Quenching Oils

Servoquench 11: Servoquench 11 is quenching oil blended from high viscosity index base stocks having good oxidation stability, good fluidity and low volatility. This oil is used for normal quenching operations on a wide variety of steel parts such as nuts and bolts, ball bearings, certain types of brake drums and meets IS: 2664–1980 straight mineral type, grade medium specifications.

Significant property of quenching oil is that it has high flash point. This means that it will not catch fire even if a part at 1100°C is quenched. The other property is that it does not oxidize rapidly. Heat transfer property should be good. It should not produce harmful fumes when a heated part is quenched.

Front door horizontal furnaces are generally used for small to medium size tool parts such as cutting ring of a cut and cup tool, draw cum blanking punch, extrusion punches and a number of such parts. Draw die and draw punch of a kitchen pot may be quite large and heavy. A horizontal furnace may not be suitable due to difficulty in putting the part inside the furnace and removal. A vertical furnace may be suitable for such parts. An overhead crane may be used for handling large and heavy parts. Same consideration applies to tempering furnace as well.

A variety of tongs and rods with wooden grips are used to handle parts. Placement and removal of heated parts from the furnace is an 'art'. Placement of parts in the furnace chamber should be such that quick holding and taking out the part from the furnace is possible. Let a cutting ring of about 18 cm diameter is to be placed in the chamber of the furnace. It can be placed just on the plain surface of heating chamber. When it comes to taking out, how can part be held by a tong? If cutting ring is placed over two blocks then there would be space to hold the cutting ring by means of a tong.

A convenient arrangement for placing a furnace and oil tank is shown in Fig. 2.5.

It is a brick or RCC construction. Top of this construction is flat on which furnace is placed. Height of construction is kept so much that it is convenient for operator (a tool maker or heat treatment man) to open the front door of furnace and place part inside furnace. There is a knob which is attached to door for lifting and lowering. The door has a counter spring mechanism to balance weight of door so that operator has not to really lift the weight of the door.

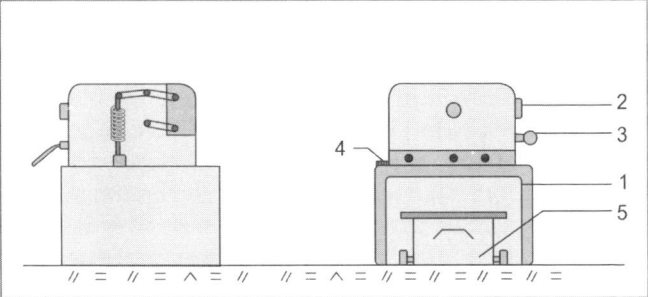

Fig. 2.5: Heat treatment furnace

Underneath the ceiling, structure has a hollow space suitably sized so that oil tank trolley is conveniently accommodated. Trolley normally remains inside. It is pulled out when **quenching** of heated part is to be done. In this arrangement, vertical distance between the furnace door and top surface of oil is minimum. Hence, while transferring heated part from furnace to oil tank, heated part is exposed to air oxygen very little. It is highly desirable that heated part is exposed to air for a very short period of time so that surface is very nominally oxidized. It can be seen in the figure that there is space on both the sides of furnace for placing various handling tools such as tongs and rods having right angled bend at the end to facilitate lifting of parts which have holes.

During hardening process, it is highly desirable that minimum possible distortion in the shape of part should take place which may affect irregular change in the dimensions. Consider hardening of a cutting ring. Bore of cutting ring is 120.00 mm after turning. This means before heat treatment, there may be a grinding allowance of 0.3 mm. Bore of cutting ring after grinding is required to be 120.39 mm. Ground bore should be completely round, not even a small portion of bore remains unground. This requirement is possible only when ovality not more than 0.1 mm develops after hardening.

Heated cutting ring may be dipped in quenching oil in two ways–one is that ring is vertically dipped and the other possibility is that the ring is dipped in horizontal position. This means that the full face of ring touches oil surface at a time and ring is allowed to dip in and then it is slowly moved around inside the oil. With this type of dipping (horizontal), chances are that there would be no distortion, hence no ovality for a cutting ring made of non-shrinking non-distorting tool steel.

Horizontal dipping of heated cutting ring is not normally possible if it is taken out of furnace by means of a tong. For the purpose of horizontal dipping, cutting ring is suitably tied with wire and a hook is formed in the center so that when it is lifted by means of a rod, face of ring remains horizontal. Another method which may be adopted is to fix a lifting plate as shown in Fig. 2.6.

There may be tool parts of shape and size which can be taken out from furnace by means of a tong. Holding heated job (part) by means of tong will cause temperature of that portion of part to go down when it is held. It may cause stresses around holding point and less hardness spot after hardening. For critical parts, there should be no stress or less hard point. To avoid this, part has to be bound by a wire with a hook formation. If there is a hole in part, a suitable iron or steel strip may be fixed only for taking out the part from furnace and then to quench the whole combination in oil.

Sometimes, a number of combination of tiny parts is needed to be hardened. In such a case, a small steel perforated bucket may be used for heating all tiny parts together and then to dip perforated bucket in quenching oil.

Some parts of press tools need to be tough at the core but hard at the surface to make the surface more wear resistant. A strip guide plate, top ejector pin or strip indexing system parts may have hard surface and tough inside. Steels selected for such parts are of low carbon percentage. For achieving hard surface, parts are placed with cast iron filings. Carbon granules or coke in a steel container with proper arrangement to put it inside the furnace and then to take out when quenching temperature is achieved and sufficient soaking time is given. Heated box is taken out, placed at suitable

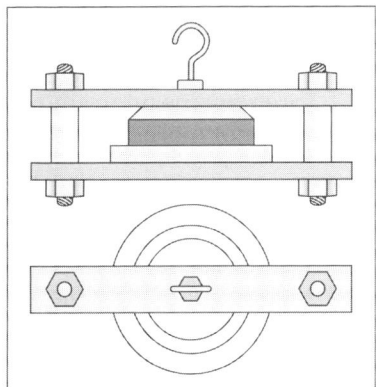

Fig. 2.6: Part lifting jig

place and wired parts are taken out quickly and quenched in oil or brine. It depends on the type of steel and recommendation of steel makers. Parts made of particular steel are also quenched in still air or by an air blast. There are steels from which parts are made and quenched in special equipment having inert gas such as nitrogen.

Generally, there is a **Rockwell hardness** testing machine, set for testing hardness on C scale (HRC). Generally, practice on shop floor is that hardness of part, when it cools down after quenching, is tested by a medium cut file. Tool makers have a 'feel' if hardness is up to the mark. In case precise hardness value is required to be known then hardness tester is used. It is worth noting that hardness is checked after the hardened part is properly cleared and scale is removed with the help of rough and fine emery paper. This precaution ensures that correct hardness is measured.

The method of testing hardness of a hardened steel part by Rockwell hardness tester is explained with the help of Fig. 2.7.

Part (1) is a platform which can be raised or lowered by rotating circular hand wheel 2.

Part (3) is a diamond point which penetrates into hardened part.

Part (4) is a dial indicator which is calibrated to read hardness directly in HRC.

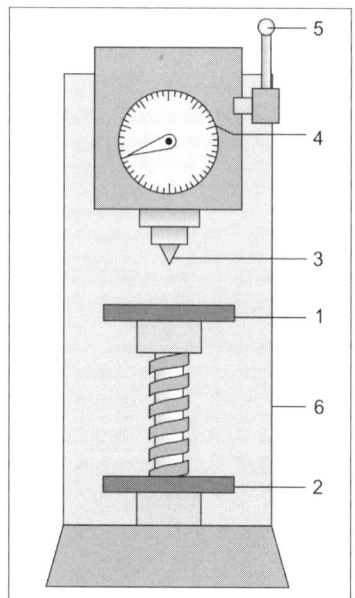

Fig 2.7: Hardness tester

Part (5) is a lever which slowly releases the transfer of loads on diamond pointer.

Part (6) is a C.I. body of tester.

As diamond is the hardest material, it is brittle also. Extreme care has to be taken that it is not subjected to jerky load.

Part of which hardness is to be tested, is first cleaned. If necessary, the location at which hardness is to be tested may be polished by a fine emery paper and then wiped out with a clean cloth. Now the part is kept on platform and raised so much that diamond point comes in contact with part surface. At this stage, further rising of platform should be very slow. Watch needle of dial indicator rotating. It moves up only so much that dial indicator pointer takes about quarter of a full round.

This process is called 'Pre-loading' of indenter (diamond point). This ensures proper seating of indenter point on steel surface. 'Pre-loading' load is generally 2–5 kg.

Now outer rotatable ring (Benzel) of dial indicator is rotated so much that zero line of dial coincides with pointer. Now pull the lever slowly. It will activate internal mechanism of load transfer to indenter. Mechanism is housed inside tester body. Load applied to ind0enter is standardized. For Rockwell C scale, it may be 150 kg for a 120° diamond indenter. On application of load, diamond indenter would penetrate into hard steel. Amount of linear penetration is indicated by dial indicator pointer. Suppose pointer has reached a position on circular scale which is marked as 58, then the hardness of steel is 58 HRC.

Case Hardening

As the name suggests, steel part is covered by a layer (Case) which is hard. Low carbon steel parts which have thick sections do not get much hard and steel in 'core' (depth of thickness) remain ductile, therefore, part does not break under normal twist or bend. Many such parts need to have hard surface to resist wear due to rubbing action between two parts under load. Crankshaft and big end bearing may be a typical example. Many agricultural implements such as harrow or disc harrow have parts which are subjected to severe tensile and compressive loads while breaking the soil since there is a continuous rubbing of parts with hard soil while work is going on. This situation needs that surface of parts should be hard with tough core, so sometimes parts are case hardened. Case hardening of steel parts may be done in the following manner:

- Flame and induction
- Carburizing
- Nitriding:
  - i. Gas nitriding
  - ii. Salt bath nitriding
  - iii. Plasma nitriding

a. **Flame and induction:** In flame case hardening, steel part is heated by acetylene excess flame. When proper temperature of steel parts is reached, carbon from acetylene is absorbed by the surface of steel part. After proper temperature is reached, say 800°C, part is quenched in water, brime or oil. Quenching in oil may be because some carbon particles from oil get cling to red hot surface of part. This further adds to the thickness of 'case' to become, say 0.2 mm.

b. **Carburizing:** Carburizing is also a type of case hardening. Steel parts where core toughness and surface hardness is needed, carburizing is done. Carbon steel parts having low carbon percentage by weight, say 0.3% its surface may be hardened by increasing percentage of carbon. It can be done by packing carbon steel parts surrounded by cast iron filings or charcoal in a steel container which can be sealed. It is to be noted that sealed box is not to be air tight. During expansion of heating air, some chemical reactions take place which causes carbon to adhere to red hot steel part in an environment free of oxygen. When quenching temperature is achieved with proper soaking time given, part is taken out and quenched in water or brine. In carburizing process, thickness of hard surface may not be uniform. It may vary from 0.05 to 0.3 mm depending on carburizing material, temperature of part and soaking time.

c. **Nitriding:** Nitriding is a sophisticated 'casing' process which provides hard and very uniform 'case'. Hardness may range from 60 to 64 HRC and uniform depth of hard surface may be achieved. Depth of hard surface may be achieved between, say 0.05–0.35 mm. This depends on the type of nitrating technique adopted and variable parameters such as temperature of tool part, amount of carbon donor, time given and number of times nitriding is repeated. A part already nitrided may be re-nitrided.

Nitriding of steel parts may be done in three ways: Gas nitriding, Salt bath nitriding and Plasma nitriding.

i. *Gas nitriding:* In gas nitriding, ammonia gas is generally passed around steel part to be nitrided. Ammonia gas (NH_3) is a nitrogen rich gas. When it comes in contact with red hot part, splits into nitrogen and hydrogen. Nitrogen defuses into the surface of steel and forms nitride which is very hard. Hardness may reach 64 HRC.

The advantage of nitriding is that many small parts may be nitrided at a time. All surfaces are equally nitrided, this process may be precisely controlled with the use of proper sensors and computerized system.

ii. *Salt bath nitriding:* In this process, small and big parts may be nitrided in bulk quantities. Parts having a previous hardness of say, 54 HRC may be nitrided by dipping in a salt bath having a nitriding temperature of about 880–950°C. A case of hardness 64 HRC is formed. This process is fast, hence, large quantities may be nitrided in a given time. It means that salt bath nitriding system has a better productivity as compared to other processes.

Salt bath is prepared of sodium cyanide. Its disadvantage is that salt is poisonous, therefore great care has to be taken. Moreover, sodium cyanide salts are quite expensive. Small organizations generally do not prefer salt bath nitriding. After nitriding, parts are thoroughly rinsed to remove any residual sodium cyanide.

iii. *Plasma nitriding*: Plasma is the name given to highly active gas with ionized molecules. Intense electric fields are used to generate ionized molecules of gas around the surface to be nitrided.

Tempering

When steel parts are heated for hardening, its grain structure undergoes change which is termed as phase. Referring to Fig. 2.4, suppose a part of carbon steels having about 0.6 percent carbon by weight is heated up to 900°C then there would be combination of ferrite and austenite in that phase of microstructure transformation. If red hot part is quickly quenched at this stage then the grain structure which has already taken shape is 'arrested' by cooling of part in quenching.

Due to sudden 'arrest' of transformation, internal stress is generated which may cause failure of parts. This difficulty is overcome by reheating the steel parts at a low temperature, typically 250°C for a time period of, say 45–60 minutes and then allowed to slowly cooled down to room temperature. This process releases stresses inside steel part, hence its failure is almost eliminated.

Sometimes it becomes necessary to anneal a portion of already tempered part. Fig. 2.8 shows pressure pins of a cut and cup sheet metal tool.

Parts (1) are pressure pins which are generally turned from carbon steel. These are then hardened and tempered to 50–60 HRC. In practice, it is found that head (3) of the pins chips off or break. To avoid this, heads of pins are tempered.

Method of tempering is as follows:

First of all, pins are polished so that steel surface becomes bright. Pressure pins are generally around 10–20 mm in diameter and 50–100 mm long. A thick piece of iron plate is heated to dull redness. Pins with heads on the plate are placed. It would be noticed that polished steel surface of pins starts getting color which keeps on changing as the tempering temperature of pins head rises. When color reaches light brown to brown then the pins should be removed from hot plate. If pins are not removed at this stage, the color becomes blue. This may be the sign of over tempering in which a case hardness of pin head goes down to say, 48–50 HRC, such a low hardness is undesirable.

Consider again a typical example of a cutting punch cum draw die made of high carbon high chromium steel. After hardening, its hardness reached up to 61 HRC. Since the weight of this part is

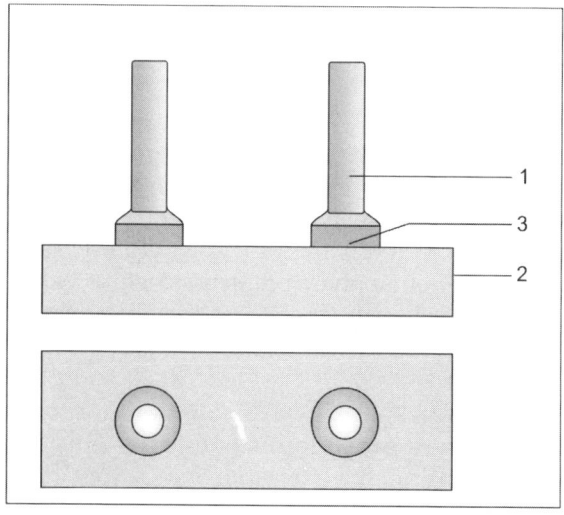

Fig. 2.8: Tempering arrangement

about two kilogram, it has to remain in tempering furnace for 45 minutes to one hour with the temperature of furnace as 180°C. After about 45 minutes, furnace is switched off and part is left inside to cool down. It may be taken out when temperature of furnace drops to say, 120°C. Part is then allowed to further cool down to say, 59 HRC, a drop of two Rockwell C.

Hardened steel part has a martensite crystalline structure which is very hard and brittle. Reheating below re-crystalline stage and gradually cooling down brings about change into crystal structure from distorted body-centered cubic crystal tetragonal to crystal somewhat body-centered cubic crystal. This change in crystalline structure reduces hardness, brittleness and internal stresses. Part then gives desired performance such as retention of sharp cutting edge for much greater number of cuttings and less wear of draw radius.

2.4 COMMERCIAL SPECIFICATIONS OF STEELS

Most of reputed steel makers provide detailed specifications of various grades or types which they produce. It includes composition, properties, standard dimensions in which steels are supplied and color codes. Since there is a large list of dimensions, it is beyond scope of this book to provide all dimensions. However, some idea is given with the help of Fig. 2.9.

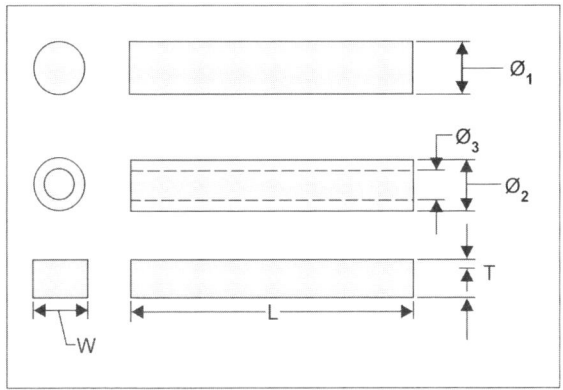

Fig. 2.9: Commercial specification of steel

Φ_1 range: 50–300 mm Φ_2 range: 50–300 mm

Φ_3 range: 20–100 mm   T range: 20–70 mm

W range: 25–100 mm   L range: 2–4 meters

Color codes are provided to visually see as to what grade/type of steel bar it is. Strips of code color are generally painted along the length of bar for identification.

Color coding of steels are not universal. Various steel manufacturers indicate their own color coding system which may consist of large number of color combination for various products. Some steel manufactures, in addition to color coding, punch code numbers on steel bars such as EN34, SVL, DF2, S5, etc.

In India, M/s Assabsripad steel is a renowned company producing a variety of steels with various trade names. It is highly recommended that the following site may be visited on internet:

www.assabsripadsteels.com

Information provided below is as on 13th April 2012. There may be some changes when reader visits the site.

- Home page of Assabsripad steels open
- Click on product tab
- Another page opens where three buttons are provided

  Cold (cold work steel)

  Plastic

  Hot
- Click button 'cold'
- Page opens where names of various steels, chemical composition, properties, recommended uses and color codes are mentioned.

  Names of various steels as mentioned are as below:

Grade	Medium		Colour	Carbon
Calmax	vac, air, oil, gas	—	white/violet	0.6
Cormo	oil, vac, air	—	red/violet	0.6
Caldie	vac, air	—	white/grey	0.7
Severker 21 (HcHCr)	vac, air	—	yellow/white	1.55
Severker 3 (HCHCr)	vac, air	—	red	2.05
Sleipner			blue/brown	0.9
Vanadis 4			green/white	1.5
Vanadis 10			green/violet	2.9
Vanadis 6			green/dark green	2.6

Calmax, caldie, serverker 21 and serverker 3 may be of more interest to tool and die makers. Since chemical compositions are provided, equivalent steels from other manufacturer may be chosen for designing and making of sheet metal press tools.

2.5 NON-METALS

In designing and making of sheet metal press tools, sometimes non-metals such as nylon, hard rubber is used. All these non-metals are used in the form of rods, hollow rods, blocks and thick sheets as starting materials. It would be advantageous for a designer and maker of tools to have a brief idea of properties of mentioned non-metals.

Nylon

Technical or chemical name of Nylon is Polyamide which is produced in a variety of types and grades in the form of rods, plates which are suitable for making some parts in the construction of a sheet metal press tool. In this context, the most important property of Polyamide is that its coefficient of friction is low and wear resistance is much better as compared to other plastics, which are non-metal materials. Parts such as small bushes, pins and actuating levers are made from Nylon.

Hard Rubber/Elastomers

It is well known that rubber has a spring like behavior. For example, if an erasing rubber is pressed, its thickness gets reduced. When released, it comes back to its original size. If this action is repeated for quite a number of times then the thickness may not regain its original thickness. This means that a permanent deformation has taken place to some extent. Therefore, such types of rubbers cannot be successfully used in tools.

Sometimes there are space constraints in sheet metal press tools, consequently, steel springs cannot be accommodated, for example, see Figs 2.10 and 2.11 where instead of spring there is **rubber/elastomer** of special grade having retention of 'spring back' property and hardness.

Description of various scales is given below:

What does 'Shore Hardness' mean?

There are different Shore Hardness Scales for measuring the hardness of different materials. These scales were invented so that people can discuss these materials and have a common point of reference.

The **Shore A00** Scale measures rubbers and gels that are very soft. The **Shore A** Hardness Scale measures the hardness of flexible mould rubbers that range in hardness from very soft and flexible to medium and somewhat flexible to hard with almost no flexibility at all. Semi-rigid plastics can also be measured on the high end of the Shore A Scale. As you can see from the chart (Visit source), there is overlap on the different scales. For example, a material with a Shore Hardness of 95 A is also a Shore 45 D. **Shore D** Hardness Scale measures the hardness of hard rubbers, semi-rigid plastics and hard plastics.

Rubber is used in block form to create a Logo impression on a thin metal sheet of about 0.25 mm (Figs 2.10 and 2.11).

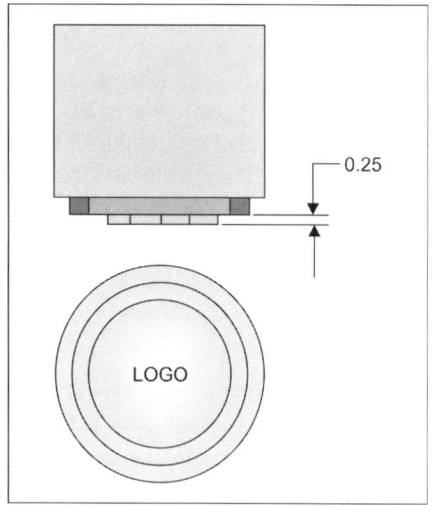

Fig. 2.10: Example of logo impression

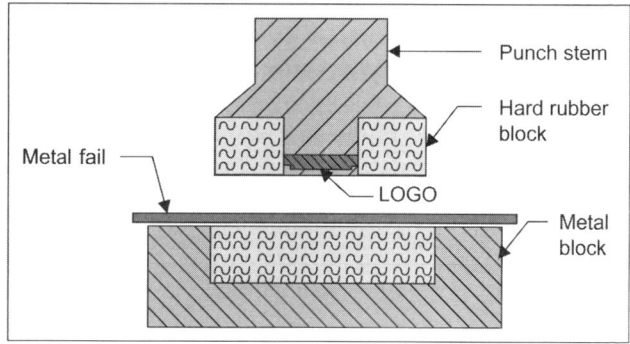

Fig. 2.11: Arrangement of creating logo impression

Apart from spring factor, this method of creating raised impression or embossing does not need a matching Logo female part insert. Consequently, tool is economical in making. It may be noted that a sharp and deep embossing cannot be generated with good results.

Another example is illustrated with the help of Fig. 2.12.

Parts (1) and (2) are two halves of die cavity. These cavity blocks can be opened and closed. A **sheet metal shell**/tube is placed inside the cavity through its opening as shown in Fig. 2.12 'a'. Part (3) is a steel punch on the face of which a hard elastomer piece is fixed. Steel punch is lowered and force is applied by press RAM, which makes elastomer to expand sheet metal shell to take shape of cavity. Once steel punch with elastomer is withdrawn, shell is left inside the closed die. Now the die is opened and the bulged sheet metal part is taken out.

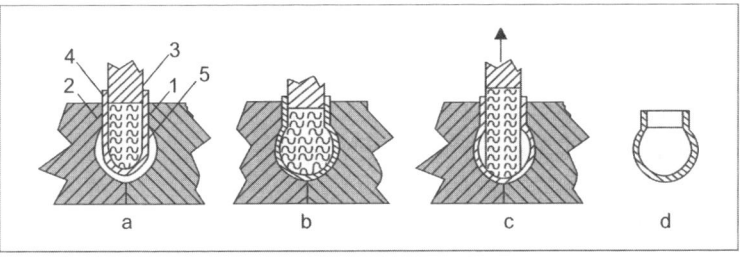

Fig. 2.12: Example of formation with the help of rubber punch

3

Machining Processes

This chapter consists of the following aspects:

3.1 CUTTING

The first steel cutting place in a tool room is steel store where a variety of steels are stored. These steels may be solid round, hollow round, rectangular sections, plates and round discs of different thickness. Purchase of round disc is expensive, therefore, it is preferred to have bars in stock and to cut pieces of desired thickness. However, sometimes procurement of standard round discs of desired thickness is necessary and economical. Suppose a particular type of steel in diameter 100, 150 and 200 mm is often needed then pieces are cut as per required thickness. In case a steel of 250 mm diameter is occasionally required then it is advisable to have round disc of desired thickness rather than going for long bars, if available at all. In case it is done then it may amount to blocking money unnecessarily. Shop floor general practice is that a requisition slip is given to steel store well in advance, say one to two days. A typical form of a demand slip is shown in Fig. 3.1.

Steel Demand Slip

Serial No: Date:

Date needed:

By, Time:

Steel Code:

Steel name:

Dimensions: $\emptyset_1$ = Diameter =

                              $\emptyset_2$ = Hollow =

Piece thickness: T =

Rectangular section:

          L (Length) =

          W (Width) =

          T (Thickness) =

Number of pieces required:

Signature of In charge Tool Room

Fig. 3.1: Steel demand slip

Sometimes steels as per demand slip are not available then steel storekeeper verbally informs I/c tool room as to what is available and what is not. They discuss as to what steel is available or by which date desired steel would be available. After discussion, I/c tool room decides if he should get alternative steel or to wait for desired steel to arrive in store, then a fresh requisition is given.

Steel stores are generally equipped with sturdy racks for holding steels bars of various sizes. In between two rows, there is a bay, sufficiently wide so that a moving hoist can lift a bar and to bring it to power hacksaw and place it at suitable 'rest' support attached to hacksaw. Bar is then adjusted for length to be cut. This adjustment and clamping of steel bar may be of pneumatic or hydraulic power to be controlled by hacksaw operator.

Figure 3.2 shows a schematic set up of steel cutting system where part (1) is robust cast iron frame of hacksaw and part (2) is a reciprocating frame which slides in guide ways. Length and depth of frame is so much that maximum diameter of steel bar may be cut. Power hacksaws are designed to take a minimum and maximum diameter of stock. These limits are specified by manufacturer of machine. There may be a number of models of various capacities.

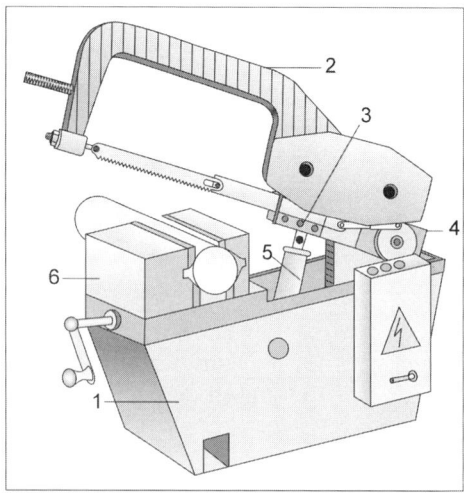

Fig. 3.2: Power hacksaw

In large tool steel stores, generally there are two power hacksaws–one for handling large diameter steel bars such as 200 mm and another is smaller model to cut small diameter bars such as 60 mm. Part (3) is a guide way in which a slide moves. The slides and frame (2) are generally integral portions and (4) is an eccentric. Its connecting rod is attached to slide of frame. To and fro movement of frame is achieved by rotation of eccentric of power unit of machine. Part (5) is another short stroke jack. The piston rod of this jack is attached to guide ways body by means of a free pin. The purpose of this jack is to lift saw a little on its return stroke. Forward motion is a cutting stroke and (6) is a clamping vice. In light duty machines, it may be a mechanical clamping by manual effort. Heavy duty machines may have hydraulic clamping system. Clamping jaws are changeable. For clamping round bars, there is a pair of jaws having V groove on both jaws. For holding rectangular section bar, another set of clamping jaws are used. On these clamping jaws, there is a big serration for strong grip on stock.

Shape and Dimensions of a Typical Hacksaw Blade

Hacksaw blades are generally manufactured from steel. After making proper dimension of width W, length L, holding holes, cutting teeth and staggering them, row of teeth are flame hardened and tempered. Ultimate hardness of teeth reaches around 64 HRC and blade body attains toughness and springiness.

Referring to Fig. 3.3, following are the typical dimensions:

$$L = 300 \text{ mm}$$

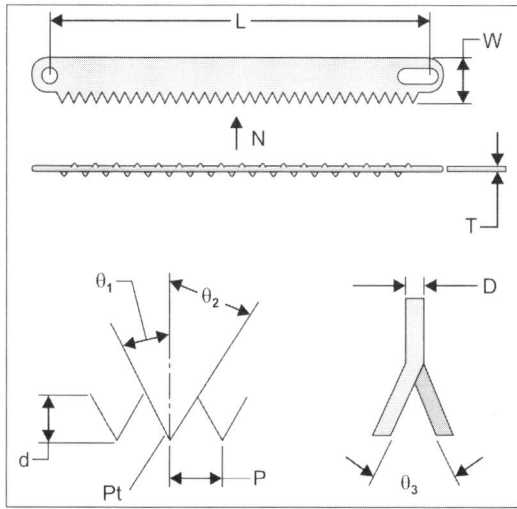

Fig. 3.3: Hacksaw blade

W = 12 mm
T = 0.7 mm
D = 1.1 mm
d = 0.7 mm
θ_3 = 3–5°
θ_1 = 5–10°
θ_2 = 45°
N = 10 per cm   (number of teeth)

On hydraulically operated saws, speed and pressure of cutting may be adjusted.

Recommended cutting speeds for various steels are given in Table 3.1.

Light duty power hacksaws have fixed speed and stroke. Setting of stock is manually done.

	Table 3.1: Cutting speeds for hacksaw blades		
S. No.	*Steel type*	*Brinell hardness number*	*Cutting speed meter per minutes*
1.	Carbon steel	170–190	1
2.	Tool steel	195–210	0.8
3.	High carbon high chromium steel	210–230	0.7
4.	Iron	150	1.5

In many makes, power hacksaws are provided with cutting fluid jet which serves as cooling cum cutting medium. This helps faster cutting and blade retains longer life of teeth.

There are a variety of power hacksaw blades. These are generally made from strips of alloy steel, high speed steel and carbon steel. Blades are generally specified in the following manner (Table 3.2):

In many tool rooms, there are circular saws. In this machine, instead of a reciprocating saw, there is a circular saw rotating at a high rpm such as 1200. It is gradually lowered to touch securely held steel bar. Piece of bar gets cut very fast as compared to hacksaw. Here plenty of coolant is used having suitable guards to contain flying coolant from rotating saw. Selection and installation of a circular saw depends on the maximum diameter of rod, round or rectangular to be cut. Manufacturer of circular saws produce a number of models having various capacities. Selection of model has to be done from available standard models. Typical specification of a circular saw is given below to give an idea as to how a circular saw is specified.

- Circular saw ———————————— Dry cutting
- Power motor ———————————— kW/HP
- Speed of circular saw at no load ——— 1250 rpm
- Diameter of circular saw ————— 14 inches
- Number of teeth ———————— 24–80
- Clamping system ———————— Quick action
- Changing circular saw —————— Spindle lock system
- Mounting hole size to suit spindle —— 25Ø mm

Figure 3.4 gives an idea of tooth shape.

The following websites may be visited for more information on circular saws:

Google —— Circular saw for steel.
A number of sites are available.

Table 3.2: Specifications of power hacksaw blades

Dimension mm			TPI	
Length	Width	Thickness	Teeth per inch	Finish
300	25	1	10, 14	Color as per makers silver
450	38	1.6	6, 10	" "
For fine work				
155	6.35	1.25	32	——

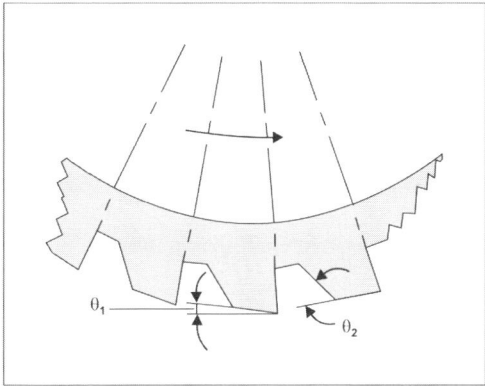

Fig. 3.4: Circular saw (tooth shape)

3.2 FILING

Suppose a rectangular section piece of steel is cut from stock by means of a power hacksaw. This piece has six faces: five are smooth and the sixth is rough, having rough 'hills and valleys' on the surface. To make the surface smooth, 'peaks' of 'hills' have to be removed. This is done by a hand tool known as FILE and the process is known as filing. There are a variety of files which are classified according to the following:

• Size
• Type of cross section
• Type of cut
• Material
• Shape

Figure 3.5 shows few types of hand files.

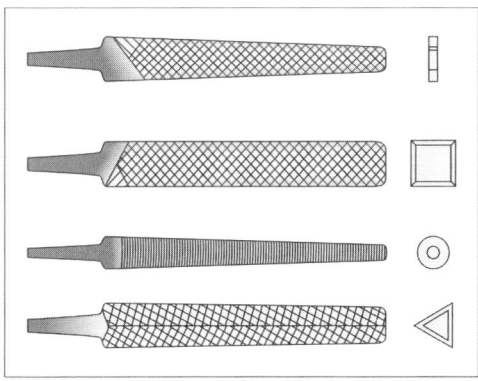

Fig. 3.5: Hand files

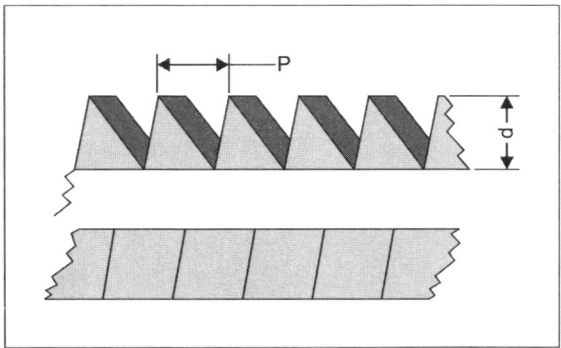

Fig. 3.6: Shape of blade cutting edge

Faces of files are provided with equally spaced rows of cutting edges, known as ridges. This is done in slant configuration with respect to axis of file. Cutting edges are generally in two sets, each on a certain angle. Figure 3.6 provides some details of shape and size of cutting edges.

Parts of a File

Figure 3.7 shows parts/portion of a file with generally used terminology. Tang is the portion on which a wooden or plastic handle is fitted. It is a friction fitting. This means that tang is hammered (lightly) inside already provided hole in handle.

The body of file contains rows of fine cutting edges, generally on all the four sides, two flat surfaces and two edges. Sometimes, one edge is kept plain and the width of body is not uniform throughout. Generally about two thirds of length have uniform width and remaining one third width becomes slightly taper towards 'points'. Differences between the widths may be approximately 2–5 mm, depending on the size, type and shape of file. Figure 3.8 may be seen.

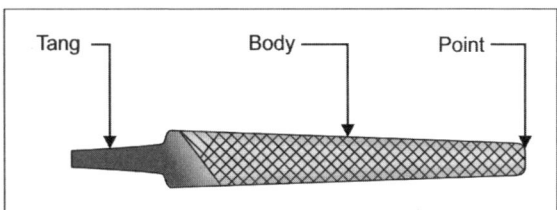

Fig. 3.7: Parts of file

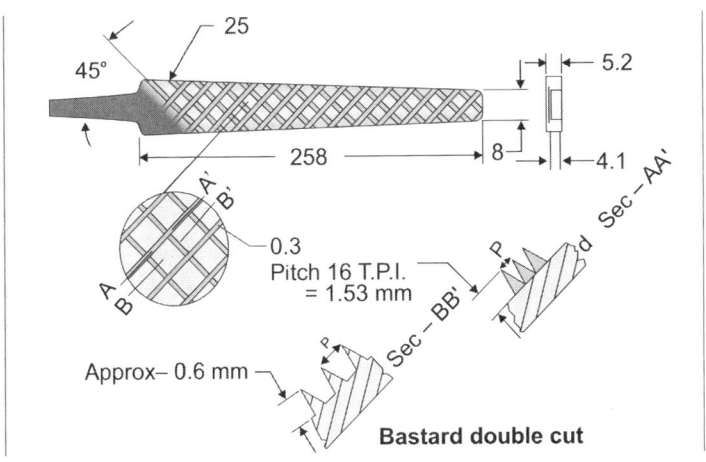

Fig. 3.8: Fine cutting edges in a file

'Point' is the end of file. This portion of file may have shapes as shown in Fig. 3.8.

Length of file varies according to shape, size, cut and standard lengths of which files are produced. Idea of this is already provided in Table 3.2.

Pitch 'P' of ridges is generally expressed as teeth per inch. If T.P.I. is 10 then the pitch would be 0.1 inch. If T.P.I. is 14 then pitch P would be 0.07 inch. As far as depth 'd' is concerned, it is not normally specified by file maker.

Figure 3.9 indicates the aspects in which files are specified.

Some typical files are shown in Figs 3.10 to 3.12.

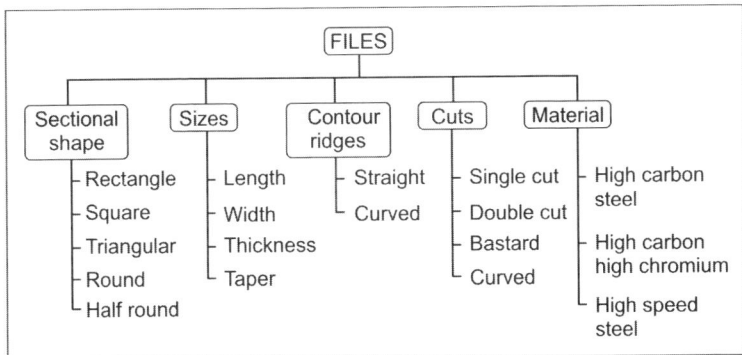

Fig. 3.9: Files classification (Schematic)

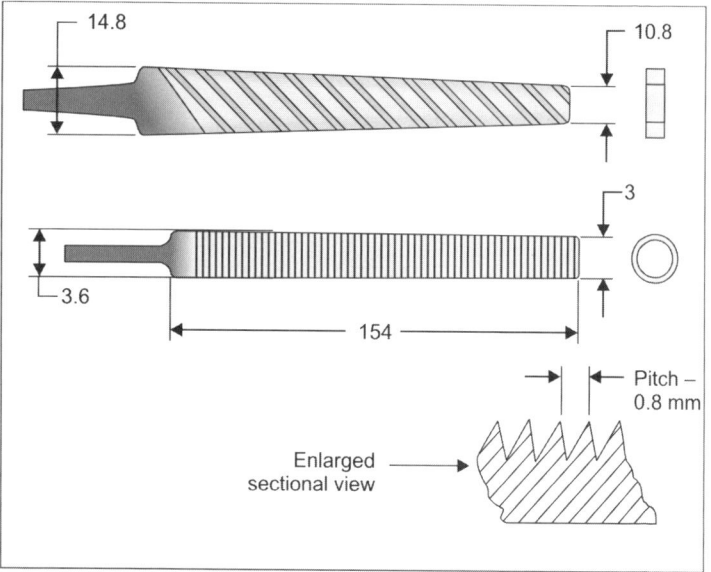

Fig. 3.10: Files classification (Smooth single cut)

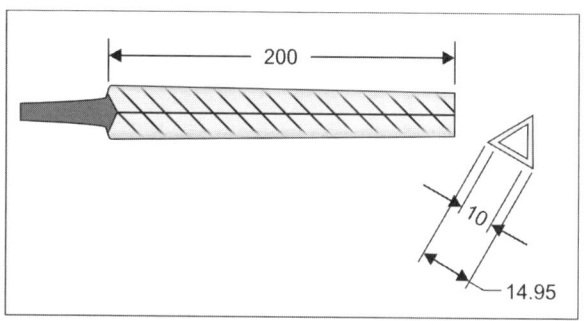

Fig. 3.11: Files classification (Single midium cut)

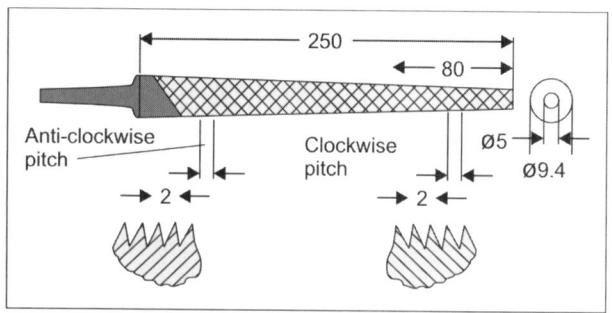

Fig. 3.12: Files classification

Manufacturers of steel files make files in various dimensions and cuts. Table 3.3 provides some typical dimensional and cut data.

Table 3.3: Typical dimensional and cut data

Length mm	Width mm	Thickness mm	Cut
152	15	4	Bastard
"	"	"	Second
"	"	"	Smooth
203	20	5.6	Bastard
"	"	"	Second
"	"	"	Smooth
254	25	6.5	Bastard
"	"	"	Second
"	"	"	Smooth
305	26	7	Bastard
"	"	"	Second
"	"	"	Smooth

For detailed study about steel files, it is recommended to visit the following **website** (see appendix) on internet as available on April 16, 2013, 8 AM (IST):

Needle Files

Very small files of various shapes, sizes and cuts are given a group name as Needle Files. These files are used for delicate and final accuracy attainment work. Some typical needle files are shown in Fig. 3.13 to give an idea of sizes and shapes.

Diamond Impregnated Needle Files

Diamond impregnated files are almost of the same sizes and shapes. The only difference is that body of these files is of bright look. It is either made of stainless steel or steel having chromium plating. It is given a layer of strong binding material and then minute particles of diamond are impregnated and bond is then turned strong so that diamond particles do not get detached/dislodged while the file is used for very delicate and fine removal of stock from hardened parts and even from tungsten carbide parts. Figure 3.14 shows a few jobs on which diamond needle files are used.

In Fig. 3.14, part (1) is of hardened steel. It has a square hole in which a square male part has to be matched. At the time of

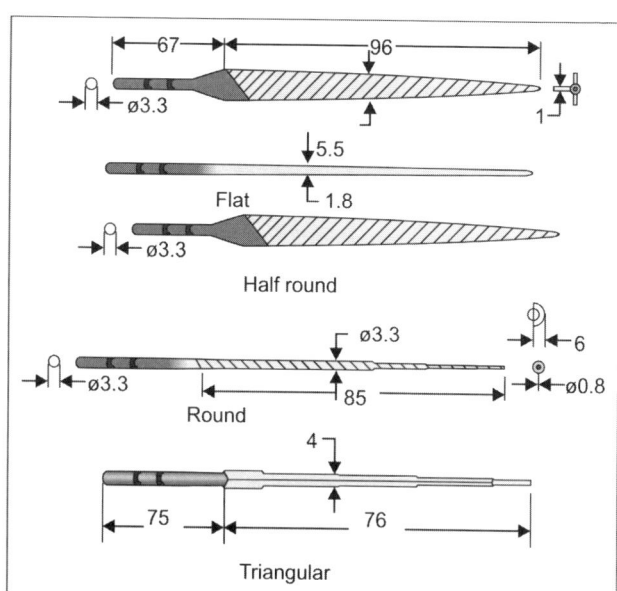

Fig. 3.13: Needle files

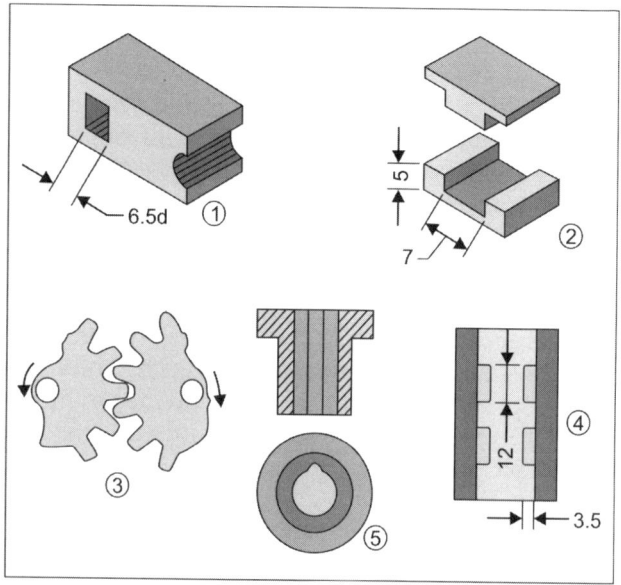

Fig. 3.14: Few jobs for filing by diamond needle files

assembly, it is found that one of the corners of square hole is not sharp, instead having a slight radius which is interfering in assembly, so to make the corner sharp, a square needle file with one face blank (having no impregnation) may be used to make the corner sharp. A flat diamond file may also be used for the purpose provided die maker is very capable and has a study hand and feeling of touch.

Part (2) in Fig. 3.14 is a hardened slide and guide way. Slide has to move with a precision fitting. It is found that it does not have enough clearance, therefore, it is sticking. There must be a 'high point' which is causing sticking. To find out on what point sticking is taking place, copying carbon paper may be rubbed to make the sliding surface black. After this, effort should be made to push slide into guide way. By doing so, high point would become visible and then these may be filed by flat or square diamond needle file.

Part (3) is a combination of small gears. At some place, meshed rotation of gears become hard due to high points. After locating high points, they can be removed by using flat or triangular diamond file. Selection of suitable file depends on the experience of die maker.

Part (4) is a slit making die, having four slits in a brass sheet component. Four punch combination has to enter in all the four slits smoothly with a clearance not more than 0.03 mm. In such a case, side plates are removed and die portion is polished by means of a diamond needle file of a triangular section or a flat file.

Part (5) is a notching die for an aluminum sheet part. On trial, it is found that material is getting burnished at one side of the notch because of lack of clearance between punch and die. Clearance may be increased by slightly removing hard steel by means of a round diamond needle file.

Uses of Steel Files

There may be innumerable situations where removal of material may become necessary and that too by filing. Two typical examples of material removal are briefly described here.

Figure 3.15 shows C.I. casted die block just after casting, removal, cooling down and cleaning. Surfaces S1, S2, S3 and S4 are made smooth by means of a 254 mm long bastard file.

Slots are filed to get proper shape and width, 245 and 203 mm long bastard and medium cut files are used. In addition to this, round and medium cut file is also used to give a smooth finish to rounded end of slot.

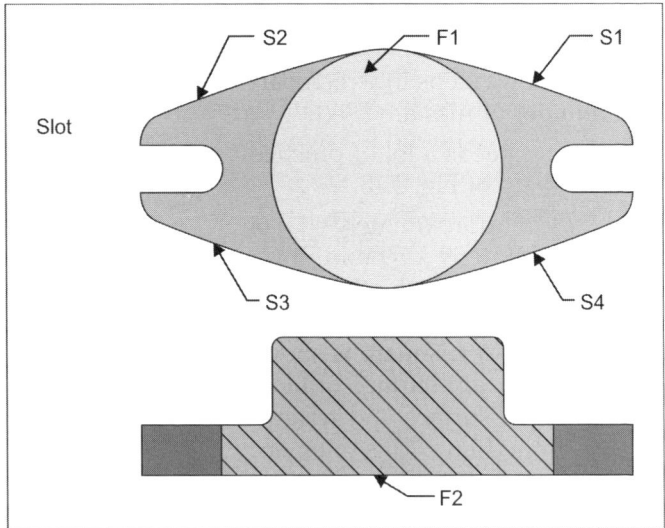

Fig. 3.15: C.I. die block

Figure 3.16 shows an idler gear arm of a bench lathe which is to be made from a steel plate of 12 mm thickness and by using drill

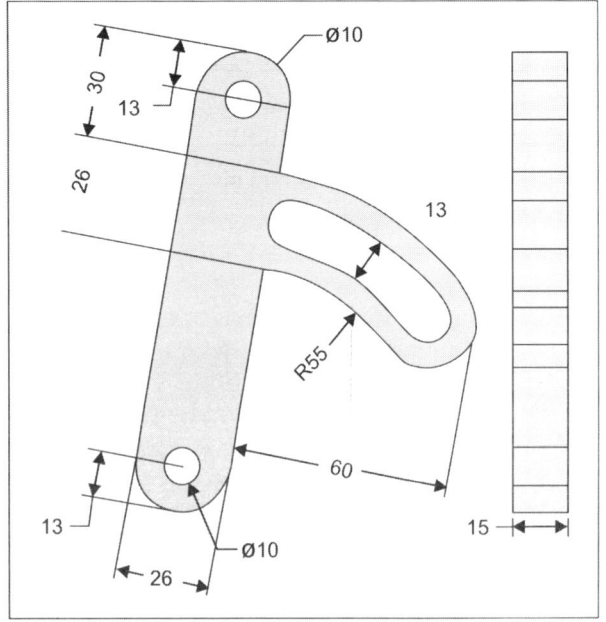

Fig. 3.16: Idler gear arm

press and files. First of all, a piece of appropriate size is cut from plate and then marking is done.

Figure 3.17 shows steps in which part is to be made. In making process, a number of files are used.

Accurate filing needs a lot of practice. Idea of accurate filing is given with the help of Fig. 3.18.

A steel block of approximately 60 × 60 × 60 mm is to be filed to achieve a size of 59 × 59 × 59 mm. It is required that all the sides should be right angle to each other. Fig. 3.18 'A' is an orthographic view of the block whereas 'C' is isometric elevation of the same block. In this view, a trisquare is shown to check right angles. It can be noticed that top surface of block is not straight, hence, a gap between block surface and trisquare blade is visible. It is a result of inaccurate filing. This often occurs if die maker is not an expert filer. Figures 3.18 'B' and 'D' show accurate filing where

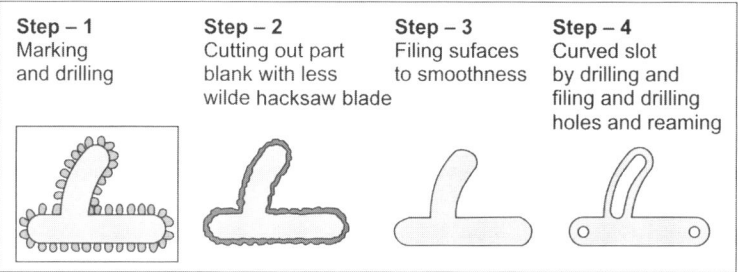

Step – 1	Step – 2	Step – 3	Step – 4
Marking and drilling	Cutting out part blank with less wilde hacksaw blade	Filing sufaces to smoothness	Curved slot by drilling and filing and drilling holes and reaming

Fig. 3.17: Idler gear arm making steps

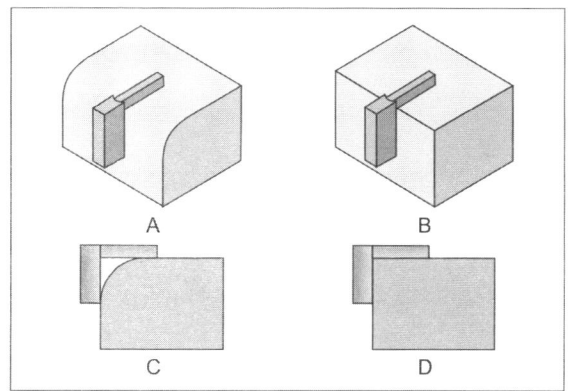

Fig. 3.18: Idea of accurate filing

trisquare is sitting on two surfaces without showing any gap between trisquare surface and block surface. It is the result of good filing.

In the beginning, a bastard file may be used to remove some stock quickly from all the three sides with the aim to bring six surfaces as near to right angle as possible; at the same time, to leave some stock for finishing and to achieve desired dimension of 59 × 59 × 59 mm. For this purpose, medium and smooth file may be used.

3.3 DRILLING

Drills are basically cutting tools. These are available in various standardized diameters and lengths. Drills having straight shanks are held in the chuck of drill press. Big diameter drills, say 25 mm may have a taper shank for direct fitting into drill press spindle. Drills can also be loaded on lathe journal with the help of reduction sleeves. Small size drills may also be loaded on other machines such as milling machine or a radial drill press.

3.4 REAMING

Reamers are basically machine and hand tools. The purpose of reaming is to finish holes to very accurate size and smoothness. Generally, two types of reamers are available: one group has reamers of fixed diameters, another group has arrangement for adjusting diameters. Extent of adjustment is generally very small, say one millimeter. Holes to be reamed are provided with reaming allowance which may range from 0.1 to 0.5 mm.

Reaming may be accomplished by hand or machine. Choice depends on nature of job and accuracies required.

3.5 TURNING

Turning is a machining process through which a non-circular object is made circular.

The most primitive form of turning was that a man used to rotate a wooden stock by hand and the other person used to appropriately touch a steel or stone shape edge to rotating wood. Consequently, wooden stock used to become roughly round. With passing times, the arrangement of turning took shape of a machine which is called lathe. Lathe is considered to be the 'mother of machine tools' because a variety of operations could be performed such as key way making, spleen cutting, grinding, cutting of non-circular recesses, etc. With the advancement in machine tools technology and development, other machine tools such as shaper, milling and grinder came into existence.

The present day lathes are quite sophisticated from the view point of accuracies, speeds, sturdiness, versatility, safety and automation with the use of programmable logic control (PLC).

Modern machines can perform the following operations:

- All types of internal or external turning
- Internal and external taper turning
- Thread cutting of various types and pitches
- Semi-auto or auto chucking of stock
- Drilling, reaming and taping
- Knurling on diameter of workpieces

Most of the tool rooms engaged in making sheet metal press tools have good quality conventional lathes because tool parts are not required to be mass produced. PLC controlled lathes are generally used in big workshops handling heavy turning work such as shafts, flywheels, connecting rods, cylinders, pistons, axils, rail wheels, etc.

Figure 3.19 shows basic construction of a typical engine lathe.

Part (1) is bed of the lathe, which may be considered as a 'chassis' because other parts are assembled on it. Bed has flat and V guide ways for carriage (10) to slide along the length of bed precisely.

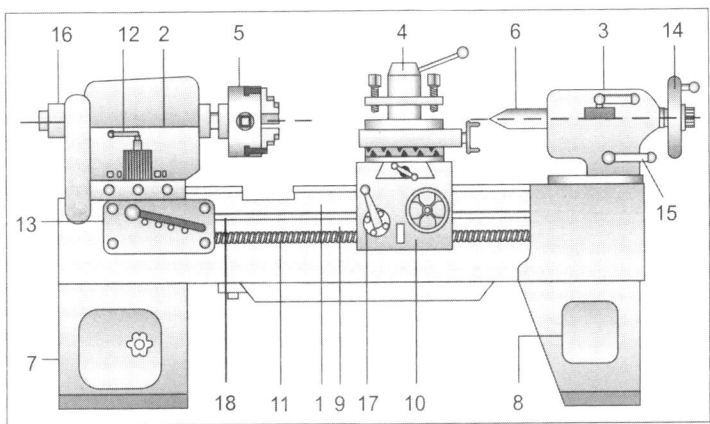

Fig. 3.19: Engine lathe

Beds are casted in cast iron. Castings are generally left in the open for a long period of time, say twelve to eighteen months. Whatever warping has to take place happen during 'aging'. After aging period, beds are machined. Result of this process is that beds generally do not warp during bed's service life. This ensures smooth and accurate movement of carriage with accuracy maintained. Bed body is supported on box legs (7) and (8). Part (8) being on tailstock side and (7) on headstock side, (2) is the headstock body in which hollow journal (16) is fitted. Journal carries crown gears, belt pulleys, ball bearing and thrust bearings. There is also gearing, all housed inside headstock (2). Journal end of headstock towards tailstock has external threads of strong section and wide pitch. A 4-jaw independent or 3-jaw self centering chucks are loaded. Once chucks are completely tightened, holding thread ring which is on the journal, a small flange keeps the ring stay in axial direction when it is screwed on the flange of loaded chuck. This arrangement ensures that chuck will not unscrew out when motion of journal is reversed for bringing back the tool to its starting or any other positions towards tailstock.

Tailstock (3) has locking system to lock it on a particular position on bed guide ways. In unlock position, tailstock can be shifted towards chuck or away from it, depending upon need. There is a sleeve (6) inside tailstock, sleeve can be moved out and in by rotating adjustment hand wheel (14). Generally, sliding sleeve has an axial line marked with graduation to see position of sleeve in or out. Bore at the end of sleeve has a morse taper to receive a taper shank of a center, live or dead. A drill chuck having appropriate taper shank can also be loaded in tailstock sleeve. Taper shank of drill chuck or center cannot be removed by hand as there is a firm grip due to taper holding, so to remove taper shank, there is an arrangement inside. On bringing the sleeve in by rotating the hand wheel, a position would come where automatically taper shank gets released.

Refering to Fig. 3.20, there is also an arrangement to shift tailstock left or right at right angle to bed slide ways. This movement is not very much. It may be around 25 mm in each direction from central location which is generally marked at the back face of tailstock near the bottom. There is a permanent central position mark on axial side block which slides on bed guide ways.

Part (1) is bed of lathe and (2) is the C.I. sturdy block which has guide ways which precisely match with those of bed. Part (3) is a tailstock body, bottom of which has a sunken guide way exactly right angle to spindle axis. Block (2) has raised guide way properly

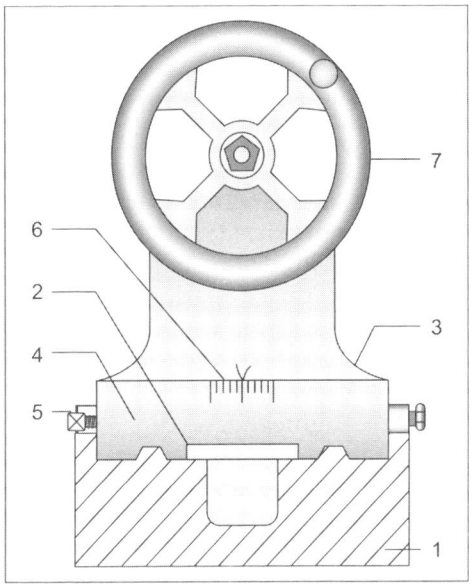

Fig. 3.20: Tailstock

matching to sunken guide way of tailstock body. There is a precision screw (5) of fine pitch. By rotating this screw, tailstock body can be shifted left or right. Amount of shift may be seen at marking (6).

It may be noted that normally two zero marking lines are kept coinciding each other. In this position, points of tailstock center and headstock center should precisely coincide. A center can also be loaded inside journal by using a reducing sleeve. Alternatively, a center may be turned on a small round piece of iron.

Carriage (10) of lathe is a part which is big in size. It can slide over bed guide ways with the help of a drive/shift wheel. Carriage carries cross slide, apron, tool post and on top of cross slide, there is facility to fix taper turning attachment. Apron (17) is the part of carriage in which mechanisms are housed to provide automatic motion to carriage either towards headstock or tailstock. This motion is picked up from horizontal rotating shaft (18) which has a rectangular sunken groove to drive gear train while carriage moves. Another shaft (11) is having threads. There is a 'half nut' arrangement inside the apron. Half nut is engaged with the threaded shaft by lowering control lever outside apron. Threaded shaft provides motion to carriage which moves to give a definite pitch in thread cutting in a job. Threaded shaft (11) gets its rotation through a gear box which is equipped with a lever to select pitch of threads on workpiece.

Drive motor of specified power is generally fitted in headstock side box leg (7). Alternatively, motor is also mounted outside at the back of box stand. From here, rotational power is transmitted on auxiliary V belt pulley. Shaft of pulley carries a three-step belt pulleys to transmit power to journal pulley. Auxiliary shaft pulley stand is designed in such a way that by pulling a lever, auxiliary shaft can be brought near to headstock. Consequently, flat belt becomes loose enough to manually shift the belt from one step to another when motor is stopped, thus, getting different journal rpm(s).

Construction of Tool Post

Figure 3.21 shows a typical tool post of a bench lathe for turning small precision parts. Part (1) is the carriage on which cross slide (2) is mounted. On the top of cross slide, there is a circular platform (5) on which tool post slide is fitted. Tool post slide may be moved backward or forward by means of hand lever (8). Tool post slide can also be rotated to any angle which is marked on circular seat and then can be locked by means of screw (9). The shape of tool post and support ring is such that height of tool tip may be adjusted to match center. This arrangement is shown in Fig. 3.22 which is self explanatory.

Another type of tool post is shown in Fig. 3.23. This type of tool post is generally provided on heavy duty lathes.

Part (1) is cross slide on which tool post (2) is mounted. Square shaped tool post is made of tough steel with case hardened surface. It carries eight bolts having square head. Upper and lower flanges

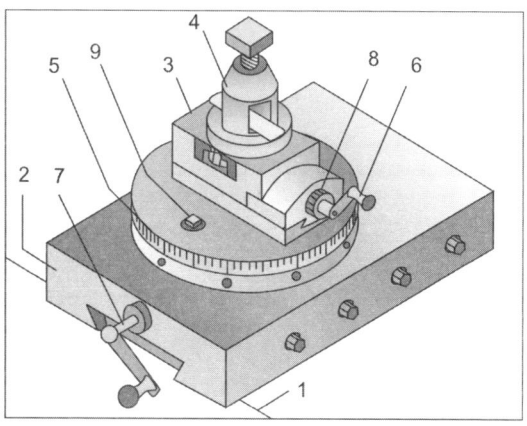

Fig. 3.21: Tool post

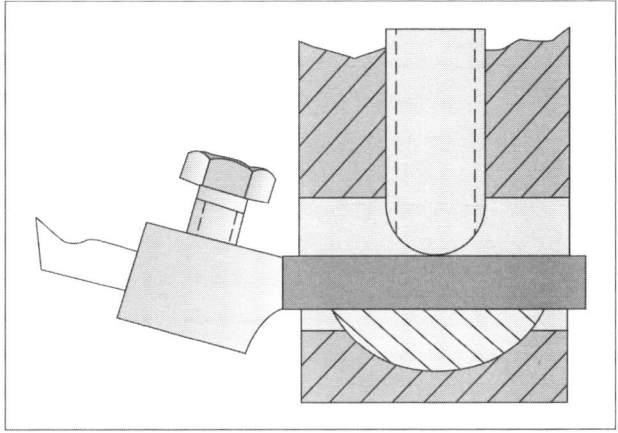

Fig. 3.22: Tool tip adjustment arrangement

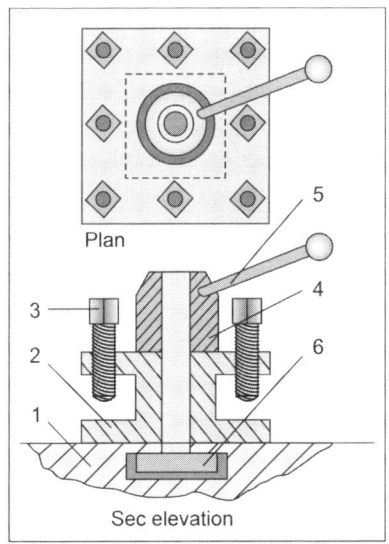

Fig. 3.23: Heavy duty tool post

of tool post are integral part, upper flange having eight threaded holes to carry bolts. These bolts are used to tighten tool bits, generally of 12 mm square section. In fact one, two or three different turning or parting tools can be fixed at a time (see Fig. 3.24). While setting tool bits, care has to be taken that cutting edge or tip of tool coincides with tip of center. This can be achieved by putting spacer strip of suitable thickness under tool bit and then two or three bolts

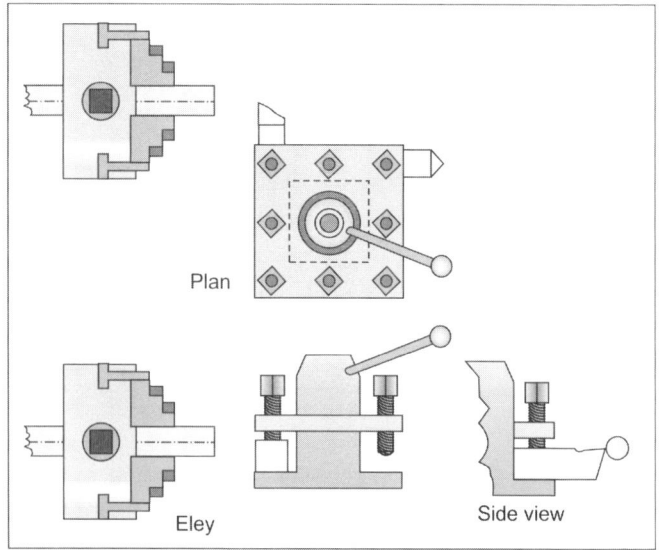

Plan

Eley

Side view

Fig. 3.24: Multi-tool holding is shown

are tightened. Matching of tool tip and center point is checked once again. Part (6) is a bolt with T head with suitable dimensions so that it may slide in T slot of tool post. Holding nut may be loosened by rotating it in anti-clockwise direction with the help of lever (5). Now tool post block may be rotated to attain a suitable position of tool with respect to job already located on chuck for carrying out turning operation.

During description of tool post (Fig. 3.23), it is mentioned that height of tool tip is brought in line with the center of job by using spacer of appropriate thickness. This system is time consuming as a number of spacers may have to be tried out.

A more sophisticated tool holder is described with the help of Fig. 3.25. The advantage of this tool holder is that height of tool tip may be adjusted by lowering or raising tool holder by means of a bolt.

Tool holder block (1) has a dovetail guide way in vertical direction, assembled on a vertical member of cross slide. (It has a vertical dovetail guide on which tool holder block precisely moved up and down.) Vertical position of block may be adjusted with the help of bolt (3), which has a locking ring (4). Tool bit (6) is tightened by set screws (5). A gauge may be used to bring the tool tip at the center. This type of tool post is generally available for heavy duty modern lathe. Arrangement on tool post slide is such that conventional square tool post can also be fitted if so desired.

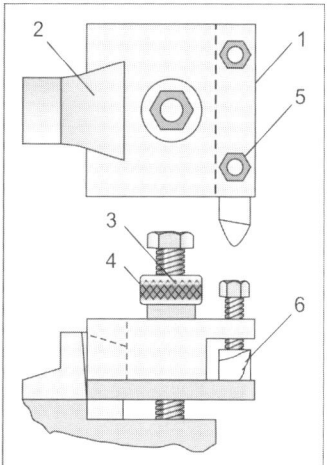

Fig. 3.25: Sophisticated tool holder

Taper Turning Attachment

The movement of cross slide is achieved by rotating long screw by hand or automatically. There is a nut fitted from underneath in cross slide body. This nut is tightened to cross slide body by means of bolt from the top of cross slide. If this bolt is removed then cross slide becomes free and it can be slided in forward or backward direction even by hand.

A mechanism known as taper turning attachment is connected to cross slide. When carriage is moved towards headstock, cross slide also moves away from chuck. This movement with respect to longitudinal movement of carriage creates a taper movement of tool which turns diameter of stock. Figure 3.26. shows basic arrangement of taper turning.

Taper turning arrangement shown in Fig. 3.26 is a view from top. Part (1) is backside guide way of bed of lathe, (2) is the carriage, (3) is cross slide which is free by removing securing bolt (9). Part (8) is an extension column fitted to cross slide by means of bolts. Hanging end of this extension column has a slot. Part (4) is the base of taper turning mechanism. It is rigidly fixed to back side of bed by means of bolts. Manufacturer of lathe provides thread holes and machined surface to tighten base (4). There is a long slide way (male) (5), say about 30 cm long, on to this slide way is a sliding block (6) which can slide towards both ends of slide ways. This motion is very smooth without any looseness or tightness. Slide way (5) is hinged in the middle on base. Consequently, it is possible

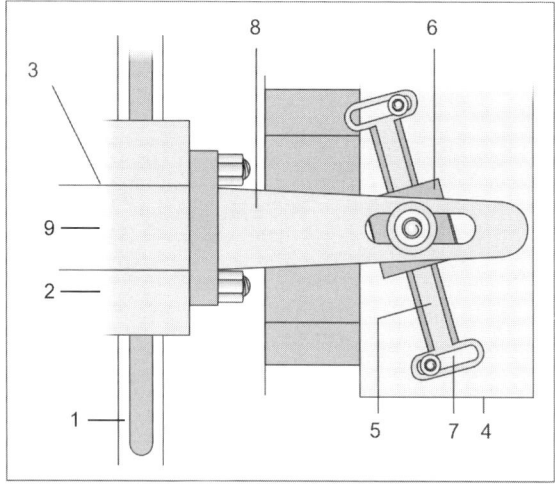

Fig. 3.26: Taper turning attachment

to swing slide way to any desired angle marked at the end of slide way. Extent of swing is around 15° both sides from center. Once angle is adjusted, slide way is tightened on base. Now, when carriage moves, say towards headstock, sliding block (6) also slides along slide way, thus, shifting cross slide towards operator as it moves towards headstock. In this way, a taper is turned on job. First, one or two cuts may be trial cuts. After cut, diameters of both ends of long job are measured to ensure if angle set is correct. If not, slight re-adjustment may be done in angle setting of slide way.

Figure 3.27 shows two typical jobs with dimensions.

Job A shown in Fig. 3.27 has a taper portion. It may be turned by tool post slide by setting it at an angle. After one or two trials,

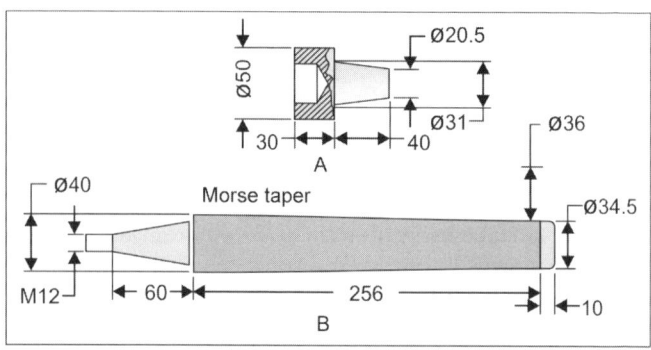

Fig. 3.27: Typical taper turning jobs

desired diameters 31 and 20.5 mm may be achieved. This could be possible because length of taper portion is 30, under limit of tool post slide range.

Job B is quite long and has a fine taper of only 4 mm on diameter on a length of 256 mm. Obviously, it cannot be turned by tool post slide. For such cases of taper turning, taper turning attachment is used. In this case in between centers, job is rotated by means of a 'dog'.

This job may also be turned without taper turning attachment. Tailstock may be shifted towards operator by a pre-calculate amount to achieve a taper of 4 mm on diameter.

Taper Calculations

Taper angle calculation for Fig. 3.28 considers right angled triangle CDE where θ is the angle of taper.

(**Note:** Ignore dimension 60. It is 30 mm.)

In triangle CDE,

$$\tan \theta = \frac{ED}{CD} = \frac{(31 - 29.5) \div 2}{30}$$

$$= \frac{1.5 \div 2}{30} = \frac{0.75}{30}$$

$$= 0.025$$

$$\theta = 1.40°$$

Taper angle calculation for Fig. 3.27 B

Difference in diameter = $(40 - 36) = 4$

Half of difference = 2

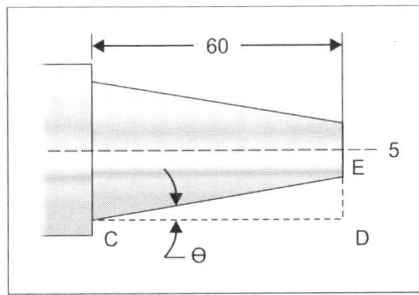

Fig. 3.28: Taper calculation

Hence,

$$\tan \theta = \frac{2}{2.56} = 0.0078$$
$$\theta = 0.3°$$

Turning Tool Bit

Turning tools are given various shapes and sizes by grinding tool bits of 6 mm square, 12 mm square. High speed steel bars of square section are given various rough shapes by forging. Cutting tips are then hardened to maximum attainable hardness, say 64 HRC. Rough shape of cutting portion is given final shape and sharpness by grinding.

Tool bits are generally made from alloy steel having the following alloying elements:

• Cobalt
• Chromium
• Tungsten

High speed steel typically has the following composition:

Tungsten ——— 16%
Chromium ——— 5%
Vanadium ——— 0.4 – 0.7%

Tungsten carbide tips are used after braze or screw mounted over a holder. Tungsten carbide is quite hard as compared to HSS and more wear resistant. Consequently, it retains sharpness for much more cutting as compared to alloy steel tool bits. Carbide tips can well tolerate high temperature, therefore, higher cutting speeds may be adopted.

Figure 3.29 shows a few turning tool bits.

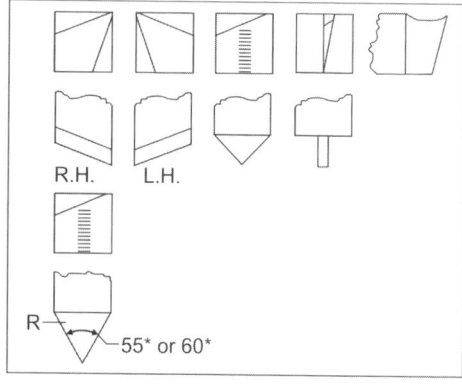

Fig. 3.29: Turning tool bits

Tool Shapes and Angles (Refer Figs 3.30 and 3.31)

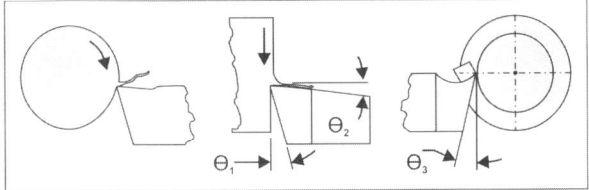

Fig. 3.30: Tool shapes and angles

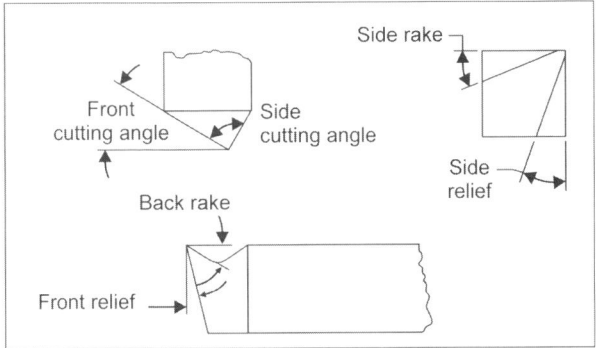

Fig. 3.31: Tool shapes and angles

Typically, the following values of angles are found to work satisfactorily for tool steels.

Side cutting angle	8°–12°
Front cutting angle	5°
Side rake	15°–30°
Front relief	5°–9°
Back relief	5°–9°

In case side rake is more then there is a likelihood of formation of curled chips. Less side rake angle tends to produce broken chips.

Cutting speed of various metals is different for best cutting performance. Cutting speed is the distance travelled by a cutting tool in a minute (see Fig. 3.32).

Let (1) be a metal strip on which tool (2) moves for one minute and traversing a distance D meters.

The cutting speed = D per minute

Or mpm (meter per minute)

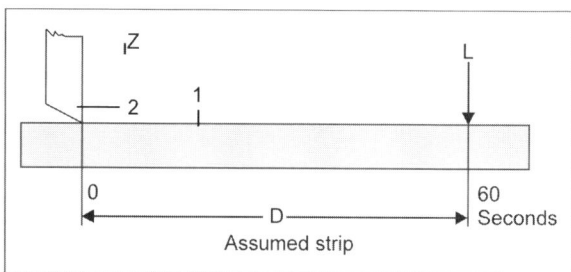

Fig. 3.32: Cutting speed representation

It is all the same if tool remains stationary and strip moves. Suppose assumed strip is accurately wrapped around a circular piece on lathe chuck. Zero is the point where two ends of strip touch each other. If circular piece takes complete one round then it can be said that surface length of strip passed through tool point in one complete round in meters. (see Fig. 3.33).

Suppose round piece makes 100 complete rounds in one minute then total surface length passed through tool point would be 100 meters.

Conclusion may be drawn that rpm would have to be calculated for turning a tool steel rod of a particular diameter. Hence, recommended cutting speeds for various types of steels must be known.

Cutting speed for few steels is given below which is determined by practical experiences:

Free cutting steel ——————— 30–60 meters per minute

Carbon steel ——————— 20–39 meters per minute

Alloy steel ——————— 18–36 meters per minute

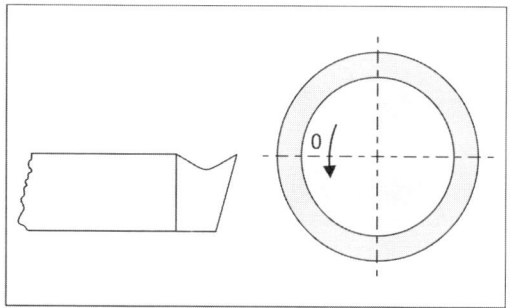

Fig. 3.33: To explain cutting speed

High carbon steel ———————— 14–17 meters per minute

Cast iron ———————— 16–22 meters per minute

Formula for determining rpm

Cutting speeds in meter per minute

rpm = Π × diameter of jobs to be turned

Example: An HCHCr bar of 15 cm diameter is to be turned. Recommended cutting speed is 15 mpm. Determine by what rpm job should rotate.

$$\text{rpm} = \frac{\text{Speed}}{\Pi D} = \frac{15}{3.14 \times 0.15}$$
$$= 32 \text{ rev per minute}$$

Thread Cutting

Metric threads have an angle of 60° and British threads 55°.

There may be two ways of giving feed of tool to job. It is explained with the help of Fig. 3.34.

Method of tool feed as shown in Fig. 3.34 A is prone to produce chattering in thread surface as flow of material on tool surface is from two directions, hence interfering with each other. In case of B method, flow of material is at one edge of tool. Feeding of tool at 30° may be done by setting tool post slide at 30° from zero setting. For BSW threads, an angle of 27.5° may be set. In this case, a back rake of 5–9° may be given.

Accuracy of Lathe

Accurate turning of a job very much depends on the accuracy of lathe. Many lathe manufacturers provide accuracy test certificate

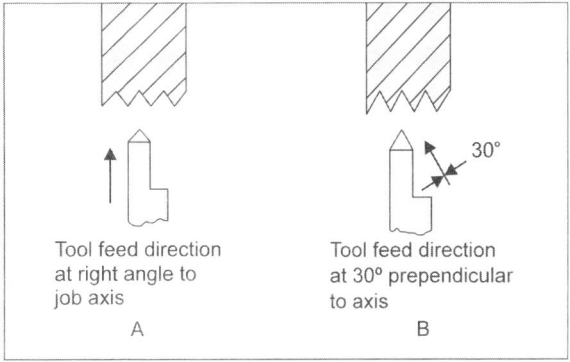

Fig. 3.34: Methods of thread cutting

to customer when a lathe is to be purchased. In case lathe supplier is a reputed party, one may rely on the certificate. It may be a good practice to develop an understanding that delivery of lathe would be permitted after accuracy checks are performed in presence of purchaser or his representative and found to be okay.

Accuracy test may also be performed on a lathe which is in service for a long time and there are signs of inaccuracy in turning a job. Generally, accuracy of a lathe has two aspects—one is dimensional accuracy and the other is surface finish and uniformity in pitch of cut threads.

Uniformity in Diameter or Bore (Refer Fig. 3.35)

Hold a mild steel bar of about 50 mm diameter and 300 mm long. Use a round nose turning tool. Set a rough movement of tool so that turning lines are visible. Turn only so much that no spot on bar remain unturned. Now measure Ø1 and Ø2. Difference in diameter should not be more than permissible limit of 0.012 mm.

Now turn again with a very fine carriage movement, resulting in a very smooth turned surface. Check diameters of test pieces at various points. Variations would point to worm out carriage guide ways and bed guide ways.

Variation in Ø1 and Ø2 may also be due to non-parallelism of journal axis and bed guide ways. This is a defect which crops up during lathe manufacturing. If it is OK then it is expected to remain OK unless headstock is dismantled for any reason and reassembled with some unintentional error.

Next test may be by turning a test piece held from center to center. Make sure that 0, 0 for side ways shifting of tailstock is matching, with this setting Ø1 and Ø2 should be equal. If it is not

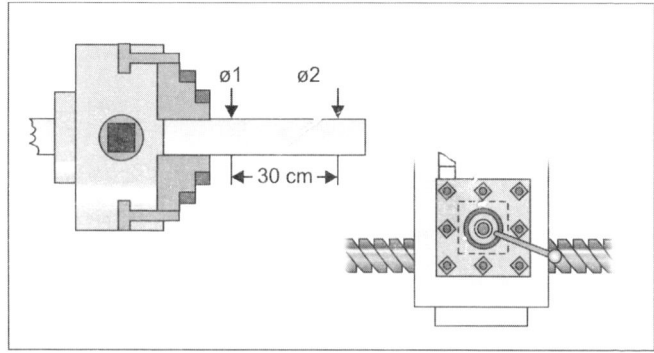

Fig. 3.35: Uniformity in diameter or bore

so, then it indicates that 0, 0 marking is not accurate. It needs shifting of tailstock to achieve Ø1 equal to Ø2.

Accuracy can also be tested without actual turning of a test piece. In this case, a standard hardened and ground test bar is used. Test bar has a morse taper at one end to suit taper sleeve and journal taper hole. It also has proper center holes to hold test bar between centers.

Referring to Fig. 3.36, a dial indicator is held on the tool post and set to touch test bar. By moving the carriage by hand wheel, reading on dial test indicator may be read at both ends. No deflection would show that axis of journal is parallel to bed guide way in both horizontal and vertical plane. Same test may be repeated by setting the test bar, center to center.

Uniformity in Thread Cutting

Irregular wearing of screw rod of lathe may cause non-uniform pitch in threads cut on a job. This defect may not be felt by merely seeing it. Pitch of thread at various points may be checked on Profile projector.

Surface Finish

Surface finish of a turned job very much depends on the condition of bearing of journal. In case bearings are worn out then there may be a play in journal, in both horizontal and vertical directions. This may cause chattering marks on turned surface of job. It may also cause non-uniformity in thread diameter. The effect of looseness in journal is more prominent when carrying out parting cut operation on a job.

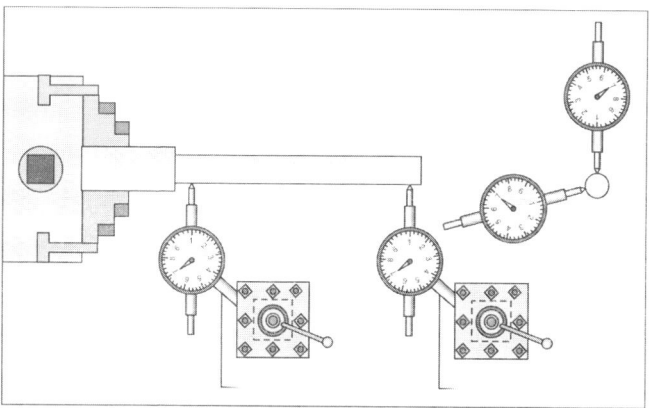

Fig. 3.36: Accuracy test by dial indicators

3.6 SHAPING

Shaping is a machining operation where straight surfaces are machined. Straight surfaces do not necessarily mean that it is just a straight plate of cast iron or steel. Job to be shaped may have any configuration of shapes.

Figure 3.37 shows a few typical shapes of work to be shaped.

Machine which shapes the job is called Shaper. All the shapers are designed to handle work up to specified value of height, width, maximum depth of cut, maximum degree of slant surface and length of stroke.

Figure 3.38 shows schematic diagram of a typical **shaper**.

The basic working principle of typical shaper is explained with the help of Fig. 3.38 where (1) is sturdy C.I. body having suitable height, width, length and base. C.I. bodies after proper seasoning are machined to have guide ways for sliding of RAM (2). There is a table (3) which can be raised or lowered according to need. Once height of table is adjusted, it is supported by a sturdy jack (10). Table can also be traversed left and right at right angle to up and down movement guide ways.

RAM (2) is a part of shaper which moves forward and backward. RAM carries tool holder block (9). Block can be tilted away from vertical by say, 35° on both sides. Tool holder block is so designed that on forward stroke of RAM, tool holder back face remains

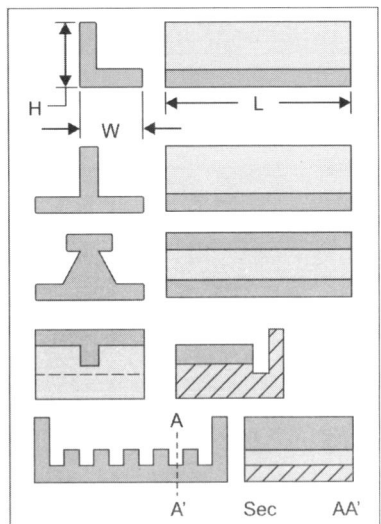

Fig. 3.37: Typically shaped jobs

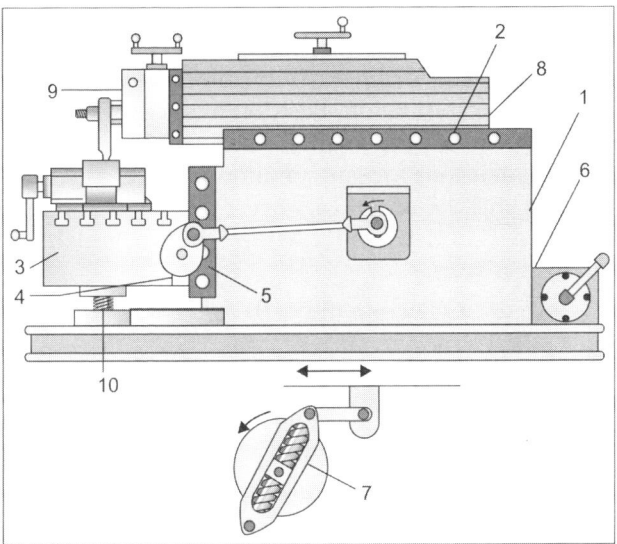

Fig. 3.38: Typical shaper

supported by counter face of RAM. On return stroke of RAM, tool holder is free to tilt so that cutting tool block just slides back with RAM without any obstruction from job. It is shown in Fig. 3.39.

Complete travel of RAM from back position to forward position of RAM is its limit. Amount of this movement of RAM is called 'Stroke'.

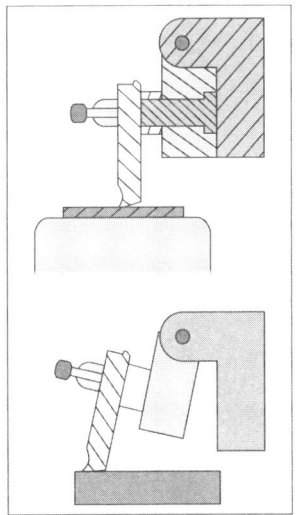

Fig. 3.39: Tool holder block

Stroke of RAM can be adjusted between its minimum and maximum limit by mean of 'Quick return mechanism' (7) which is shown in Fig. 3.38. The location of slide block in swinging arm slot may be adjusted from outside. Position of slide block from center of bull gear determines the length of stroke. Direction of rotation of bull gear is such that forward motion is affected when sliding block is on top and the arc of forward movement is much bigger than the return arc. Consequently, return stroke is much faster than forward cutting stroke.

$$X/R = \sin (90° - \theta)$$

Referring to Fig. 3.40,

R = Distance between center of bull gear and center of sliding block hole, so it is the radius of circle on which block center moves.

D = Distance between center of bull gear and center of fulcrum pin. Swinging slotted arm swings around pin (F). It is a fixed dimension as per design and construction of machine.

Figure 3.41 shows how stroke of RAM changes by changing radius of circular path on which center of sliding block rotates. Let r_1 be the minimum possible radius and r_3 the maximum possible radius. Radius r_2 may be at any place in between r_1 and r_3. Stroke S_1 relates to r_1, stroke S_2 relates to r_2 and stroke S_3 relates to r_3.

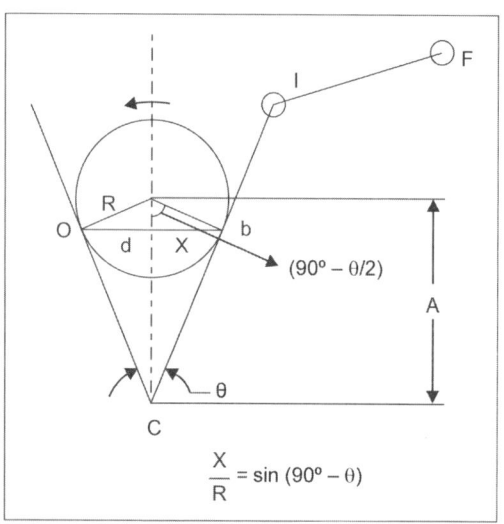

Fig. 3.40: Explaining shaper RAM stroke length

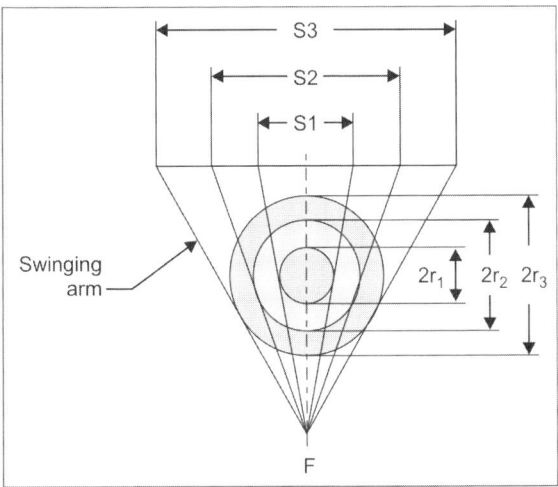

Fig. 3.41: Stroke changing explained

Now it remains to be explained as to how value of radius is changed from outside of machine. It is explained with the help of Fig. 3.42.

Referring to Fig. 3.42, Part (1) is bull gear which is driven by a pinion extended from a gear box (6). On the face of bull gear, a guide way (2) is built in which sliding block (3) slides. Part (4) is a screw rod which passes from threaded hole in sliding block. At the end of screw rod towards center, a bevel gear (5) is fitted which matches bevel gear (6) attached to shaft (7) that passes through hole in the bull gear shaft. End of shaft (7) outside bull gear shaft is

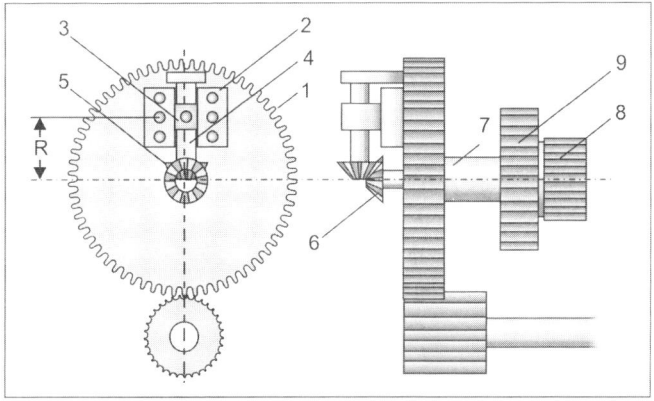

Fig. 3.42: Arrangement for changing value of radius

fitted with a hand knob (8) which should be rotated when shaper is not running. Rotation of shaft (7) shifts the position of sliding block (3). Consequently, the value of radius R can be changed. In heavy duty shaper, instead of a hand knob, there is a square shape at the outer end of shaft (7). A key spanner having a square hole is generally used to rotate the shaft. Once the desired position is reached, shaft is locked with bull gear shaft by means of a locking circular ring having serrations on diameter or provision of slots for using a 'C' spanner.

Positioning of RAM

Referring to Fig. 3.43, length S of stroke may be needed at any place on the job. The position of tool T is at a distance 'd', from reference line XX, while the length of stroke is S. There may be a situation as shown in 'b' where desired position of tool is just at reference line XX' and length 'S' of stroke remains the same. This change of position of stroke may be adjusted by loosening locking nut 'A' and pushing the RAM so that tool approximately reaches near vertical reference line XX'. In heavy duty machines, shifting of RAM position is achieved by screw mechanism at the back of RAM. By rotating screw, RAM may be moved forward or backward with respect to locking bolt 'A', hence, adjustment of stroke length and its position is possible.

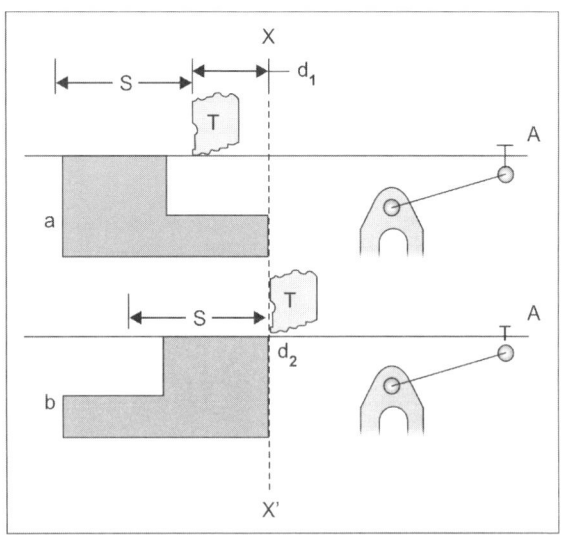

Fig. 3.43: Positioning of length of stroke

It is necessary that it should be possible to vary the speed of RAM as cutting speed for different metals vary. Recommended cutting speeds for various metals are already given in Table 3.1, so generally there is a provision of changing cutting speeds by means of a change gears box lever. In old and simple models of shapers, change in speed was affected by shifting flat driving belt from one step to another step pulley.

Secure Holding of Job

Generally, there are two ways of holding a job securely on table of shaper. One is to tighten the job directly on the table and the other is to hold the job in a vice which is mounted on the table. Table is provided with tee slots for clamping the job directly. In case job is to be clamped in a vice then vice is first clamped on to the table. During cutting stroke, a big force acts on the job to either slide it forward or turn it around. Both the cases are dangerous from safety point of view of operator, machines and job. Hence, utmost care should be taken to clamp the job.

Excessive feed and depth of cutting tool overloads the machine and even tool may break. How much feed and depth may be given to a particular job set up? There is no formula to work out exact answer. It needs a long practical experience to decide as to how much feed and depth may be safely given.

There are small versions of shaper which may be fixed on a table of size, say 60 cm × 40 cm. These are called bench shapers. Such

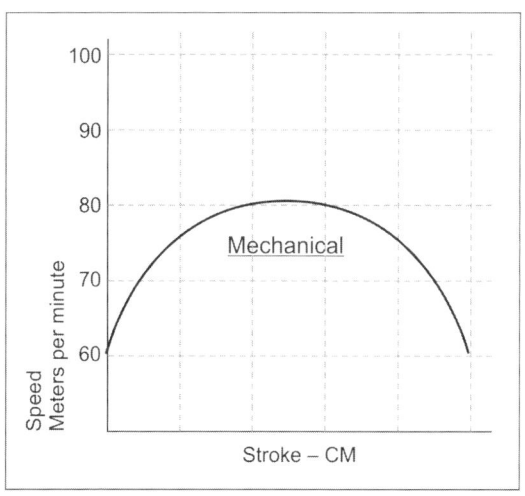

Fig. 3.44: RAM speed profile

type of shapers are used for machining small parts of sheet metal press tool or parts of other mechanism. These shapers are also used for training purposes. There may be hydraulically operated shapers. These are not commonly used. The advantage of hydraulic shaper is the constant speed of the cutting stroke. In case of mechanical shapers, cutting stroke speed rises from start to maximum in the middle and again zero at the end of stroke (see Fig. 3.44).

3.7 MILLING

Milling is a name given to a group of various operations which are performed on parts of tools, dies, machine parts and many others. Operations are planning of surface, slot cutting, key way, tee slots, holes, gear cutting, holes and other operations. Machine which performs these operations is called milling machine.

Broad classification of milling machines is vertical and horizontal. It signifies that the main spindle of machine on which cutting tools are fixed, is either vertically placed or horizontally fixed. The basic classification of milling machines is as under:

- **Horizontal Milling Machines**
  Horizontal arbor
  Vertical head attachment
  Manual x, y and z axes of tables
  Vertical slotting attachment
  Indexing head, table mounted, center stock
- **Vertical Milling Machines**
  Swivel head in vertical plane
  Swivel or no swivel table, x and y axes
  Manual x, y and z axes of table
  Slotting attachment
  Indexing head, table mounted, center stock

Both types of milling machines are also manufactured with power movement of table in x and y axes. Spindle speeds may be varied according to circumferential speeds required. Spindle speed may range from 70 to 300 rpm.

There may be many occasions when job requires both vertical as well as horizontal operations in one setting of job on machine table, so to cope with such needs, a vertical spindle head attachment is available for fixing on to the horizontal spindle supported on a boss protruding out of mechanical machine column. Axis of operations, such as hole, performed in vertical as well as horizontal position is exactly at right angle as the machine is so precisely designed and manufactured. A vertical machine normally does not have a horizontal head attachment.

The basic construction of various types of milling machine is briefly described with the help of figures.

Figure 3.45 shows a horizontal milling machine. Part (1) is cast iron body which is machined after due seasoning and Part (2) is horizontal arbor which normally rotates in clockwise direction when seeing towards arbor. Arbor has a standard diameter with a key throughout its length. Circular cutter (5) of various specifications are loaded on to the arbor. The cutter has standard key way which slides over key of arbor. There are spacers (7) to locate cutter at desired position on arbor and to tighten cutter with the help of nut (2). Part (8) is spacer cum arbor support. There is a sturdy rod (3) on which an arm (9) is mounted. Arm is so designed and dimensioned that bush bearing (11) precisely slides over spacer cum arbor support (8). Once arm is in place, it is tightened on rod (3) with the help of bolt (10).

Work table (12) of machine can be moved in three directions, viz. longitudinal, traverse and vertical. All the three movements are performed by hand wheel. Milling machine which is described

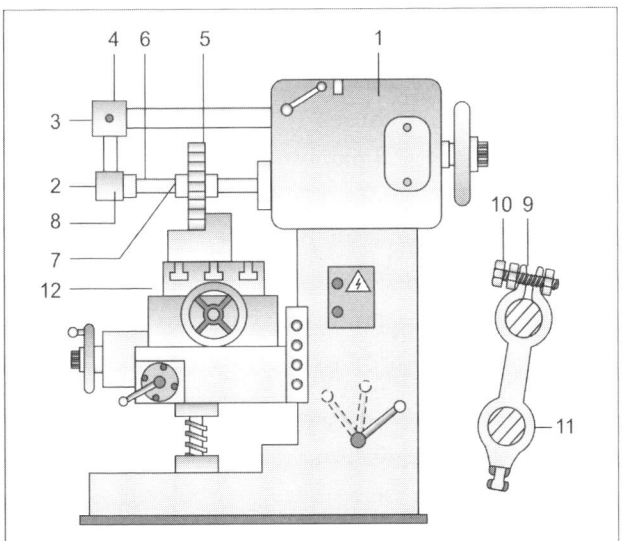

Fig. 3.45: Typical horizontal milling machine

above is a simplest one, yet useful. Three attachments are normally used on such type of machines. These are vice, center block and dividing head. Many tool parts are held between dividing head 'dog' plate at one end and by center block at the other end. This set up is generally used for gear cutting; it may also be used for machining equally spaced slots or lines on a circular job as shown in Fig. 3.46.

Dimension of job as shown in Fig. 3.46 are finished dimensions and are given below:

$$L = 220 \text{ mm}$$
$$I = 185 \text{ mm}$$
$$\varnothing 1 = 42.50 \text{ mm}$$
$$\varnothing 2 = 42.80 \text{ mm}$$
$$\theta = 30°$$
$$d_1 = 0.8 \text{ mm}$$
$$d_2 = 0.6 \text{ mm}$$

Part shown above is a die rod to produce lines on a sheet metal thin tube (say, 0.35 mm thick sheet) of brass. Die rod is to be made of steel having wear resistance and hardness. Blank size would be as shown in Fig. 3.47.

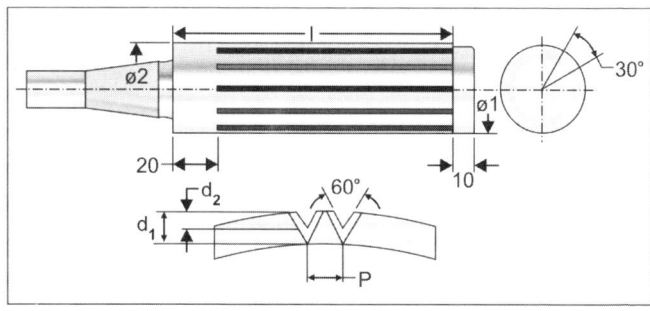

Fig. 3.46: Typical job of line milling

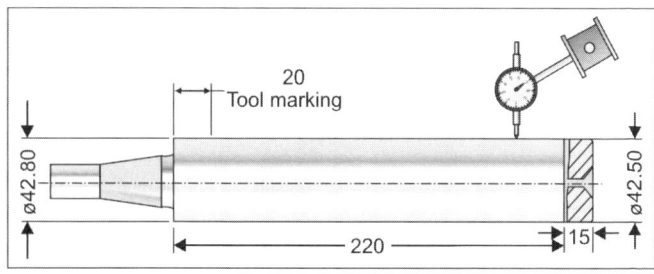

Fig. 3.47: Die rod blank

Length L of job has a taper of 0.15 mm on diameter at one side of axis. This blank rod has to be set on milling machine so that surface of rod on the top is perfectly parallel to longitudinal movement of table. This can be checked by means of dial test indicator and moving the tables longitudinally. There should not be any movement in the needle of dial test indicator. This setting may be achieved by adjusting the height of center block. Generally, there is a provision of height setting.

Please note that if a vee groove of 60° is cut by a 60° cutter then the depth of vee groove would be the same throughout the length whereas it is desired that depth of groove at 42.50 Ø end should be 0.8 mm and 0.6 mm at the other end. This can be achieved by raising center block by 0.2 mm. Hence on cutting, depth of groove would be 0.8 mm at 42.50 Ø end and 0.6 mm at 42.80 Ø end. Annular distance between two pairs of line is 30°, therefore, index head is to be set for turning (indexing) the job by 30° by rotation of indexing crank according to setting.

CNC Milling Machines

Modern tool room engaged in making sheet metal press tools and plastic moulds are equipped with Computerized Numerical Control (CNC) machines. These machines are highly accurate and having least movement of 0.005 mm. It is controlled by PLC (programmable logic control). Once various machining parameters are entered, machine may be operated on automatic mode.

The type of work performed by CNC machines necessarily is explained with the help of illustrations.

Indexing Head

Indexing head is an attachment used on milling machine for the purpose of rotating a job for a number of equal annular divisions. The basic construction of an indexing head is explained with the help of Fig. 3.48.

Part (1) is the body of indexing head, (2) is the center on which job having a counter sink is supported and (3) is a plate fixed to the center spindle. It has a slot to accommodate 'dog' protrusion so that job indexes with center spindle. Fixing of 'dog' protrusion in the slot is such that there is no relative play between 'dog' protrusion and plate slot. Part (4) is an indexing plate having a number of circles on which there are different number of holes equally placed. Typically there may be plate having three circles of holes. Outer circle is having 30 holes, middle circle 27 holes and the innermost circle has 13 holes. There may be a number of plates

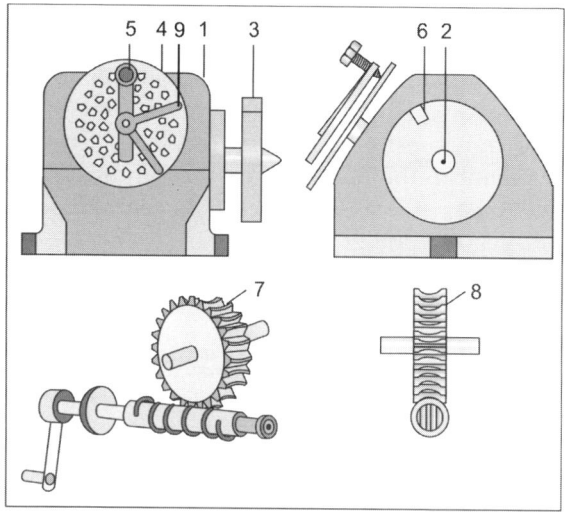

Fig. 3.48: Indexing head

having different number of circles and holes. These indexing plates are interchangeable. Part (5) is a selector lock arm called crank handle with a spring loaded pin holder. Any desired hole may be engaged by pulling the crank handle up, rotating the crank to reach the desired hole on a selected circle of holes. Pin handle is then released which enters into selected hole. Part (9) is a pair of stopper arms. Both the arms can be moved away or closer to each other. The purpose of these arms is to show range of holes 15, 20, 25, 30, etc. for repeated indexing till required number of indexing of job is accomplished.

Figure 3.49 shows combination of two plates of a compound 'follow on' sheet metal press tool. The final component to be produced and its sequence of operation is shown in Fig. 3.50.

Die sinking operations have to be very accurate for bore size and pitches of holes. An accuracy of + 0.01, – 0.00 is necessary in bores. Pitches of holes lengthwise and widthwise have to be with an accuracy of + 0.015, – 0.015. In Fig. 3.49, it is clear that there are two plates, one over the other with a gap. Both the plates are dowel pinned so that there is no chance of relative shifting. Furthermore, during assembly of tools, the alignment of both plates remain undisturbed. In upper plate, hardened and ground die inserts are fitted during assembly. Lower plate carries punch inserts of suitable design. It is worth noting that last station would be trimming. In this particular case, trimming punch would be on top and die insert in lower plate.

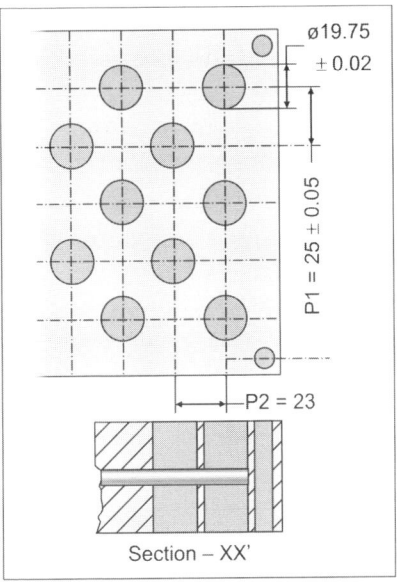

Fig. 3.49: Plates combination for die sinking operation

In case such a job is handled by manual setting of coordinates then there is a possibility of human error in setting, rendering the plates useless. It will also take quite a long time mostly due to manual setting. Due to all above reasons, CNC milling machines are used. There are two possibilities of carrying out milling operation which are as below:

- Coordinates of hole axis is entered into PLC console.
- Preparing drawing on AutoCAD in such a way that computer aided machining is possible.

In modern tool rooms, most of the tool parts are machined through the use of CAD and CAM.

Another example of milling operation:

Figure 3.50 (b) shows a typically shaped trimming punch. It is to be milled from steel blank shown in Fig. 3.50 (a).

Referring to Fig. 3.51, first of all, the job has to be loaded on an indexing head and center block for support. Ensure that diametric surface on top is exactly parallel to machine table. Indexing head is so adjusted that the job can be rotated by 120°. Manual sequence of operation would be as follows:

- Bring the cutting edge of cutter tooth just in the level of blank diameter on top.
- Shift the table manually so that cutter is a little away from job.

- Set the depth of first cut which might be, say 0.3 mm.
- Set the longitudinal travel of table in the direction of arrow so that length of cut is x mm.
- Move the table back so that job is clear of cutter.
- Rotate job with a small angle, say 3°.
- Take next cut.
- Process has to be repeated till full arc 'c' is machined.
- Move table so that job is out of touch from cutter.
- Index the job so much that 4.5 mm of circumference rotates.

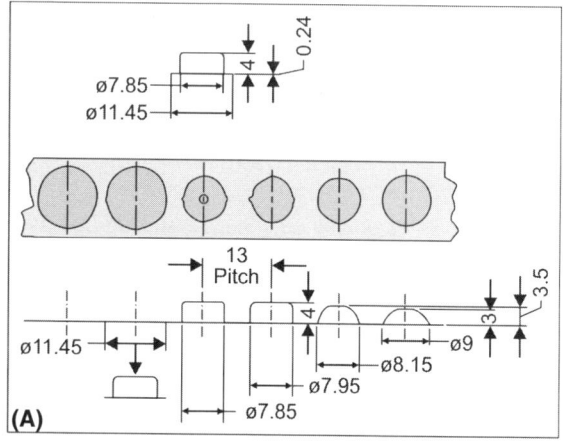

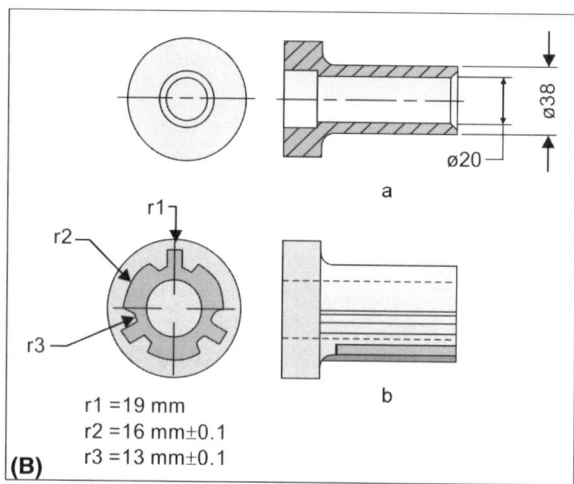

Fig. 3.50 (A) and (B): Typical indexing jobs

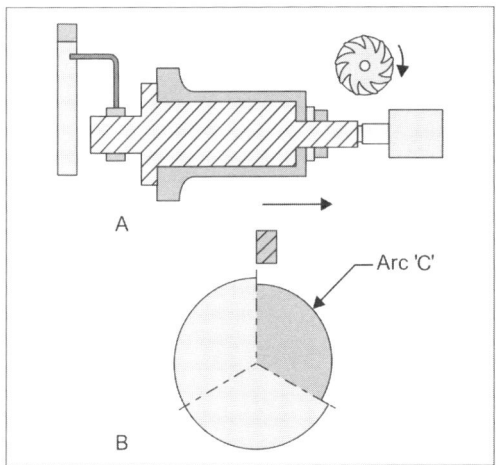

Fig. 3.51: Loading of job blank

- Resume machining for this sector also.
- Repeat process for remaining sector.

It may be well appreciated that manual operation will take quite some time and there is a likelihood of error in doing various settings.

This job may be machined on a CNC milling machine quite fast and accurately. First of all, a drawing is prepared on a CAD with the aim of computer aided machining. Data storage device may be taken out from computer and loaded on CNC milling. Various commands in the form of electronic bits are translated into physical actions of table, indexing head and machine spindle. It is accomplished by servomechanism which precisely rotate x, y and z actuating screw spindles to which hand wheels are also attached for manual operation if need be.

Universal Milling Machine

A universal milling machine may have a horizontal spindle with a vertical head attachment. This head may be swiveled clockwise or anti-clockwise up to a certain angle. It may be up to say, 30°. It depends on design of machine. Table of a universal milling machine may also be tilted along longitudinal as well as traverse axis. Figure 3.52 shows typical design, machining of which may need a universal milling machine.

Referring to Fig. 3.52, four holes are to be drilled on a cylindrical part. Starting point of hole would be located at a point on vertical line passing through center of hole. Hole to be drilled is tilted in

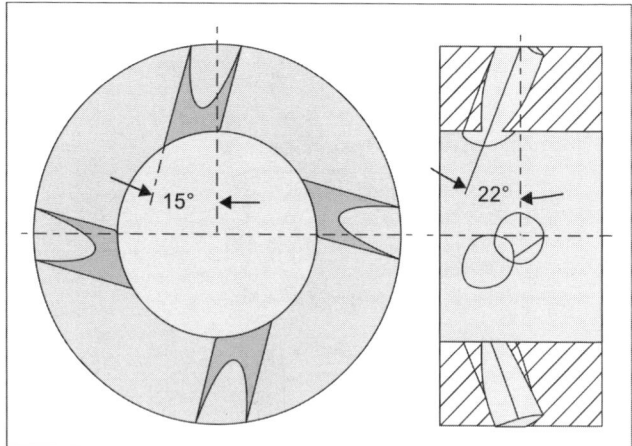

Fig. 3.52: Typical job for universal milling machine

two directions, one at an angle of 15° from vertical line and 22° from vertical line toward face of part. Since there are four holes, job had to be equally indexed for performing drilling of each hole. It is worth noting that due to tilted holes, a torque would be generated on the job which should not be sustained by dividing head. There should be a suitable arrangement for securely holding the job on turret after the job is indexed.

A vertical swiveling head may be swiveled 22° and angle 15° may be set by swiveling the universal type swiveling table.

Milling Cutters

There are three main classification of milling cutters according to holding method (Fig. 3.53). These are as follows:

• Cutters which have holes suitable for precision entry over machines arbor. Examples of such cutters are slitting cutters, side and face cutters, involute cutters, etc. These cutters generally have keyways to suit machines arbor key. In fact manufacturers of arbor maintain standardized diameter and key. Similarly, cutter manufacturers maintain standardized bore and keyway.

• A cutter having straight shank which is held in a chuck or collets. Diameters and length of shank are standardized.

• Big size planning or facing cutter has a standard taper shank suitable for internal taper in machine spindle. In addition to taper shank, there is a locking extension which fits into slots provided on the face of machine journal spindle. This ensures

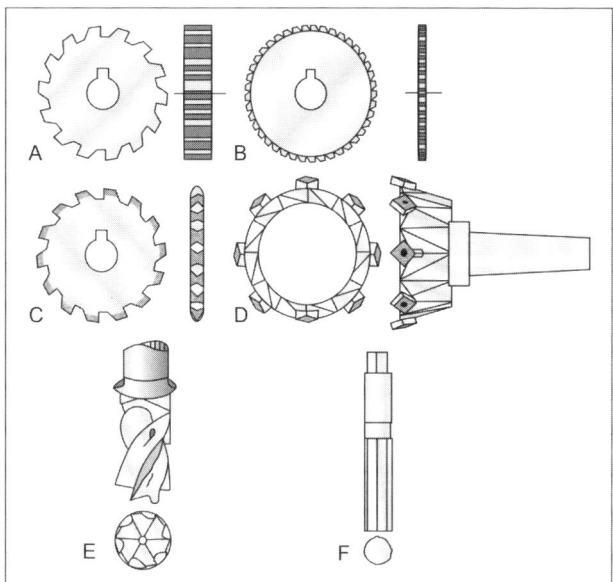

Fig. 3.53: Various milling cutters

that all torque loads are sustained by extension slot combination rather than taper matching of shank taper and machines journal taper.

Big size planning, facing, radius cutting, milling cutters are generally fitted with tungsten carbide inserts. Due to high hardness, around 64 HRC and wear resistance, higher cutting speeds may be adopted as compared to steel cutters.

Milling cutter is generally produced from cobalt steel which has higher hardness as compared to high speed steels (HSS). Milling cutter cutting teeth are generally coated with high wear resistant material such as titanium.

Figure 3.54 shows slitting operation in a job which has a step. Slit is to be provided on lower face in such a way that face of slitting cutter is about 0.02 mm away from face of the wall joining two faces of the job/workpiece. It can be seen in the figure that cutter is rotating in anti-clockwise direction when seeing towards machine column. Table with job is moving from right to left. In this way, the shape of material removal would be as shown in figure.

Figure 3.55 shows a workpiece where a wide slot is to be machined throughout its length. A side and face cutter can be used for this purpose. By shifting the position of cutter, width of slot may be cut to obtain desired dimension of width and depth.

Examples of milling operations using various cutters:

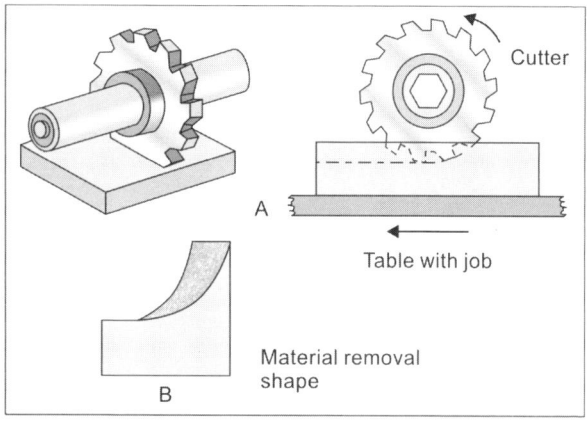

Fig. 3.54: Example of milling operation

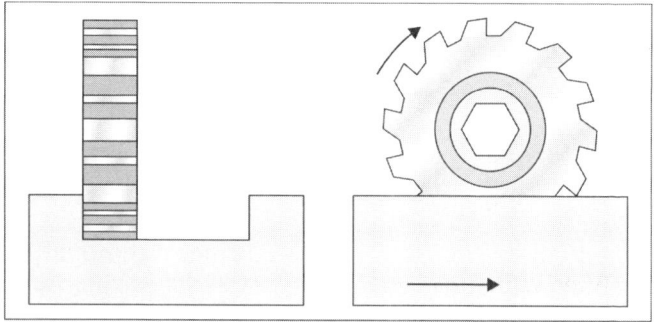

Fig. 3.55: Wide slot machining

Figure 3.56 shows milling of a form profile. Milling cutters of some standard profile, such as segment of a circle may be available in the market. Alternatively, milling cutter manufacturer may be approached for getting cutter/s made of desired profile designs.

Figure 3.57 shows machining of dovetail in a part/workpiece, already having a wide slot as shown in Fig. 3.55. In case dovetail machining is required on both walls of slot, the cutter would have to be shifted on the other side. Before taking a cut, it has to be decided that cut should start from which end of the workpiece. Rotation of cutter cannot be changed. If cut is taken from the same side then cutter would have a tendency to pull the job. Therefore, back lash in screw and nut of table may create a problem, so the alternative is to take cut from the other end of workpiece. Now the table would move from left to right.

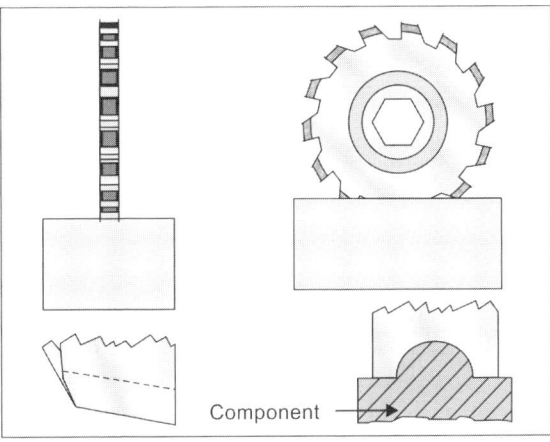

Fig. 3.56: Form profile milling

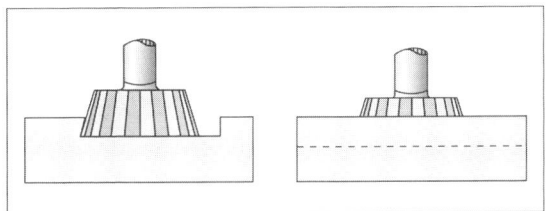

Fig. 3.57: Dovetail machining

Figure 3.58 shows a tee slot being cut by a woodruff milling cutter. While cutting one leg of tee on left hand wall of slot, table is moving from right hand to left hand, when cut would be taken on the other wall then table would have to move from left hand to right hand.

In Fig. 3.59, a pair of 60° channel is being machined. This cutting is performed by a standard 60° cutter which has sharp cutting point. After machining of one channel of desired depth, table is shifted

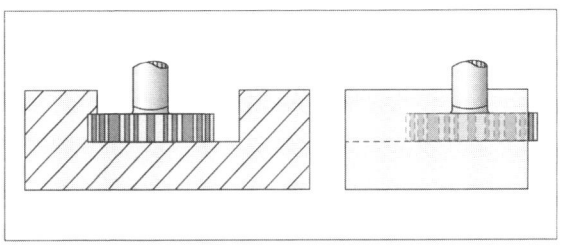

Fig. 3.58: T-slot machining

to required value to create a desired pitch between two channels of the pair.

In Fig. 3.60, 'A' is a workpiece on which a recess is to be machined up to a certain length L and depth D and corner radius 'R' as shown in 'B'. It may be done by using an end mill cutter of suitable diameter.

Reamers are like end mill cutters with the difference that these are made to produce a definite accurate size of a hole. Suppose a hole of 25 mm ± 0.01 is to be produced in a number of pieces then first of all, hole is to be drilled and machined by an adjustable boring head. A negative margin of 0.02–0.05 mm is left for finishing by a

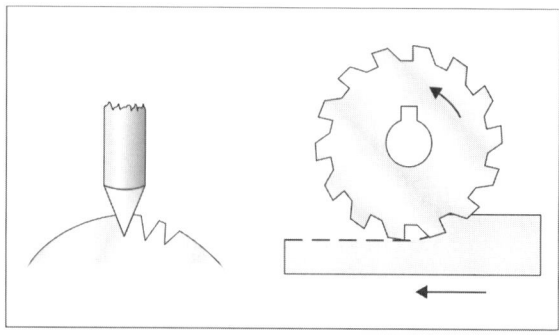

Fig. 3.59: Channel machining

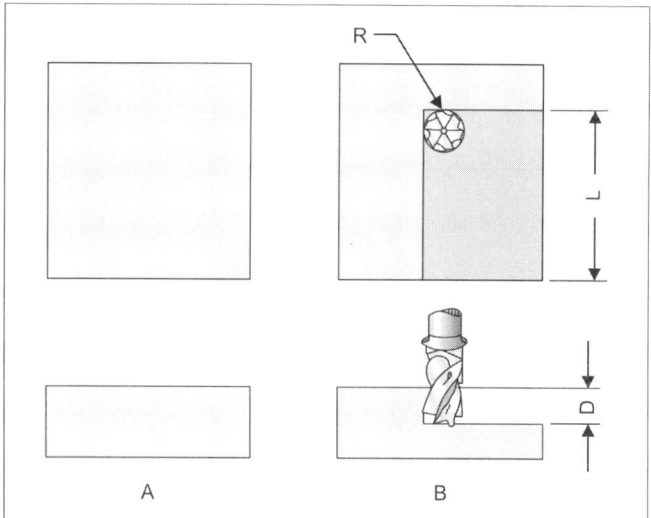

Fig. 3.60: Example of recess machining

reamer. The advantage of using a reamer is that holes generated would have almost the same dimension, hence, interchangeability of parts is possible. By merely boring, interchangeability may not be generated.

3.8 HOBBING

Hobbing is a process by which gears of various varieties such as spur, helical, involutes, etc. are machined, generally on commercial level. Since this machining process is not directly involved in designing and making of sheet metal press tools, therefore, hobbing is not dealt herewith. Use of gears may be there in the design of a sheet metal press tool or plastics mould but then a gear has to be a bought out item.

3.9 3-D PANTOGRAPH MACHINING

There are certain sheet metal components on which some alphabets, a geometric or artistic design is embossed. Sometimes these are big in size (say within a circle of 6 cm), others may be as small as within a circle of 1.5 cm and even smaller.

To get these embossing, matching die and punch inserts are to be machined. Suppose desired embossed impression is of an eagle as shown in Fig. 3.61.

Raised impression on sheet metal has different depths at different points of eagle pattern and hence dies and punch have also to be like this. Before a pantograph came into existence, say about a century back, such type of engraving on die inserts used to be done by expert engravers. No matter how expert an engraver was, there had to be defects. Exact repetition of a design was not possible.

Fig. 3.61: Embossing impression

Now what requires is to make a model/pattern of 2 to 3 times larger than the desired size on die insert. For example, Eagle shape pattern (Fig. 3.61) is prepared on a brass or copper sheet in 3-dimensions by a sheet artist. A sheet is placed over a plane surface of a wax like material and artist creates the image by skillfully and delicately hammering the sheet by small chisel like tools and small hammer. Once prepared, it is used as pattern on a 3-D pantograph. (A 2-D pantograph cannot be used to produce variable depths of artistic design).

The most suitable machine for machining an impression die is a 3-D pantograph. This machine is briefly explained with the help of Fig. 3.62.

Referring to Fig. 3.62, machine has four main arms (1), (2), (3) and (4). These arms are generally of I section. The main arm is mounted on a joint (11) on which it can swing and tilt up and down, making an angle with horizontal. Part (5) is a high rpm motor mounted on arm (1). A round belt (6) connects cutter pulley with motor spindle and the other on fulcrum (11). Pulley on fulcrum (11) has two steps—one for motor belt and the other to join cutter pulley (7) by means of round belt. Part (8) is a grip to move the stylus (12) all around. Position of stylus can be changed on arm (3) as it is mounted on a sliding block. This arrangement is made to set the ratio of size of pattern and engraving on die insert. Ratio

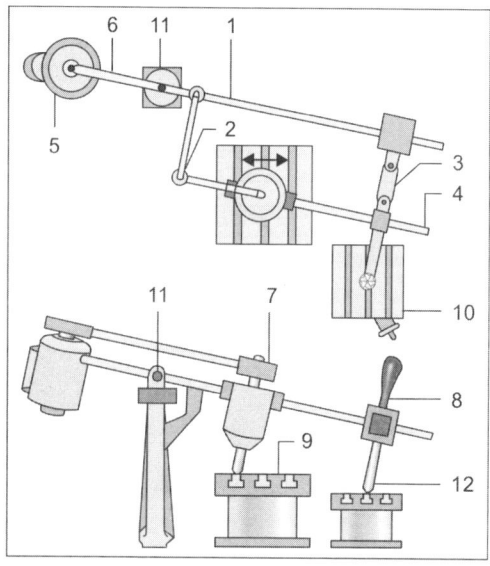

Fig. 3.62: 3-D pantograph

may be of any proportion under the machine specification which is generally 1: 1 to 1 : 3. For example, if the length of eagle from tail to beak is required to be 20 mm then the same dimension on pattern has to be approximately 60 mm.

For carrying out the machining, first of all workpiece is fixed on table (9). Now the pattern is fixed on table (10). Both the tables are kept down. Stylus (12) is then fitted to chuck. Stylus is a rod, its diameter being three times the diameter of cutter. End of stylus may be round or having pencil like shape. It depends on the contour of design to be engraved. For example, details of wings of an eagle may be achieved with a stylus having a sharp radius pencil like shape. Now table (10) of pattern is raised so much that stylus touches the surface of model and arm (1) is almost horizontal. Care is taken that table is raised only so much that arm (1) is approximately horizontal. Now try to move the stylus over the model/pattern surface. If there is a lack of smoothness in the movement of stylus then a fine layer of petroleum jelly may be smeared on the surface of pattern. Now fix a suitably shaped and sharpened cutter in cutter holder spindle. Note that workpiece is still much below the cutter end. Run the motor to see if cutter is rotating perfectly. Fix a trial piece of flat plate on work table (9). Raise the table to give a trial cut. Once satisfied, re-fix the workpiece and raise the table to bring the surface near the cutter end, leaving a little gap so that it does not touch the surface of steel workpiece. Now the machine is ready to carry out complete engraving. When first cut impression is completed, gradually raise the table of workpiece, say 0.05 mm each time till engraving is completed, thus eagle shape cavity/core in die block or punch is made. For machining die cavity, depth side of pattern is towards stylus. If punch is to be machined then raised side of pattern is towards stylus. Suitable clearance between die and punch is maintained by suitably changing the diameter (cutting point) of cutter or stylus diameter.

Latest design of pantograph may have high speed electric motor directly rotating the cutter. No belt is needed in this type of machine.

Quality of finish of machined surface depends on robustness of machine, cutter sharpness, cutting angles and skill of operator.

3.10 GRINDING

Grinding is also a material removal operation. Unlike single point cutting operation in lathe or milling, there are multiple fine cutting points which carry out cutting process with very fine cuts in terms of feed and speeds. Normally depths of cut do not exceed 0.01 to 0.02 mm. Grinding is performed by means of a grinding wheel.

Figure 3.63 shows a grinding wheel made of hard material which is a minute cluster of grains. Grits are closely bounded by means of suitable binding material. If peripheral surface 'S' is seen with sufficient magnification then the structure would look like the one as shown in Fig. 3.64.

There are three views, A, B and C. In all the three, Part (1) denotes the hard grits which act as cutting points. Part (2) is the binding

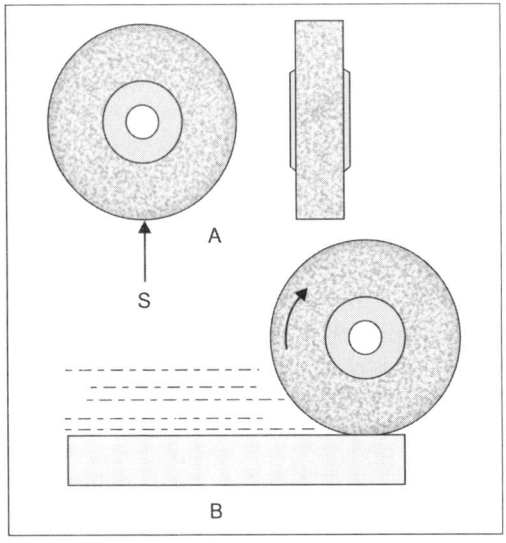

Fig. 3.63: Grinding wheel

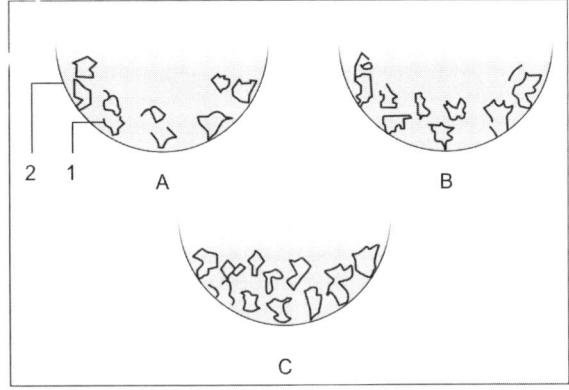

Fig. 3.64: Grains/grits in a grinding wheel

material which strongly holds the grits in place. In Fig. 3.64 A, hard grits have more space between them as compared to Fig. 3.64 B. In Fig. 3.64 C grits are more closely placed.

Wheels with coarse grits or grains spacing are used to grind surface of unhardened steel plates or parts having Brinell hardness number (BHN) around 110 or less. The advantage of using coarsely shaped grits wheel is that it does not get clogged rapidly as compared to 'B' and 'C' wheels with fine grain spacing; provides rough surface both on non-hardened and hardened steel surface. Hardened (50–60 HRC) steel surface is grounded by a grinding wheel having closely placed grits of fine grain, hence, a fine surface finish may be obtained.

Grinding Wheel Specification

Generally, grinding wheels are specified under the following headings:

- Material
- Grain size
- Bond type
- Spacing
- Wheel grade

There are a number of materials used in the manufacture of grinding wheels. Mainly there are two materials from which grits are made from particles/grains. The grains may be aluminum oxide, silicon carbide, tungsten carbide, diamond and cubic boron nitride (B). Wheels made of aluminum oxide grains are suitable for unhardened steels. Silicon carbide grinding wheels are generally used for grinding cast iron and non-ferrous material. Diamond impregnated wheels are mostly used for grinding tool components made of tungsten carbide and ceramics.

Grain size of any of the above materials differs according to end use. This means how fine a finished surface is required and what grinding wheels peripheral speeds are required. Grain size varies from 8 (coarsest) to 1200 (finest) and ultra fine grain for precision finish.

Properties of bonding materials are such that they strongly hold the bulk of grits in the shape of wheel. High revolution (approx 6600 rpm) of grinding wheels generates tremendous centrifugal forces which may explode the wheel. That is why, bonding is made very strong. When the space between the sharp points of wheel is clogged with steel particles then it becomes necessary to remove dull grinding surface of wheel by dressing it with a diamond dresser. Property of bonding material is such that during wheel dressing, bonding material is also removed to allow fresh cutting points to appear when top dull surface is dressed out.

Generally, the following materials are used as bonding materials by manufacturer of grinding wheels:

• Resinoid   • Rubber   • Shellac

WHEEL GRADES

Figure 3.65 shows a typical grinding wheel on the plain surfaces of which are pasted circular 'name plate'. Specification of grinding wheel is written in the following manner in Fig. 3.66 (photo is pasted below).

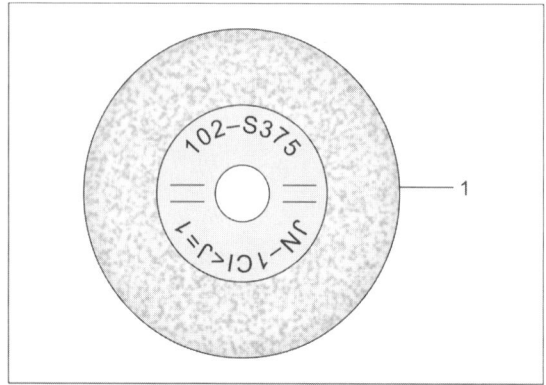

Fig. 3.65: Grinding wheel with pasted 'name plate'

Fig. 3.66: View showing 'name plate'

(*Source*: *www.wikipedia.com* as on September 4, 2013)

Safety Precautions

Safety precaution is associated with selection of grinding wheel and its use. Select grinding wheel produced by a reputed manufacturer who is expected to maintain high quality of wheel, especially bonding material and bonding process so that it is highly unlikely that wheel would explode while it is rotating free or grinding is being done.

Sometimes due to miss handling, grinding wheels develop hair cracks. It may not be visible casually. General practices on shop floor are to tap the wheel lightly. There is a difference in tapping sound. It needs some practices to recognize by sound if wheel has a hair crack or no crack.

Grinding wheel should rotate 'True' when loaded on machine spindle. Even slight eccentricity may cause bad surface finish and even dislodging lightly held jobs on vice or magnetic chuck.

It should be ensured that safety guards are in place. It includes guards around wheel and guard plate fitted at the end of work table of the machine.

This prevents injury to a person if happens to be there and job gets dislodged and thrown out by grinding wheel. While working on grinding machines, safety glasses should be used to prevent any particles entering into the eye. In case of wet grinding, direction of fluid nozzle should be kept in such a way that the liquid fall over grinding point and does not splash over the operator.

Before starting grinding machine, check if lowering of vertical column is not jerky or sticky. Sometimes, it is found that due to improper maintenance of grinding machine, grinding wheel (Spindle is attached to vertical column) shifts down suddenly on the job surface resulting in either damage to job or grinding wheel.

3.11 ELECTRIC DISCHARGE MACHINING (EDM)

Electric discharge machining (EDM) is a process where removal of metal (steel) does not take place like hack sawing, filling, turning, milling, grinding, etc. In EDM process, micro particles of metal are removed/dislodged.

The basic principle of EDM process may be explained with an example of spark plug of a petrol engine of a motorcycle or a car. It is a general observation that a new spark plug does not have any pit on its electrode. After long use, a pit is developed as shown in Fig. 3.67.

Why does pitting take place? Answer to this question is that minute spark of electric current has a very high temperature

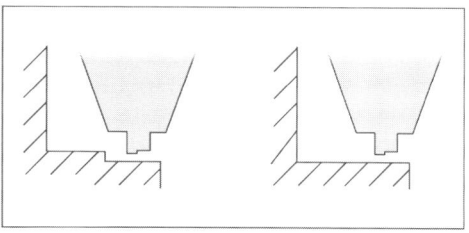

Fig. 3.67: Electroed of spark plug

enough to melt micro particles from the body of steel. Suppose if a high intensity electric sparking continuously take place then in a short time, steel particles removal would be quite fast.

There are many factors such as strength of spark, dielectric medium through which spark has to 'jump', spark gap, electric current and potential difference (voltage), arrangement for flushing out dislodged steel particles, arrangement that spark takes place at the least gap points, set least gap to be automatically maintained. Sparks should be in the form of 'storm', auto stop in case there is a short circuit current between electrode and steel workpiece, etc.

All the above mentioned requisites and even more are provided in a spark erosion machine.

The basic construction of a typical spark erosion machine is explained with the help of Fig. 3.68.

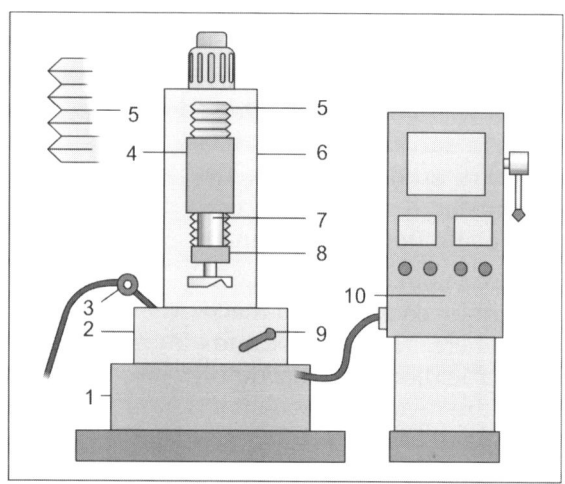

Fig. 3.68: Spark erosion machine

Part (1) is the box type base of machine. Hydraulic pumps, electrical control gadgets, dielectric liquid tank, etc. are housed in the hollow base. Part (2) is the dielectric work tank in which workpiece holding table with T slots is provided. There are four walls of the work tank. Out of these, front and right hand side panels (walls) can be wide open to facilitate loading and fixing of workpiece. These two panels are so designed that they can be seal locked with the help of lever (9). There are very accurate and effective rubber seals to completely make the tank leak free. Tank can now be filled by dielectric liquid by operating pump from control panel (10).

Part (6) is machine column on which electrode work head (4) is mounted. Work head carries electrode holding device (7). Up and down movement of electrode holding device is achieved by a stepper motor which is controlled automatically from control panel (10). Control panel contains a number of printed circuit boards (PCB) with mounted electronic components. There may be six to eight boards to control various operational actions such as spark gap, spark power, time, safety device control, dielectric fill control, flushing dielectric jet control, fine coarse spark erosion, etc. Part (5) are with flexible bellows to cover guide way of electrode head. Part (8) is electrode which actually does spark erosion.

Necessity of Electric Discharge Machining

Generally, there are two situations when electric discharge machining is resorted to. First situation is that many times, it becomes necessary to carry out some additional operation on a hardened steel part. A simple example is that a hole for a M-10 Ellen screw head was to be drilled. Due to oversight, counter sink was not machined and the part was hardened. Now the counter sink may be spark eroded by means of a circular electrode of appropriate diameter. It might be necessary to use two electrodes, one for roughing and the other for final finishing.

Second situation is that certain shapes and dimensions may be machined or built by conventional machining, but it takes a lot of time and effort whereas same shape and dimensions may be easily achieved by electric discharge machining. Two examples of such workpiece/part are given below:

Part shown in Fig. 3.69 cannot be made by conventional machining in one piece blank. It may be made by redesigning the part to be made by joining two pieces.

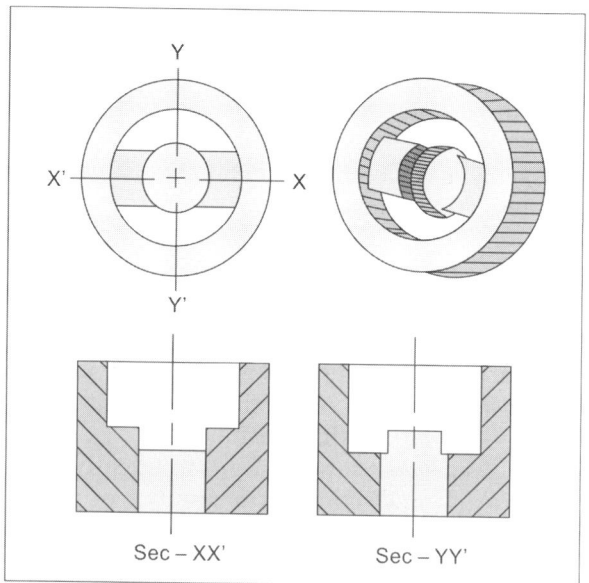

Fig. 3.69: Part for which EDM is suitable

It is much easier to machine by electric discharge machining. Figure 3.70 shows machining set up.

In Fig. 3.70, part (1) is the hardened steel workpiece in which ribs are to be machined. Part (2) is a high purity copper electrode which has a slot. Width of slot is equal to desired width of rib plus widening allowance due to sparking. This allowance is normally 0.02–0.10 mm, depending on how coarse or fine EDM is to be carried out.

Another example is of a plastics injection mould cavity which can be made either by hobbing or EDM (Fig. 3.71).

Hobbing is a process in which a hardened steel hob (punch) having desired profile, serrations and dimensions is forced by a hydraulic press in a block of unhardened steel, having its soft condition hardness, approx below 80 BHN. Cavity made by this process has very high internal stresses and are prone to cracking during hardening even if stresses are removed by heat treatment before hardening. It is a cumbersome process and rarely restored to. EDM is the best method. A copper electrode has to be made to desired shapes and dimensions. Cavity block may be completed with all fittings and machining operation. Even the cavity may be rough machined, leaving margin for spark erosion machining. Block can then be hardened and tempered to make it for EDM.

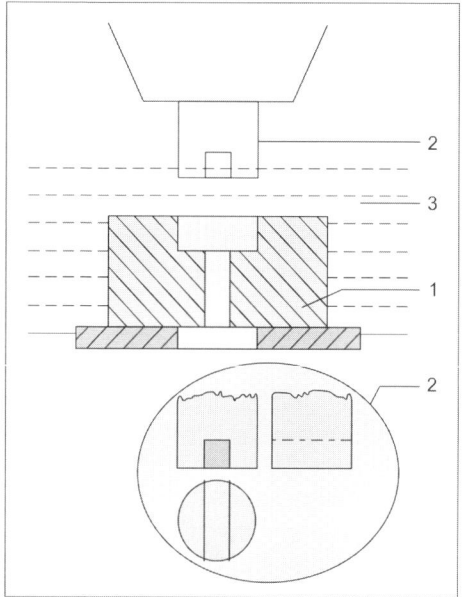

Fig. 3.70: EDM process shown

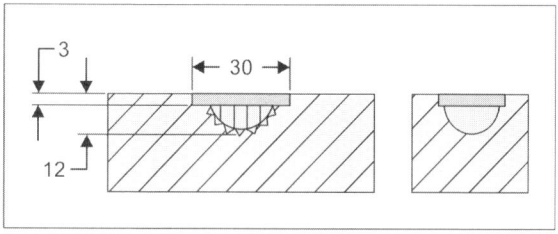

Fig. 3.71: Another example for EDM

Automatic Control of Process

Modern machines are equipped with programmable logic control (PLC). This enables operator to select code numbers of a set of various process parameters control. It may be a previously fed process parameter data with a particular code. It may be 'called' by pressing appropriate button on control panel. Alternatively, all desired process data may be entered into the system through PLC. Typically, data to be fed are as below:

- Electrode code numbers. First time it has to be written on display screen of PLC.
- Desired current intensity of spark.
- Minimum spark gap between electrode point and workpiece.

- Number of times electrode to go up and down in a minute for facilitating flushing of eroded particles.
- Operation of flushing jet (yes or no).
- Setting command or auto start command.
- Feeding value of depth in millimeters up to which electrode should go down after which spark erosion process would stop.

Electrodes

Electrodes can be made out of graphite block, copper and even steel. Performance of graphite electrode is better as compared to copper and steel. The reason is that graphite is light weight, can withstand higher temperature and wear is less. The disadvantage is that it creates black dusty environment when machined. Further, maintenance of dimension is difficult due to its softness. Mostly copper of high purity (96–98%) is used for making electrodes. It is because of the fact that copper is a good conductor of electricity and heat. Wear of electrode is moderate.

There may be instances where male or female parts of steel are available and the opposite is to be spark eroded. Under such circumstances, first of all a rough erosion is done by means of a copper electrode, leaving margin for fine finishing by steel part, to be used as electrode.

3.12 WIRE CUT

'Wire cut' EDM is a very useful machining process by which complicated blanking or piercing die profiles can be machined. This process is explained with the help of Fig. 3.72 given below.

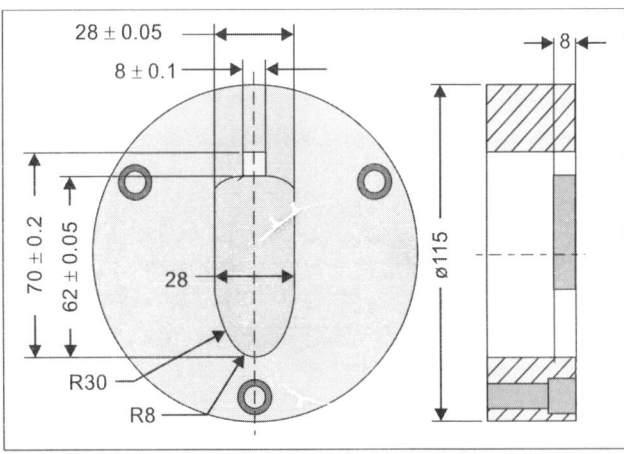

Fig. 3.72: Tool part for 'wire cut' EDM

A trimming die ring is shown in Fig. 3.72. In case it is to be machined in conventional way then maximum possible material has to be drilled and filed out. An electrode has to be made to give shapes and dimensions. Fitting operations may be carried out for fitting die ring into die set, then the die ring may be hardened and tempered. Now the desired shapes and sizes may be obtained by EDM. It is worth noting that making of an electrode, removal of excess material and filing is quite time consuming.

In case of 'wire cut' EDM, the following would be the sequence of operations:

- Prepare CAD drawing with coordinates of profile to suit computer aided machining (CAM).
- Do fitting work of die ring into die block.
- Drill a small hole into die ring as starting station, as shown in Fig. 3.74.
- During set up of machine, first of all die blank is set on the machine in such a way that job holding device does not touch wire, rather it remains away from wire.
- Machine wire is then threaded through small hole and then set in perfect vertical position and wire remains tight and vertically straight while it travels when machine is started.

Figure 3.73 shows schematic set up of workpiece for machining a trimming die profile. Part (1) is spool of brass wire of thin gauge.

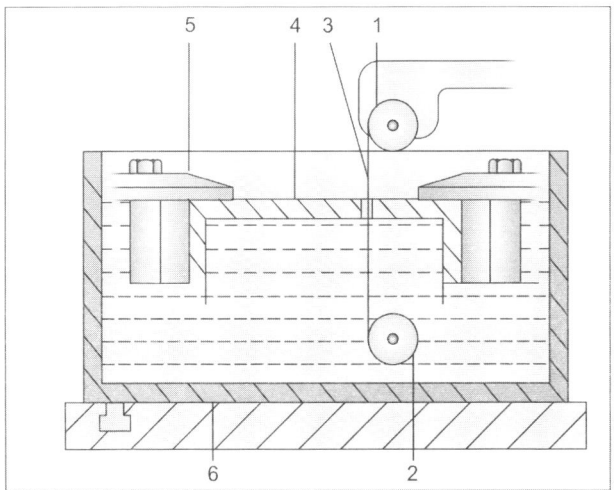

Fig. 3.73: Schematic set up for 'wire cut' EDM

Generally, it is 0.100–0.350 mm in diameter. Part (2) is the winding spool. When machine is started, spool (2) starts pulling the wire down. Both the spools are so devised that electrode wire (3) always remains in certain tension so that wire definitely remain straight and vertical. Generally, tensile strength of wire is 55 MP_a. Table (6) of machine moves precisely in xx' and yy' direction depending upon the coordinates fed to PLC by AutoCAD generated drawing suitable for computer aided machining (CAM).

Figure 3.74 shows a small hole in die ring through which wire is threaded. This hole is about 4 mm away from contour of die. As soon as machine is started, table moves in x' direction while wire is eroding its way till circumference of wire reaches to the contour line of die. From this point, table movement is in such a way that wire keeps on spark eroding the contour till it is about to reach exactly at starting point (1) of contour. As soon as cutting of contour is completed, piece (2) of steel inside the contour gets free and is likely to damage the wire. Therefore, to avoid this to happen, steel piece (2) is provided with a support before completion of cutting. It may be done automatically if such a mechanism is provided in machine otherwise machine has to be stopped for this purpose.

In case of wire cut EDM, dielectric is de-ionized water which is conductive for electricity. Like EDM in this case, tank is filled with de-ionized water to such a level that workpiece is under water.

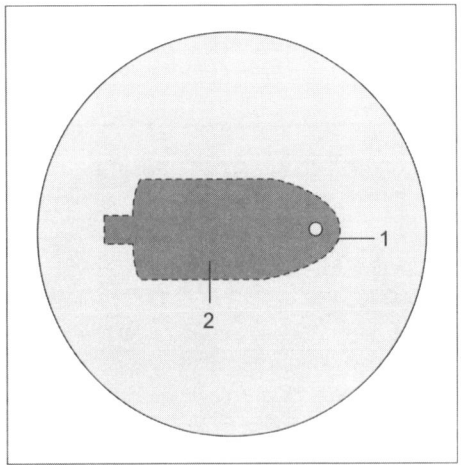

Fig. 3.74: Starting hole in a die blank for 'wire cut' EDM

3.13 LAPPING

All solid materials have surface being rough or smooth. The words 'rough' or 'smooth' are attributes to quality of surface. In technological and industrial environment, it is highly desirable to quantify surface finish. Before describing lapping and polishing, it is necessary to review surface finish and its measurement.

Figure 3.75 shows a typical magnified surface of a steel plate which is being shaped. Portion A is a certain type of roughness. B and C are the repetitions of A. There is a sudden defect between A and B, B and C and so on. It may be due to some machine defect or a vibration generated in the vicinity of shaping machine. It appears that vibration is in the form of jerk. This type of surface roughness is called waviness. If this surface roughness is measured by a mechanical measuring instrument, its stylus has a very fine rounded off point. It is moved across the 'lay' which is typical characteristic for a particular machining operation. It may be straight, radial, circular, etc. While a diamond point of stylus moves in straight line, it also moves up and down in the humps and valleys of surface roughness. Sensitive electronic sensor and circuit translate up, down and traverse movement into form of a graph. Values of humps and valleys are recorded in nanometer.

If the surface is lapped by an abrasive stone (lapping stone) then the humps above line AA', as shown in Fig. 3.76 would be lapped away and thus the surface would become smoother. If surface is further lapped up to BB' then it would become further smooth. Depending on lap material and lapping paste, further smoothness may not be possible due to scratches generated by particles of lapping paste or lapping stone. Lapping process has a great influence on the achievement of surface finish.

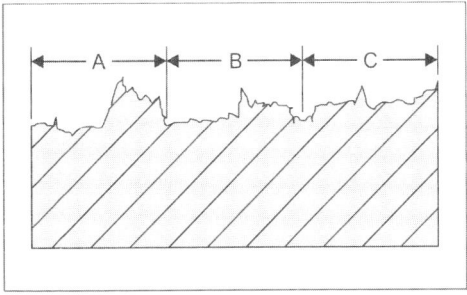

Fig. 3.75: Magnified surface of steel plate

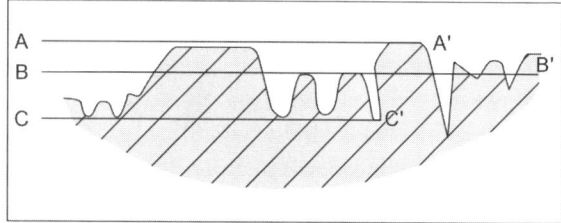

Fig. 3.76: Hump and vallieys in a surface

3.14 DE-BURRING AND POLISHING

De-burring is the term used for removing minute protrusion of metal due to a machining process. This protrusion is called burr.

Figure 3.77 shows a steel or brass ring which is faced on a lathe. Cutting tool has moved from circumference to center while the tool tip is crossing into already machined hole, a minute bit of material is pushed towards hole. This effect is shown by an enlarged view in Fig. 3.77. Depending on mechanical properties of metal, condition and setting of cutting tool, turning speed and depth of cut, dimension of burr may vary from 0.05 to 0.5 mm or even more. Presence of burr in any part of press tool interferes with assembly and relative motion between two parts. It is therefore necessary to remove burrs. This process is called de-burring.

Polishing is a process by which **surface** is made so much smooth that light reflected from polished surface gives a shine, which may be called '**mirror bright**' polish. Complete successful process of achieving mirror bright polish is described with the help of typical example actually handled by author.

There are two very important pre-requisites for achieving mirror bright polish. Firstly, steel should be free from microstructure fault.

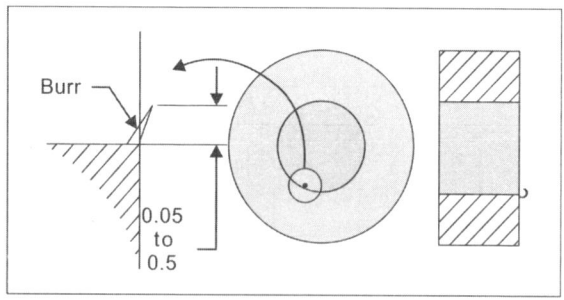

Fig. 3.77: Burr is shown in a steel job

The composition of steel should be favorable for achieving high quality polish. Typical composition of steel may be as under:

Mirrax ESR	C	Cr	Mo	V	Mn	Si	Ni
	0.24	1.3	0.35	0.35	0.5	0.3	1.4

(*Source*: Uddeholm AB Sweden, High performance steel)

It is highly desirable that this steel should have been produced by Electro Slag Refining (ESR) process. Such steels produced by Assabsripad are Stavax ESR and Mirrax ESR. Secondly, polishing equipment, paste and paper should be at hand. There should be good arrangement for flushing particles and nano-particles while polishing is going on. Polishing person/turner should have good amount of patience. After achieving certain smoothness, if excessive pressure or further amount of same paste is applied then scratches would reappear (would be generated).

Figure 3.78 shows the core of plastics (ABS) reflector mould. The parabolic surface is required to be **polished to mirror brightness** without distorting parabolic profile. To achieve this, the following may be the steps:

- Turn steps as shown, according to coordinates given in Table form.
- Remove humps of steps carefully by maneuvering cutting point of tool.
- When valleys appear like lines, stop removal of steel.
- Apply copper sulfate on the machined parabolic surface. Full surface would take colour of copper. This includes valleys.
- File parabolic surface with smooth flat file in such a way that valley lines remain visible (with copper colour) of equal width.
- Continue to delicately file the whole surface till all valleys are faintly visible.
- Fix strips of smooth emery paper on a wooden flat piece, such as a measuring scale.

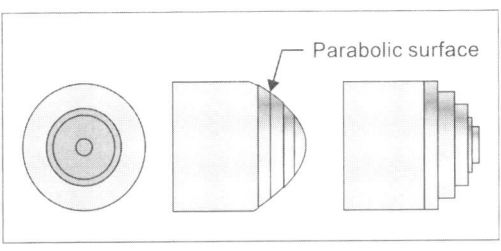

Fig. 3.78: Core of plastics reflector mould

- Fix another set of 'oo' polishing emery on wooden sticks.
- Polish surface with smooth emery delicately for about one hour.
- Now clean the surface with flannel cloth (soft cloth). Do not apply much pressure.
- Repeat the process with 'oo' ultra smooth emery paper sticks. Once again it has to be done delicately all over the parabolic surface.
- Care has to be taken from very beginning that valley lines still faintly visible at the end of polishing.
- From this stage, shine would start appearing.
- Stop lathe on which core is loaded for polishing.
- Apply green polishing compound (these polishing compounds are available in the form of cakes and bars for polishing brass articles).
- Run the job at high rpm and polish the surface with very clean soft cloth. Cloth may be fixed on a wooden stick for ease in handling and safety. Take care that hand or stick does not touch rotating chuck. This process might take half to one hour.
- Now polish is near mirror brightness.
- Finally polish the surface with calcium powder or Talcum face powder, using fresh piece of soft cloth with very light pressure of palm. Take safety care that hand does not slip to touch high speed rotating chuck. For the sake of safety, it is highly recommended that core (workpiece) may be loaded on a mandrel directly on machine journal, hence, no question of risk due to chuck as it is not there.
- Now the surface should be mirror bright. Valley lines are almost not visible. Prepared surface is now be safe guarded by keeping the core wrapped in a very clean soft cloth and kept in a plastics container till the core is about to be hardened.
- Harden and temper the core, taking precaution to avoid any damage to polished surface.
- Once the core is hardened and tempered, it has to be re-loaded on lathe mandrel.
- This time parabolic surface has to be polished to mirror brightness by extra fine diamond paste. Diamond pastes of various fineness are available duly filled in very small applicators which are like injection syringes.
- This time surface should have better mirror brightness with out trace of any scratch.

- Process is not yet completed. Steel surface is prone to rusting. Although polished surface gets rusted with difficulty but still there are chances of rusting.

- Get the reflective surface of core protected by protective chromium plating. This will almost protect surface completely from rusting for a long time.

Note: Polishing of steel and achieving mirror bright surface is an art. It needs a lot of practice. Process of generating parabolic surface may be easily and accurately accomplished on a CNC lathe.

4

Power Presses and Thread Rolling Machines

This chapter consists of the following aspects:

4.1 INTRODUCTION TO POWER PRESSES

Power press is a machine which provides dynamic force to perform many types of operations such as bending, drawing, stamping, etc. Power presses may be categorized as follows:

- All mechanical
- All hydraulics
- All pneumatics
- Combination

All mechanical power presses are those in which a heavy flywheel is rotated by an electric motor. Inertial energy of rotating flywheel is mechanically transferred to move a RAM up and down to provide force required to perform an operation and up and down motion by eccentric crank or cams.

All hydraulic power presses are those in which strong hydraulic pumps are run by electric motor. Pressure generated by hydraulic pump is transmitted to big diameter jack. A RAM attached to jack moves up and down by means of hydraulic circuitry and controls. Hydraulic presses are normally used where it is needed that material is kept pressed for a while or a large component such as metallic covers of an earth moving machine engines are drawn and stamped. Such components may be, say two meters long and one and a half meter wide. All pneumatic presses are used for light work such as riveting of assembly parts or some light bending and draw operations. Pneumatic presses are those in which compressed air is applied to a pneumatic jack for creating force in the RAM while moving down to perform an operation.

Mechanical power presses may be

- Geared inclinable
- Geared uninclinable
- Ungeared inclinable
- Ungeared uninclinable

Most of the presses are opened from the back. Body of power presses may be of cast steel, fabricated in steel and cast iron. Selection of material of construction depends on force to be applied, size of dies to be loaded. Tonnage of power presses may range from 20 to 200 tons and even more.

Construction of a Typical Power Press

Photograph of a Power Press is given below to give an idea of shape and construction (Fig. 4.1).

Details of construction of a typical power press is explained with the help of Fig. 4.2, where Part (1) is the cast iron sturdy frame. Since overall shape is like 'C', therefore, it is given the name of 'C' Frame. Sections and ribs of frame are so designed that it can sustain accidental overloads without failure of cast iron frame in which tensile stresses are generated somewhere in the middle of 'C'. Cast iron of special grade having much better tensile strength is used which is generally closed grain C.I. Press bodies after casting are left in open for natural seasoning. It might take even a year. After seasoning, necessary machining is carried out. Part (2) denotes legs of power press on which frame is mounted with the help of strong pins (10). Part (8) is a mechanical jack which is used to adjust the

Fig. 4.1: Photograph of a power press

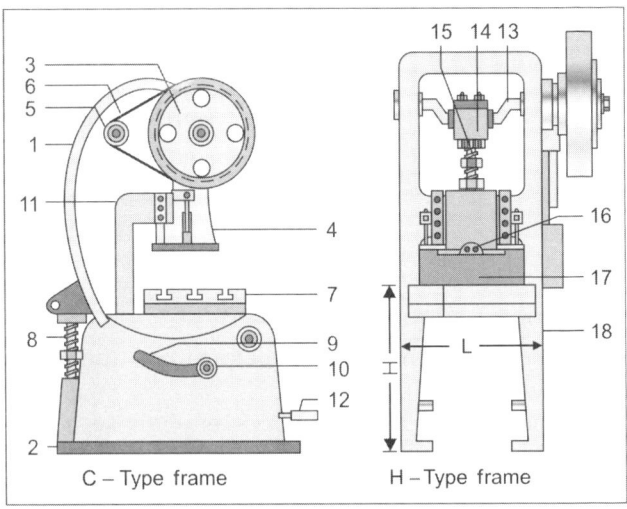

Fig. 4.2: Constructional details of power press

angle of tilt. Part (4) is a heavy and sturdy RAM of cast iron which moves up and down in guide way of frame. Part (3) is the **fly wheel** to provide inertial force for RAM to move down with a force. Part (5) is a driver pinion which drives fly wheel with the help of belts (6). Part (7) is a sturdy C.I. plate which is called bolsters plate on which tool is clamped with the help of T bolts. Part (13) is the crankshaft which rotates with flywheel when both are connected through a rolling key. It is generally operated by foot.

From operator's safety point of view, rolling key can be operated only when two (left and right) levers or push buttons are operated by press operator. This ensures that operator's hands are away from 'danger zone' that is between punch and die.

Part (14) is a connecting rod which joins crankshaft with RAM (4). Normally crank remains towards top, hence RAM is also at top dead center (TDC). As soon as stroke is given, RAM moves down towards bottom dead center (BDC), crosses it and moves back to TDC. Part (15) is a heavy duty Pitman screw which can be screwed in and out of connecting rod and locked at desired position. This arrangement facilitates setting of relative distance between punch and die. Normally punch is fixed in the RAM and die is fixed on the bolster plate with accurate alignment of punch and die. Typical example of accuracy of such an alignment between blanking punch and ring is 0.02 mm in case of a 0.42 mm thick brass sheet, blank diameter being 140 mm. Part (16) is a tool holding device. After placing the stem of tool, it is tightened by this device.

The purpose of making the body tiltable is to create ease of removal of component towards back of power press.

Latest trend does not require tiltable press because of innovative component removal systems such as air jet, pick and place mechanism and compound tools where ejection of component generally takes place towards left or right hand side of press.

Vertical column power presses are much sturdier as compared to 'C' type bodies. 'H' type bodies are generally used for high tonnage presses and where size of component is large.

Rolling Key verses Pneumatic Clutch

Joining of flywheel to crankshaft may be achieved either by rolling key or pneumatic clutch. Rolling key arrangement undergoes jerky and impact engagement. Rolling key and its housing gets damaged frequently. The function of rolling key is explained with the help of Fig. 4.3.

The basic principle of working of a rolling key in a power press is to mechanically and temporarily join crankshaft (1) (in View 1) and rotating flywheel (3) when stroking paddle is pressed by operator. Referring to Fig. 4.3, View 1 shows the position of rolling

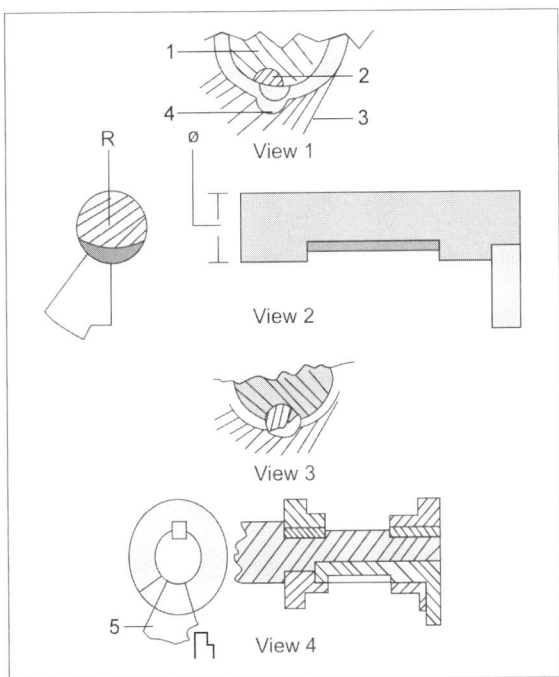

Fig. 4.3: Rolling key arrangement

key when stroking paddle is NOT pressed, hence, recess (4) 'fly pass' over the rolling key engaging section (2). View 2 shows the situation when stroke paddle is pressed by operator, freeing the rolling key to come to partially rotating mode. As soon as recess (4) approaches rolling key, rolling key gets space to partially rotate and hence it forms a 'wedge' between crankshaft extension and rotating flywheel. Consequently, crankshaft rotates with flywheel. If paddle is immediately released then after rotation for about 355°, rolling key arm (5) (in View 4) gets obstructed for further rotation. Consequently, rolling key rotates back to a non-engagement position, hence rotating flywheel gets disengaged while rotating. If the stroke paddle is kept pressed by operator, crankshaft will keep on rotating, thus making the RAM move up and down in guide ways continuously.

It may be noted that rolling key is constantly subjected to impacts and severe rubbing of rolling key recess in crankshaft extension and side collars. Due to all these happenings, life of rolling key is short, may be from few days to months. It all depends on what steel is used for making rolling key, what heat treatment is given. A rolling key is expected to have long life if its core is tough, surface hard and surface matching of rolling key on its housing in crankshaft extension rod is good.

It is worth noting that over loading of power press creates severe stresses on rolling key, which may reduce its life. The wrong setting of sheet metal press tool may lead to serious breakdown and even accident.

Figure 4.4 'B' shows position of punch and cutting ring well set. When stroke is given, punch would enter cutting ring. Figure A shows the shifted position of lower half of the tool (die block). In

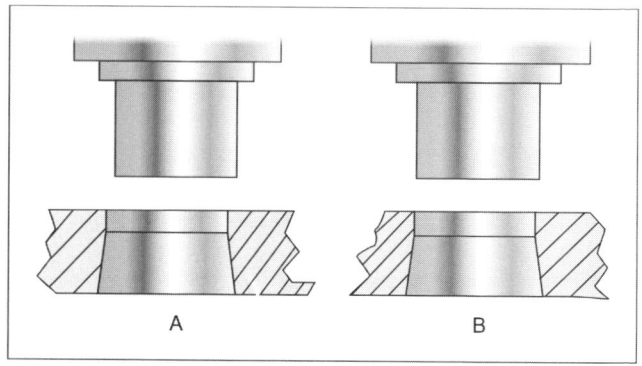

A   B

Fig. 4.4: Punch and cutting ring position

such a situation, it may be very well visualized that what would happen if a stroke is given. RAM cannot move down. Either the tool will get exploded or severe damage would take place to crankshaft or flywheel/rolling key system or even power press body.

With the development of pneumatic clutch cum brake system, use of rolling key is much reduced especially in high tonnage presses such as 100 tons. Pneumatic clutches are comparatively safe. In case of severe obstruction to the motion of RAM, slippage in clutch plate takes place, thus severe damage to die or system is eliminated.

Pneumatic clutches are so designed that brake is applied to the motion of crankshaft as soon as clutch is disengaged. Please note that brake is NOT applied to rotating flywheel. It is applied to a disc which is an integral part of crankshaft having nominal inertia. The design of pneumatic clutch cum brake is quite complicated and it is not necessary to provide engineering drawing and detailed description. However, the basic principle of its working is explained with the help of Fig. 4.5.

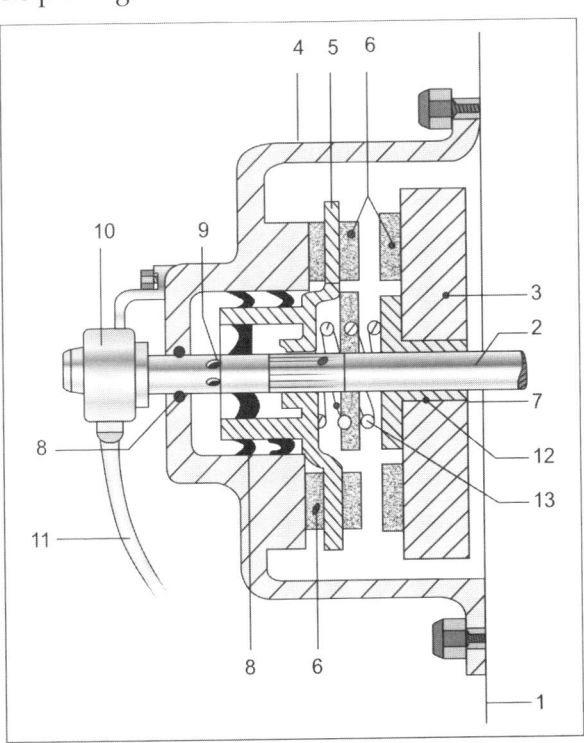

Fig. 4.5: Pneumatic clutch system

1. The body of press
2. Extension of crankshaft
3. Flywheel
4. Clutch housing
5. Friction plate
6. Friction pads
7. Collar of shaft
8. Pneumatic seals
9. Air holes
10. High pressure air injection unit
11. High pressure air hose

Flywheel freely rotates over pressure bush (12) which is fixed to crankshaft extension (2). Friction plate (5) can slide over the spleens on shaft (just for explanation) and it rotates with crankshaft extension (2). Hub of friction plate has pneumatic seal (8), which seals air gap between hub diameter, bore and clutch housing. Part (10) is an arrangement whereby it does not rotate while shaft rotates. At the same time, high pressure air can be injected through hole inside the shaft and create pressure in the space between friction plate and clutch housing (4). Due to generation of pneumatic pressure, axial thrust on friction plate (5) is created, thus, it axially moves towards flywheel against springs (13). Friction pads (6) come in contact with a force. Thus, due to friction, rotating flywheel makes the friction plate to rotate with it. Consequently, crankshaft is making the RAM movement up and down. Friction holding of flywheel and friction plates is so strong that torque created by rotating flywheel is transmitted to crankshaft.

When air pressure is released, instantly friction plate gets shifted towards clutch housing (4) due to presence of compression springs (13). This causes instant braking action to crankshaft.

The operation of pneumatic clutch cum brake is achieved by electro-pneumatic valve. Generally, there is a selector switch to set one of the following functions:

- No operation
- Inching mode
- Continuous mode

Please note that any position of selector switch starts the operation. For starting the operation, there is a button to inch the movement of RAM. It is set by operator during setting or observation. If selector switch is on 'continuous mode' then pressing of button once will start power press to run continuously. Normally, a big red color button is provided for 'EMERGENCY' stop of press. For normal stoppage of press, there is another red button near starting button.

Pneumatic clutching is quite reliable. Movement of RAM can be stopped at any stage of forward or return stroke. Further, braking of crankshaft automatically takes place when clutch is disengaged just by giving a push to 'RED' button. Electro pneumatic circuitry does the job of clutch engagement or disengagement.

Another advantage of pneumatic clutch is that an electro-pneumatic safety system can easily be incorporated whereby clutch can only be activated when operator presses two buttons, one for right hand and the other for left hand and then presses the paddle by foot. In this way, hands of operator remain away from danger zone.

While designing a tool, it is taken into account as to in what shape material is to be fed in the tool. It may be manual feeding of strip, blank, etc. It may also be automatic feeding of continuous strip, that means a roll. For roll feeding, special equipment such as roll stand, tension and loop controller, pitch feeding system for feeding specified length of roll strip into the die while tool is in open condition. This action is automatically synchronized with up and down movement of RAM. There is also a scrap 'take off' equipment after the tool. Scrap roll strip coming out of tool is taken over by a 'take off' equipment which keeps on taking scrap and cutting into pieces for easy removal.

Power Press Specifications

Power press specifications generally cover the following aspects:
- Tonnage
- Adjustability in stroke
- Length and width of bed
- Tool stem hole size in RAM
- Thickness of bolster plate
- Opening through bolster plate
- RAM adjustment (up and down)
- Opening in bed
- Opening in back of body/C frame
- Maximum distance between bed and RAM face ('day light')
- Flywheel diameter and thickness
- Number of strokes per minute
- Distance from floor to top of bed
- Motor horse power (HP)
- Motor RPM (1440 or 960)
- No. of V belts with section
- Weight of press
- Height, width and depth

Tonnage

In a mechanical press, heavy RAM moves down from TDC to BDC (bottom dead center) by means of a crankshaft or an eccentric in case of cross shaft construction.

Force attained by RAM is mainly due to two sources:

• Dynamic force due to weight of RAM moving down at a certain speed at different angles of connecting rod with vertical. This depends on the position of crank or eccentric in its circular path (See Fig. 4.6.)

It implies that dynamic force would be maximum when crank pin is at 90° with respect to top dead center (TDC) and minimum when RAM is at (BDC).

• Other source of attaining force by RAM is the transfer of force from heavy rotating flywheel through crankshaft pin. In this case, this force keeps on increasing when crank pin is moving from 90°–180° angle which is BDC.

Variation in generation of force in RAM due to flywheel torque is explained with the help of Fig. 4.6.

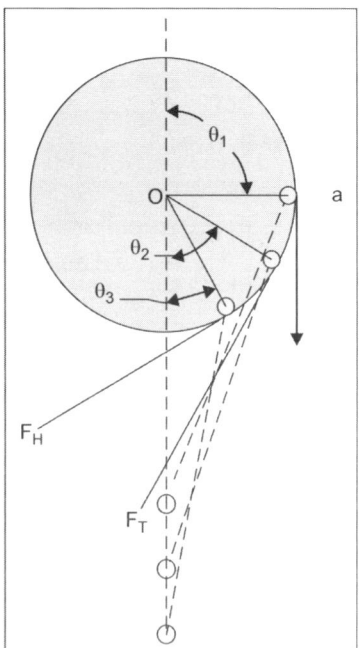

Fig. 4.6: Dynamic force generation

Let 'oa' be the radius of circular path on which PIN of crankshaft is rotating in clockwise direction. Application of force by crankshaft PIN will always be tangential to circular rotational path of crankshaft pin irrespective of the fact that at what angle radius line 'oa' is with respect to vertical.

Force F_T has two components—one in vertical direction and the other in horizontal direction. Position of radial line 'oa' is considered at angles θ_1, θ_2 and θ_3.

Let the horizontal component be F_H.

Therefore, values of F_H in terms of F_T would be

at
$$\theta_1 = 90°$$
$$F_H = F_T \cos \theta_1$$
$$= F_T \cos 90°$$
$$= F_T \times 0$$
$$= 0$$

at
$$\theta_2 = 45°$$
$$F_H = F_T \cos 45$$
$$= F_T \times 0.7071$$

at
$$\theta_3 = 30°$$
$$F_H = F_T \cos 30$$
$$= F_T 0.8668$$

at
$$\theta_4 = 0°$$
$$F_H = F_T \cos 0$$
$$= F_T \times 1$$
$$= F_T$$

Now let us consider Fig. 4.7 in which knuckle action is treated.

J_1 is the free joint between crank shaft and crankshaft PIN.

J_2 is the free joint between crankshaft PIN and J_3 which is free joint between connecting rod and RAM.

When F_N at joint J_2 moves towards left, line $J_1 J_2$ and $J_2 J_3$ tend to become straight in one vertical line $J_1 J_3$. This tendency shifts (moves) RAM 'R' down.

Now let us examine effects of F_N on force which gets transferred to RAM. Let us call force on RAM as F_R. Force F_R in terms of F_N depends on θ_{RM} (Connecting rod making angle with vertical).

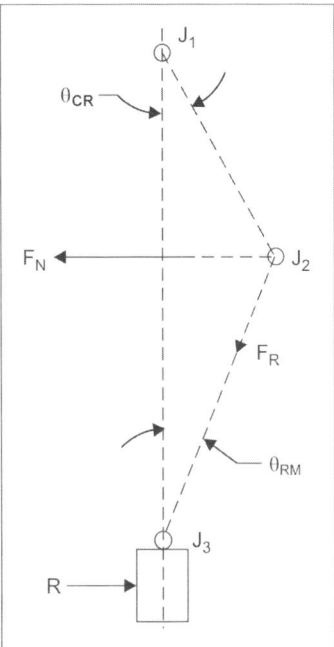

Fig. 4.7: Force generation analysis

$$\sin \theta_{RM} = F_N / F_R$$

Or
$$F_R = \frac{F_N}{\sin \theta_{RM}}$$

(1)   For $\sin 40 = \dfrac{F_N}{0.6248} = 1.556\ F_N$

(2)   For $\sin 30 = \dfrac{F_N}{0.5} = 2\ F_N$

(3)   For $\sin 15 = \dfrac{F_N}{0.2588} = 3.864$

(4)   For $\sin 5 = \dfrac{F_N}{0.0872} = 12.315$

(5)   For $\sin 0 = \dfrac{F_N}{0} = \infty$

In practice, generally one third down travel of RAM is 'blank', the space between faces of upper and lower tools. Remaining two thirds travel may be the working stroke. This two thirds travel undergoes changes from higher value of theta (θ) to zero.

Consequently, force exerted by RAM varies. Generally, mean value of force exerted by RAM is specified in terms of tonnes or tons.

Adjustability in Stroke

Adjustability in stroke means the length of travel of RAM from top dead center (TDC) to bottom dead center (BDC) can be increased or decreased under specified limit indicated by manufacturer of power press.

In practice, total height of press tools may vary, further, penetration of upper portion of tool into lower portion may also vary. For this reason, flexibility in the setting of stroke length is a desirable facility. All power presses do not have this facility.

For achieving adjustment in stroke length, connectivity rod is not directly fitted on PIN of crankshaft. Instead, an eccentric bush is first fitted on PIN and then connecting rod is adjustable with respect to eccentric bush. This arrangement makes it possible to increase or decrease length of stroke up to a certain limit.

Length and Width of Bed

Referring to Fig. 4.2, L is the length of bed, spanning from left to right. It is the integral portion of press body.

Tool Stem Hole Size in RAM

RAM of low tonnage presses are generally provided with a bore of standardized diameter and depth. Further, there is a provision to securely clamp the tool stem in RAM hole.

RAM face of heavy tonnage press (say 200 and above) carries T slot arrangement for clamping upper portion of tooling which is normally quite heavy. Generally, such type of heavy press is operated by hydraulic system.

Thickness of Bolster Plate

Referring to Fig. 4.2, Part (7) is the bolster plate. Its thickness depends on the tonnage of power press. It may vary from 70 mm for a 40 tons press, 100 mm for a 100 tons press 200 mm for a 500 tons power press. Length and width of bolster plate is kept approximately equal to bed size.

Opening through Bolster Plate

Most of the presses are provided with a through hole in its bed and bolster plate. Hole in bolster plate has got a counter step for fitting a cushion cylinder if needed. A pneumatic cushion system can also be fitted.

Through hole is necessary if blanks or components are to be dropped down. Diameter of hole varies with the tonnage of power press. It may vary from 145 mm for a 40 tons press to 260 mm for a 200 tons power press.

Spring and Pneumatic Cushions in Power Presses

Spring or pneumatic cushion in power press is needed if a tool is designed and made to have pressure pins. Example of such a tool is 'cut and cup'. It is shown below to give an idea of combination of cushion and cut and cup tool.

Referring to Fig. 4.8, (7) is bolster plate in which there is a through hole with a counter to accommodate cushion housing. In cushion housing, there is a strong compression spring (10) having two end discs, one on top and the other below. This provides cushion to movement of pressure pins (5). Normally, there are three pressure pins of equal lengths. When cutting punch cum draw die moves down to draw cut blank into a cup, then there is a need to hold

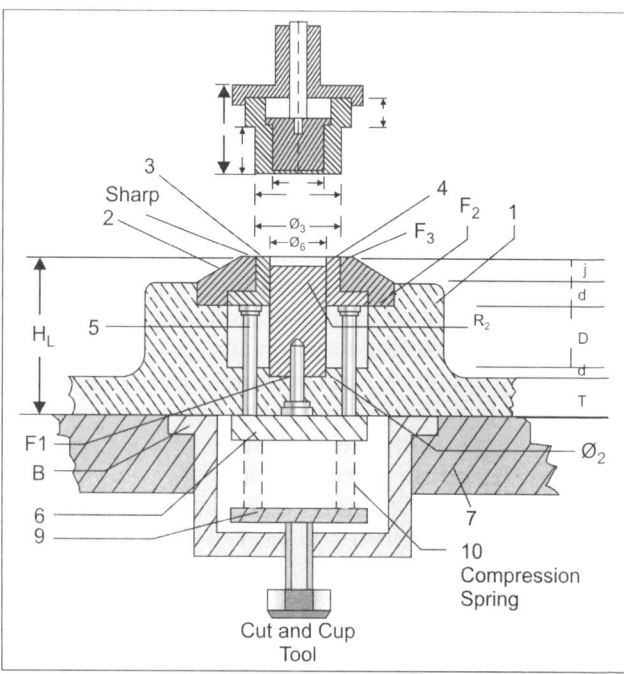

Fig. 4.8: Cut and cup tool

blank with pressure between faces of draw die and pressure pad (3). This pressure is provided by cushion spring through pressure pins. Amount of under holding pressure may vary from product to product, depending on thickness and hardness of stock (sheet) and also some other factors such as surface finish of stock, speed of draw, use of lubricant if used.

To adjust cushioning pressure, a bolt is provided. By screwing the bolt in, spring compression increases. It is increased so much that wrinkle free cups are produced. Please note that dimensions of various parts are so maintained that in no case spring gets completely closed while pressure pins are descending. This is necessary to avoid damage to tool.

Drawback with spring cushion system is that under holding pressure on sheet metal, blank is not uniform throughout its draw. It keeps on increasing while the spring keeps on getting compressed. Another drawback is that adjustment of pressure through rotating the bolt (8) is very inconvenient.

Spring cushions are still used in low tonnage presses such as 25 to 40 tons power press. Please note that cushion housing has to be removed when component or blank is to be dropped down from the tool. There are examples of tool construction that, in spite of the fact that cushion system remains in its place, dropping component or blank is ejected by air blast or mechanical system towards back side.

RAM Adjustment Up and Down

Most of the time, it is necessary that upper portion of tool (say, a punch) should precisely reach up to a certain distance with respect to lower tool. Let us take the example of a circular blanking tool which has a cutting punch on the top and cutting ring on the lower half of the tool. From cutting function and pushing the blank down, it is required that 2.5 mm length of punch should penetrate into the cutting ring. Suppose cutting punch remains. 0.5 mm above cutting ring face then what to do. Such precise adjustments are not advisable to achieve by changing the stroke length so to overcome this difficulty, connecting rod (connecting crankshaft PIN and RAM top) is provided with Acme threaded rod. This arrangement is typically illustrated with the help of Fig. 4.9.

Referring to Fig. 4.9, Part (1) is a connecting rod, (2) is Pitman screw with ball formation at the end, to be fitted to RAM (6) and (3) is the clamping strip housing. After adjustment, Pitman screw is clamped with the help of clamping strip. This strip also has thread

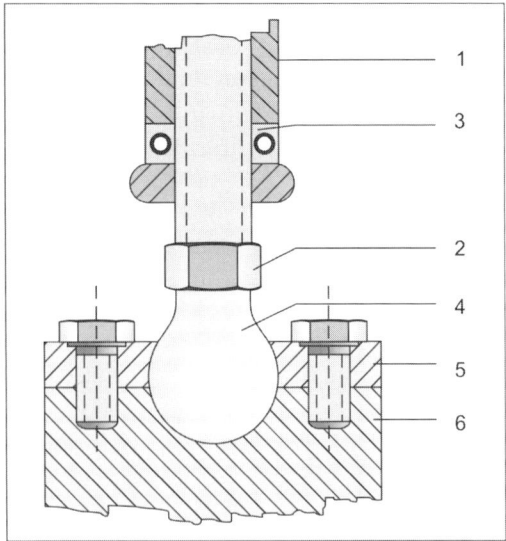

Fig. 4.9: Connecting rod and RAM joint

formation matching to screw formation in connecting rod. Ball formation allows annular (swinging) motion of connecting rod with respect to RAM.

Opening in Bed

It is already described under sub-heading 'Opening through Bolster Plate'.

Opening in Back of Press Body

Referring to Fig. 4.2, space (17) is the opening between vertical body walls (18). This opening facilitates ejection of components by means of air blast. Conveyor belt system may also be installed through the opening.

'Daylight', Maximum Distance between Bed and RAM Face

"Daylight" is the maximum distance between bed and RAM face when Pitman screw is completely inside the connecting rod and eccentric of crankshaft is so adjusted that TDC is at its maximum.

Flywheel Diameter and Thickness

The following parameters determine the energy stored in a flywheel which is transferred to crankshaft in terms of torque:

- Mass of flywheel
- Diameter of flywheel
- Weight of flywheel (kg)
- rpm of flywheel
- Calculated circumference of center of mass

Power press manufacturers normally indicate diameter of flywheel and rpm in technical literature. Generally, flywheel diameter varies 800 mm for a 40 tons press to approximately 1200 mm for a 250 tons capacity power press.

Power press designer designs diameter and thickness of flywheel to provide desired tonnage at a certain rpm of flywheel.

Number of Strokes Per Minute

One round of flywheel creates one stroke of RAM. One stroke means starting from TDC and again stopping at TDC. Number of strokes per minute is mostly a fixed value. It may be 110 rpm for a 40 tons press to 80 for a 250 tons press.

Distance between Floor and Top of Press Bed

Referring to Fig. 4.2, 'H' is the distance between the floor and the top of press bed.

Motor Horse Power

Generally, motor operates on the following:

- 3 – Phase
- 50 Hz cycles
- 440 Volts AC

Horse power of motor varies according to tonnage of power press. A 40 tons power press may have a 5 HP motor and a power press of 250 tons may have a 24 HP motor. This all depends on design and capacity of power press. Press manufacturers normally provide all relevant technical information to their customers.

Weight of Press

Knowledge of gross weight and net weight of a power press is essential for customer to know because he has to arrange construction of suitable foundation and unloading arrangements.

Height, Width and Depth

Knowledge of width and depth is necessary for designing length and width of foundation.

How high the press would be is essential to know to ensure overhead clearance during transportation, installation and statutory clearance between machine top and roof (shed).

4.2 PNEUMATIC POWER PRESSES

Pneumatic power presses are used for light duty work such as riveting. It is also used for pressing work. RAM of a pneumatic power press is actuated by means of pneumatic jack (piston in a cylinder). Generally, air is used for exerting force on piston of jack. Supply of air is obtained from air compressor. Pressure and volume of air is controlled by pressure control valve and volume control valve. Increase or decrease of air pressure increases or decreases force available at RAM face. Increase or decrease in flow of air increases or decrease speed of RAM.

Presses having control on both force and speed of RAM are uncommon. Pneumatic presses generally have pressure control valve to adjust force applied by RAM.

In riveting operation, force is needed with an impact which is achieved with the combination of RAM weight, speed and force applied on the piston of jack.

Most of the presses are table mounted. These are generally used on assembly line. Pneumatic press body construction is generally standardized by press manufacturer. Body construction may be of cast iron or of welded steel structure.

Definite length of stroke and set uniform force cannot be obtained instantaneously as air is compressible. It does not behave like hydraulic system because liquids are practically non-compressible.

4.3 HYDRO–MECHANICAL POWER PRESSES

Hydro-mechanical power presses are generally high capacity presses. These are normally used for carrying out bending, drawing, trimming and piercing of big sheet metal parts around two meters by one and a half meter in size and weighing many kilograms. Examples of such parts are cover plates of huge diesel engine, ship construction, guards and panels of earth moving equipments and many more. These may be four meters or more. Many a times, there are two robust hydraulic jacks as the RAM width is quite wide, say two meters.

Hydraulic presses are generally equipped with hydraulic pressure and volume control system.

Apart from it, mechanized holding and adjustment of location of blank sheets is also provided. Generally, there are SAFETY systems to ensure safety of operator, tools and press.

4.4 ACCURACY TESTING AND ADJUSTMENTS

Performance and **life of a press tool** very much depends on accuracy of a power press. Various factors of accuracy of a power press are explained with the help of Fig. 4.10.

RAM of a press has a standard hole to hold stem of tool and slide wedges in both sides of RAM. These slide wedges are housed inside power press body slide ways. Movement of RAM is with precision sliding fit.

Face F_1 of RAM should be perfectly at right angle to slide wedges. Perfectly means that angle should be 90° plus or minus two seconds. This is maintained while RAM is machined during manufacturing. Similarly, stem hole in RAM should be at right angle to face F_1. RAM face F_1 and bolster plate face F_2 should be precisely parallel. Distance between F_1 and F_2 at left and right of RAM face should not vary more than say, 0.02 mm. In fact, this variation depends on the capacity and size of power press. All these accuracies are maintained at the time of manufacturing of power press. Now the line of up and down movement of RAM should be at right angle to bolster face. This accuracy depends on the assembly and adjustments of guide ways blocks.

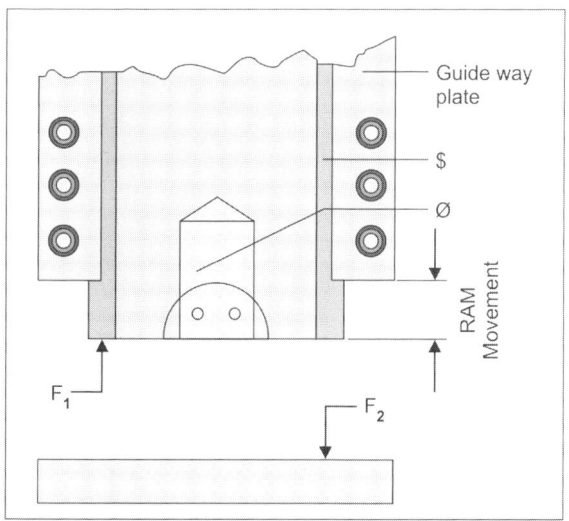

Fig. 4.10: Accuracy checking locations

The following accuracies are required to be tested:

• Stem hole right angle to RAM face
• RAM face right angle to slide wedges
• Parallelism of faces F_1 and F_2
• Axis of RAM movement to be at right angle to bolster face

How to Carry out Accuracy Test

Stem Hole Right Angle to RAM Face: Fix a ground and chromium plated test bar to the RAM hole. A precision trisquare may be used to check the accuracy. Three possible situations are shown in Fig. 4.11. In 'A', try to see through trisquare edge against white illuminated paper. In this case, test bar is perfectly at right angle to RAM face. Almost no light could be seen. In 'B', some light could be seen between the trisquare blade and test bar is little tilted towards left. Figure 'C' shows slight gap between trisquare blade and test bar. This indicates that test bar is a little tilted towards right.

Actually this accuracy test is performed at the works where RAM is machined. Accuracy test with numerical values may be carried out by putting the RAM with test bar on a co-ordinate table such as a milling machine table. A dial indicator may be used to set face of RAM so that dial indicator does not show any variation. Now the tip of dial indicator is located at test bar at diameter point. Now the table is given movement at y axis (which is right angle to x axis). Deviation shown by dial indicator is the outness of RAM hole axis. With sophisticated computerized system of testing, angle of tilt of hole may be read directly on display screen.

Same test may be performed in front side of RAM that means at right to previous test.

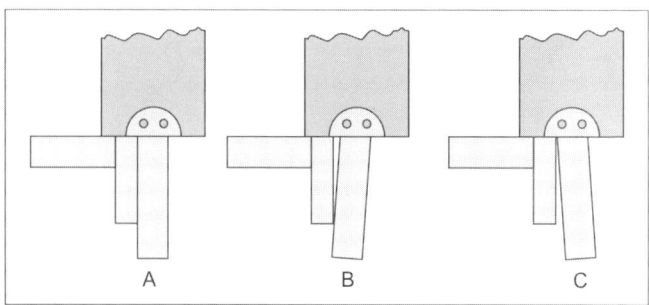

Fig. 4.11: RAM alignment checking

RAM Faces Right Angle to Sliding Wedges: It is an inbuilt quality which is maintained at the time of manufacturer of RAM. In case it is desirable to retest then the RAM has to be dismantled and loaded on the table of a milling machine. One surface, say RAM face has to be adjusted parallel to the movement of table. This means that dial indicator will not show any deviation. Now, dial indicator would have to be set to touch slide wedges. Table is now moved in other axis (right angle of previous). If dial indicator does not show any change in reading then it may be concluded that RAM wedge is right angle to RAM face.

Parallelism of Face F_1 and F_2: A well ground, lapped and polished plate of steel is placed over the bolster plate. This plate should be precisely equal in thickness at all points. Before placing on bolster plate, examine the surface of bolster plate to make sure that there is no high point and the surface is smooth. Size of plate should be bigger than RAM face. Now fix a dial gauge to a stand which has a large base area. Adjust the total height in such a way that a deflection of about quarter of a round of needle takes place when point of dial indicator is brought under RAM. Dial gauge is now slided over the plate so that left, right, back and front positions of RAM face is covered. If both the faces F_1 and F_2 are parallel, there would be no change in reading. In case faces are not parallel then reading on dial indicator would be different.

It may be noted that instead of using mechanical dial indicator, electronic instrument may be used where digital numeric reading may be read.

Axis of RAM Movement to be at Right to Bolster Plate Face: This test can be performed by fixing sensor probe on a rod fixed to RAM hole or face with so much room that tip of sensor touches trisquare blade edge or a specially made block for this purpose. If the axis of movement of RAM is at right angle to bolster plate, then there would be no change in reading. If axis of movement is not perfectly right angle to bolster plate surface then there would be a change in reading. Amount of deviation may easily be read on the screen of computer attached to sensor.

Why is it Necessary to Ensure Accuracy of Press?

It is because accurate functioning of tool depends on it. Further, life of tool gets enhanced. This is explained with the help of Fig. 4.12.

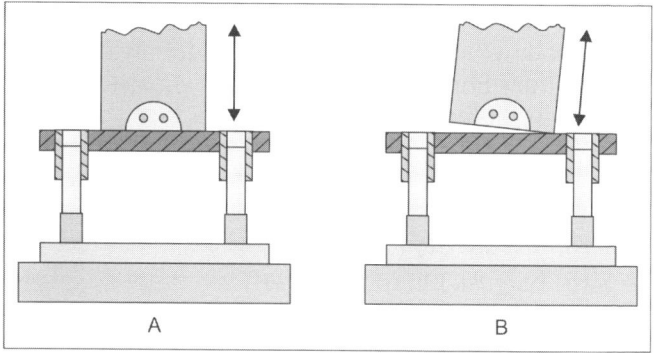

Fig. 4.12: Accuracy of press RAM

Hand Presses

Use of hand presses is quite common in small and cottage industries for carrying out sheet metal press tools operations. All sorts of operation may be carried out for small tiny size components.

Generally, size of hand presses is indicated rather than force because it depends on operator as to how much force is applied. Moreover, it also depends on the size and weight of flywheel. If operator rotates the wheel with more speed, then more inertial force would be generated.

Figure 4.13 provides some idea of shapes of hand presses.

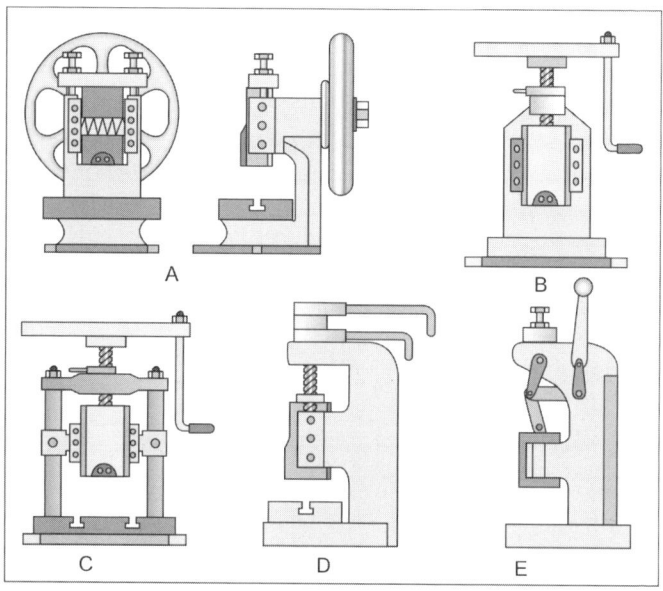

Fig. 4.13: Types of hand presses

A – Vertical wheel type
B – Horizontal wheel type
C – Heavy duty horizontal wheel type
D – Pneumatic type
E – Knuckle joint type

A - Vertical wheel type

This type of press comes with No. 2, 3 and 4. No. 2 is the smallest and No. 4 is the biggest. On rotating vertical wheel, RAM of press moves up and down. When wheel is rotated in anti-clockwise direction (seeing from wheel side), RAM moves down. Length or extent of downwards movement is adjusted by two stopper bolts fitted at the top member of RAM. Example of few components produced by these presses are shown in Fig. 4.14.

B - Horizontal wheel type

It is used to carry out operation that needs heavy force. Few examples are shown in Fig. 4.15.

C - Heavy duty horizontal wheel type

It needs much more heavy duty as compared to B type press. Example of a type of work which is carried out on C type press is shown in Fig. 4.16.

This press is manually operated by more than one person in standing posture. Sometimes two thirds to one complete round is taken by operators. Use of such type of press is almost obsolete. It is used for drawing sheet metal shells of shallow depth but large diameter, say 60 mm.

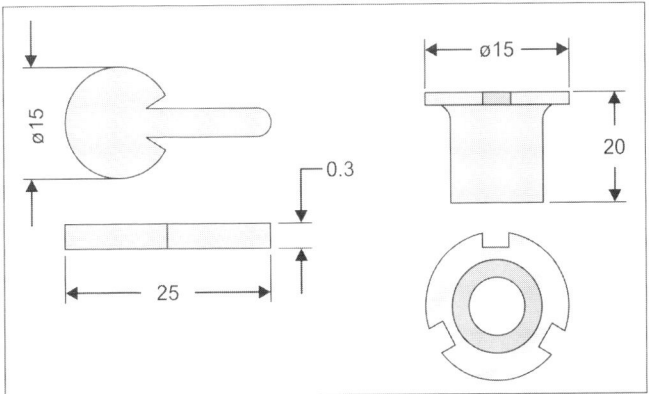

Fig. 4.14: Example of components produced (vertical wheel type)

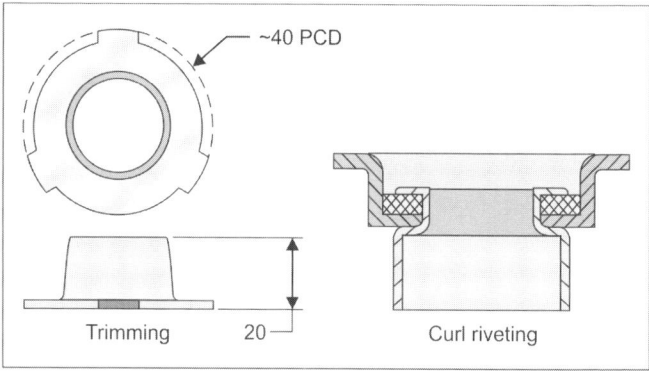

Fig. 4.15: Example of operaion requiring heavy force

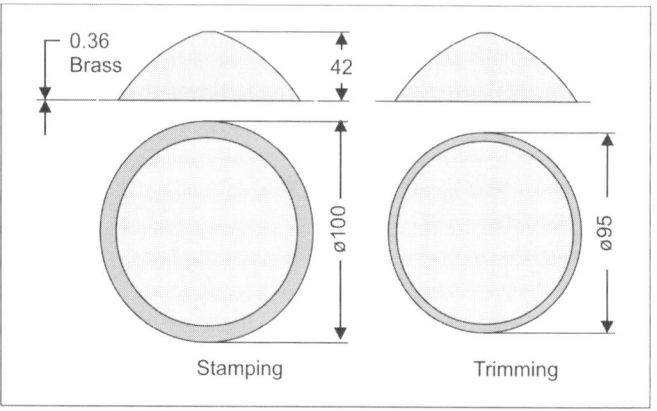

Fig. 4.16: Operation requiring heavy duty press

D – Pneumatic type

In this type of press, up and down motion of RAM is achieved by a pneumatic jack. Normally, a big diameter (say 100–140 mm) jack is used. Pressing force is adjustable by adjusting pneumatic pressure inside the jack. These presses are equipped with FRL (filter, regulator and lubricator) switch buttons and safety installation. Length of stroke in all the presses is adjusted by means of mechanical stoppers. The advantage of this press is that fatigue to operator is very much reduced.

E – Knuckle joint type

This type of press is useful where sufficiently large force is required with specified depth. This requirement is explained with the help of Fig. 4.17.

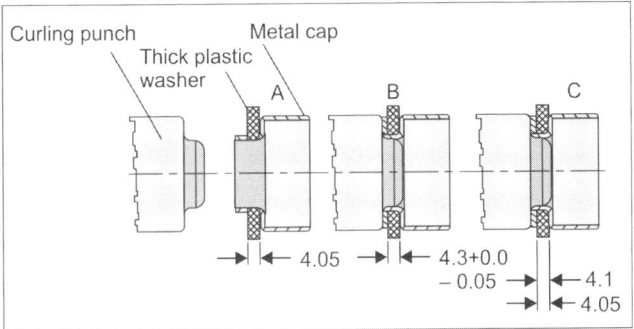

Fig. 4.17: Operation requiring knuckle joint press

Figure 4.17 shows three situations of assembly of two parts. One is a sheet metal neck cup and the other is 4.5 mm thick plastic washer. 'A' shows components in position ready for assembly by a curling punch. Fig. 4.17 'B' shows assembled components with specified dimensions 4.3 + 0, – 0.05 mm. If this assembly operation is performed by any other press then dimensions 4.3 + 0, – 0.05 cannot be ensured because variation in inertial force of RAM may create variation in this dimension.

In knuckle joint press, lower position of curling punch is set. The application of force is basically applied force by knuckle joint and NOT inertial force. In this way, dimensions 4.3 + 0, – 0.05 may be consistently achieved.

Figure 4.17 'C' shows possibility of excessive pressing by any other press than knuckle joint press.

4.5 SCHEMATIC ARRANGEMENT OF POWER PRESSES

Typical specification which is generally provided by power press manufacturers are highlighted in Table 4.1. Specimen values are provided just to give an idea of dimensions.

All the information provided in Table 4.1 is important for tool designer but the most important is one which is explained with the help of Fig. 4.18.

Referring to Figs 4.18 and 4.19, a schematic arrangement of crankshaft, connecting rod, Pitman screw, RAM and bolster plate is shown.

Let R be the radius of connecting rod pin where connecting rod is joined having a free rotating joint. When crankshaft rotates, pin joint moves from TDC (top dead centre) to BDC and then to TDC

Table 4.1: Information for tool designing

S. No.	Description	Unit	Press capacities (Tons)		
			30	100	150
1.	Force	Ton	6		
2.	Stroke length adjustment	mm	0—22		
3.	RAM position adjustment	mm	25		
4.	Distance between RAM face and bed at top dead center	mm	150		
5.	Thickness of bolster plate	mm	40		
6.	Length and width of bed	mm	320 × 180		
7.	Diameter of stem hole in RAM and depth	mm	25 × 30		
8.	Opening through bolster	mm	75		
9.	Opening in bed	mm	75		
10.	No. of strokes per minute, ungeared	mm	120		
11.	Distance of top of bed from floor	mm	690		
12.	Flywheel diameter	mm	470		
13.	Motor HP	HP	1		
14.	Motor rpm	rpm	1440		
15.	No. of V belts	Nos	2		
16.	Section of V belts	sec	B		
17.	Weight of packed press	kgs	475		

to stop there. Movement of pin from TDC to BDC produces down stroke D of RAM. Value of D does not change as value of R is fixed (assumed) which is normally with the presses having crankshafts and not adjustable eccentricity. 'd' is the amount of adjustment provided to raise or lower the RAM to change gap 'G' between RAM face and bolster plate surface. This means that there may be two values of 'G'—when RAM is at TDC and the other when RAM is at BDC. Variations in value may be stepless within specified range and depending on the adjustment of Pitman screw.

In what manner, a tool designer utilizes information (as per Table 4.1) would become clear in chapter 11 when tool designing is dealt with.

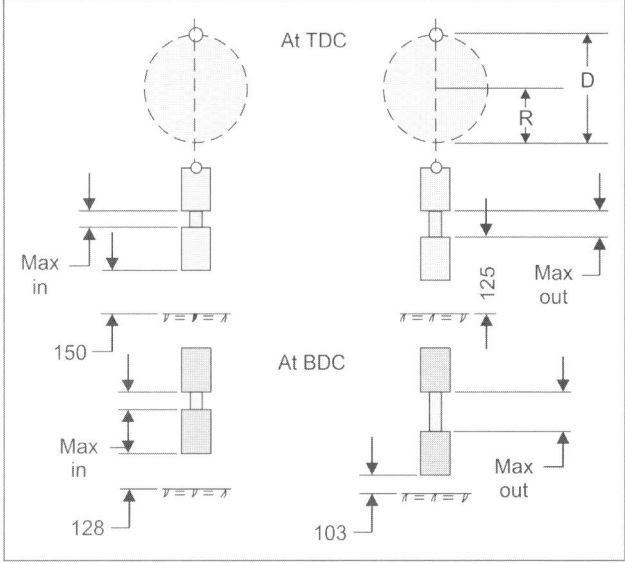

Fig. 4.18: Crankshaft and Pitman screw

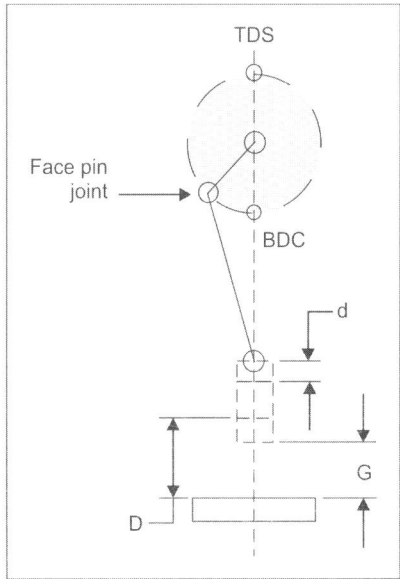

Fig. 4.19: Schematic arrangement of crankshaft, connecting rod and Pitman screw

4.6 THREAD ROLLING MACHINES

Many a sheet metal components have threads of various designs and the end of circular components trimmed to have a smooth finish edge at right angle to axis of component.

Thread rolling tools (generally discs) are also designed to suit a particular thread rolling machine and to give specified thread profile on the components and other dimensions such as start of thread, distance of a bead from thread end.

Generally, there are two types of thread rolling machines as follows:

• Hand thread rolling machines
• Automatic thread rolling machines

Both the machines are briefly explained with the help of Figs 4.20 and 4.21.

Hand Thread Rolling Machines

Hand thread rolling machines are quite inexpensive as compared to automatic thread rolling machines. These are generally used in small scale industries where accuracies in quality of threads are not demanding. Moreover, quality of rolled thread very much depends on the working of the operator.

Working of hand thread rolling machine is explained with the help of Fig. 4.20 where part (1) is the upper body of machine and part (2) is the lower main body in which all other machine parts are supported. The upper body can be swung up and down as it is hinged to the main body on free joints. The upper body carries a

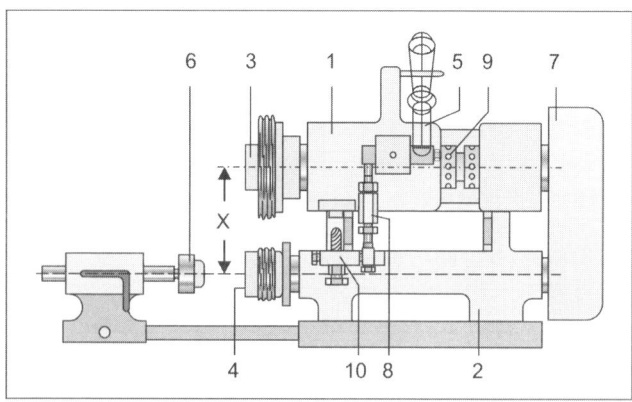

Fig. 4.20: Hand thread rolling machine

spindle (3) on which upper thread roller set is mounted. The main body also carries a spindle (4) on which lower thread rolling set is mounted. The upper body may be lowered with the help of lever (5). Distance X is maximum when lever (5) is completely pushed back. The value of X would be minimum when lever (5) is pulled and stopper (10) screwed down completely to allow upper body to come down completely by adjusting the stopper bolt. The movement of upper body may be restricted according to requirement.

Diameters of lower and upper thread rolling tools should be such that both touch each other when lever is pulled. Setting of stopper should be such that there is still a possibility of little down movement of upper body if stopper is screwed down. This precaution is taken if any of the upper or lower tool (thread roller) is found to be slightly smaller than required minimum diameter. Rotation of spindles is interconnected with the help of gear train housed in enclosure (7). Rotation of lower spindle is generally clockwise when seeing towards thread roller from back rest (6) side. The upper spindle rotates in anti-clockwise direction. The lower thread roller has right hand thread and upper thread roller left hand matching threads. Components (say, a cup) are manually placed by operator over lower thread rolling tool. Backrest (6) is moved forward to keep the cup completely in over the tool. Lever (5) is then pulled by operator which brings upper tool over the outer surface of cup and gradually pushes the material (sheet) to form threads. It takes about 2 to 5 seconds to form threads as both the spindles are rotating at high rpm, say 600.

Automatic Thread Rolling Machines

Automatic thread rolling machines are generally equipped with the following features:

- Feeder magazine to automatically bring the component in front of lower tool
- Backrest moves to push the components over the lower tool
- Thread depth adjustment by a micrometer knob
- Axial adjustment of upper spindle by means of a micrometer knob
- Ejection of component by a reciprocating ejector ring
- Facility to change rotation ratio of upper and lower spindle.

For example:

Lower spindle round – 1, 1, 1
Upper spindle round – 1, 1/2, 1/3

- Hand wheel to facilitate rotation by hand while setting the tool for proper matching
- Machine start lever
- Emergency brake stop by means of a red big button
- All movements are synchronized and achieved by various cams inside the machine body.

Schematic view of an automatic thread rolling machine is shown in Fig. 4.21.

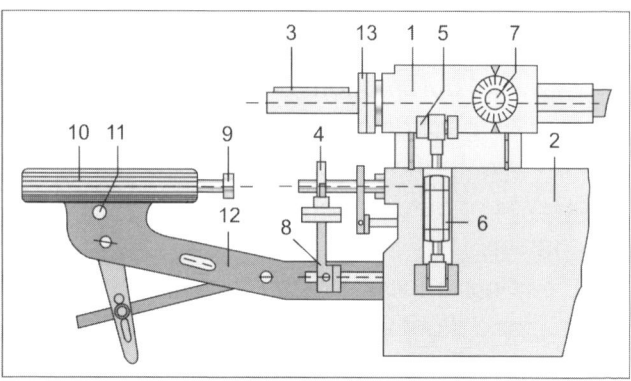

Fig. 4.21: Automatic thread rolling machine

5

Automation and Robotics

This chapter consists of the following aspects:

5.1 ELEMENTS OF AUTOMATION

It is advantageous to sheet metal press tools designer and tool maker to know under what circumstances and how a tool is designed and made to suit automation. This point would be discussed in few last pages of this chapter. First of all, various aspects of automation are described below:

Following are commonly used elements of automation:

- **Hoppers**
- Feeders
- Orientation devices
- Pick and place mechanism
- Sensors
- Electrical and electronic gadgets
- Microprocessors
- Variety of mechanical motion mechanisms
- Actuators
- Variety of mechanisms
- Computerized controls
- **Robotics**

Hoppers

Hoppers are those parts of machine or plant in which raw materials for a production are filled. Raw materials may be in the form of powder, granules, paste, liquid, balls, tablets, capsules, grains and even small components. None of the raw materials is in an orderly manner. They are just poured in. There is a large number of variety of hoppers. Their design and size depend upon the type of raw material to be filled. Quantity is retained by hopper at a time, this means its capacity.

Hoppers in which coal is unloaded from railway wagon are huge and sturdy. In cement producing plants, stones are dropped in hoppers which are heavy, strong and sturdy.

In hoppers of injection moulding machine or extruders, only few kilograms of plastic granules are dropped. These are small sized hoppers in which about 25–50 kg of plastic granules can be dropped. In dairy plant, milk is poured by vendors from their containers to receiving hopper of plant. In pharmaceutical production plants, there are hoppers fitted to various machines. These hoppers are filled with medicinal powder for producing tablets. These are only few examples which are mentioned here. There may be a large number of such examples.

Feeders

In almost all production processes, raw materials or components for further operations are fed in an orderly manner. Quantity of feeding at a time is fixed and synchronized with the movement of other mechanisms. When it comes to feeding of components, they are to be fed in a particular position. Take example of a toothpaste tube cap. In a placement and screwing machine, cap should always be fed unidirectional, either having threaded side down or up. Therefore, design and construction of feeder is done in such a way that it orientates components in a specific position before finally feeding the component to other unit or station of machine.

There are metering feeders also which are used to feed quantities of granules and powder of specific weight for each stroke of machine. There are large number of varieties and designs of feeders. Some feeders are integral part of a machine, while others are separate units attached to machine. Such type of feeders are synchronized with the machine by mechanical or electrical means.

Feeders are normally fed by small quantities of raw material or components from hopper automatically. In some cases, this feeding may be manual. Feeders are equipped with a system which stops entry of raw material or component from hopper if feeder becomes completely filled up. Generally, the speed of filling of feeder from hopper is higher than the output from feeder. In this way, feeder always remains filled up, hence, no stroke of machine goes blank.

Orientation Devices

Many components are similar from both the sides like a simple plain washer or a cylindrical spacer of metal or plastics. These types of components can be fed from any of the two directions. There is no need for orientation of one or the other side before feeding to machine head, but still there is a need for orientation of cylindrical spacer, for example, in vertical or horizontal position. If the shape of both the ends of component is different then it is necessary that component should be fed in a particular direction. For example, a sheet metal cup is to be fed in a trimming machine with open end of cup always in the direction of lower trimming roller. This is achieved by some specialized construction of chute, channel or mechanism. The whole process of bringing the component in particular direction is called orientation. Orientation mechanism may be simple to very complicated design and construction.

Pick and Place Mechanism

'Pick and place' are those mechanisms by which components or raw material are picked up from one place and placed on another location. There are a large number of varieties of pick and place mechanisms. Mechanisms may be heavy duty or very delicate to handle tiny components. Lifting of heavy steel plates from a platform, carrying it on to the bed of radial drilling machine and placing it between the guides, is an example of heavy duty 'pick and place' arrangement. On the other hand, a delicate pick and place mechanism vacuum lifts electronic chip and places it on a specialized location on PCB (Printed Circuit Board).

Pick and place mechanism may have an arm which can axially move in all the three X, Y, Z axes. Moreover, picking arm may rotate in any direction. It is not necessary that all the movements and flexibilities are available in all mechanisms. It all depends upon as to what actions are to be performed by pick and place mechanism.

In most of the cases, speed of any motion is very important. It has to be synchronized with the speed of machine. Pick and place cycle time may range from, say five minutes for heavy work like transfer of steel sheet to two seconds for transfer of tiny electronic chip. Many a times, accuracy of placement of a component is very important together with high movement speed. For such jobs, light, strong and with minimum possible vibration, mechanisms are designed and built. As already described, pick and place mechanisms are used where both picking and placing stations are fixed and at pre-determined location. In case location of item or component, which is to be picked, is not defined, it may be anywhere in a specified area then a robot may be used as it would first identify as to where target item is located. It would travel to that place and even make sure if the item being picked up is desired item or some other similar item. After picking, robot would travel upto the location where item is to be placed. It would then accurately place it after ascertaining if placing location is vacant and OK.

Sensors

Sensors in itself is a vast subject and many books may be available for detailed study of sensors. Detailed description of sensors is beyond the scope of this book, however, a brief description is provided herewith. Sensors may be classified as below:
• Mechanical sensors
• Optical sensors
• Magnetic sensors
• Proximity sensors
• Pressure sensors

Sensors are generally used to sense presence of an input/output of components, materials and products. Temperature range of a system, such as plastics injection mould, may also be controlled by sensors. Speed of a machine or process can also be measured and controlled by suitable combination of various types of sensors.

In sheet metal press working process, generally mechanical sensors are used. In this chapter, a number of mechanical sensors are briefly described.

Optical sensors are generally used for counting and sensing a missing component in a machine turret or conveyor belt. Magnetic sensors are generally used to segregate ferrous and non-ferrous objects of similar shape and size.

Proximity sensors are also used to sense presence of an object where physical touch or light exposure is undesirable.

Pressure sensors are used to measure or sense pressure and vacuum.

Electrical and Electronic Gadgets

Electrical and electronic gadgets are generally used to achieve desired action according to sensed inputs. 'Power supply' is a basic electronic gadget which supply DC power to sensor device. Associated electronic circuitry converts input signals from sensor to output signals to electrical gadget, say a solenoid to operate a mechanical prob or system on a production line. Pressure sensors are generally used to either measure or control hydro-pneumatic pressure and vacuum in a chamber such as in vacuum metalizing plant.

Microprocessors

Microprocessor is a combination of basic electronic elements such as integrated circuits, translators, resistors, capacitors, etc. The function of microprocessor is to translate input digital data into digital/analog output data.

Microprocessors are generally programmed to provide pre-decided output according to given inputs, including those from sensors.

5.2 VARIETY OF MECHANICAL MOVEMENT MECHANISMS

Various mechanical movements are required to perform different functions in innumerable machines and plants. For a given space, speed, force, torque, movement, etc. a number of machine elements are required. Size, shape, material of construction, properties, etc. of machine elements depend on design of function required, hence, there may be innumerable combinations, but the basic working principle remains the same. It is therefore necessary to understand basics of machine elements, movements and mechanisms to appreciate any automation system. Following are some essential machine elements and mechanisms:

- Reciprocating movements
- Rotating motion
- Combination of above
- Planetary movements
- Movement in a geometrical pattern like parabolic, epicycloids, etc.
- Leverage links
- Motion adoption by gear trains, rack and pinion
- Conversion of motion
- Combination of cam and follower
- Universal joints
- Pneumatic, hydraulic and electric actuators
- Precision stepping movements

Reciprocating Movements

A rod, block or spindle moving to and fro along its axis is called a reciprocating motion. It may vary in amplitude and speed according to requirement. Force available at the end of reciprocating element may or may not vary during its complete cycle of reciprocation. It completely depends upon the design of mechanism which is providing reciprocating movement to machine element, a rod for example.

Rotating Motion

Any machine element may have rotary motion. It is not necessary that rotating machine element has to be round in shape. A rod of rectangular section turning along its axis may be called a rotating rod. If the same rod is not turning along its axis, it may not be called a rotating rod.

Combination of Reciprocating and Rotating Movements

In a particular mechanism, there may be a necessity that a rod reciprocate as well as rotate. Rotation of a reciprocating rod may be a continuous rotation, partial rotation in one direction and then rotating back to its initial position. Rotary motion can be intermittent also. This means that rod rotates for specific degrees, stops and again rotates some specific degrees. In this way, it keeps on rotating. Such type of movements are required in indexing mechanisms. There may be functional need that a rod moves to one extreme end of reciprocation and stops there to rotate by 180° and then to reciprocate back to its other extreme end of reciprocation.

Planetary Motion

If a wheel, gear or rod is rotating on its axis and its axis is traveling along a circle then it is called a planetary motion. It is shown in Fig. 5.1. The planetary movements may be continuous, intermittent or backward or forward for specific degrees.

Figure 5.2 shows a point at the end of a link. The construction of linkages should be such that shown point moves along the path shown by dotted lines.

Leverage Links

The purpose of leverage links may be to transfer the movement from one mechanism to another as it is to modify movement of driving element before transferring it to driven element. Levers are also used to reduce or magnify the length of travel of a machine

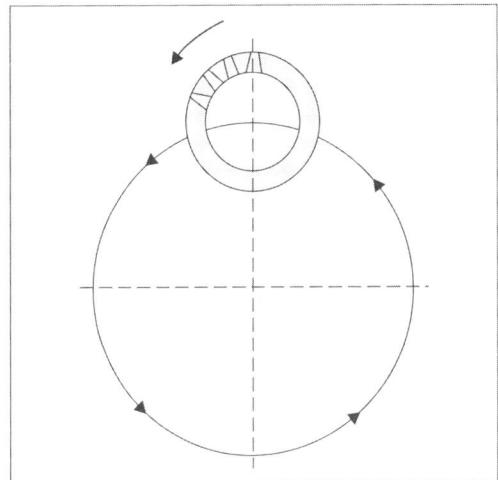

Fig. 5.1: Planetary movement

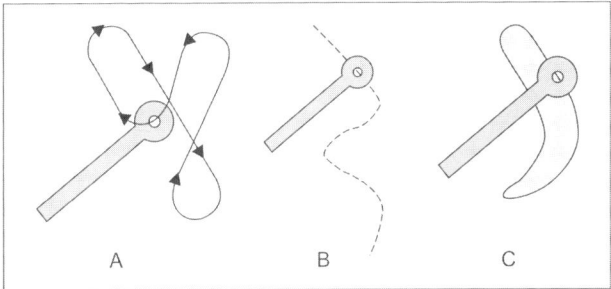

Fig. 5.2: Movement of a link point

element, say an axially driven rod. Links are also via media to convert a rotary motion into linear motion and vice versa. There are instances where links are used to increase or reduce amount of force applied to a particular machine element. There may be examples of **combination** of two or more functions of links like modifying movement of a push rod together with applied force.

Conversion of Motion

There may be the following types of motion:
- Linear continuous
- Rotary intermittent
- Linear intermittent
- Swinging motion
- Rotary continuous
- Motion of a point in a specified path

These motions may be interlinked by using suitable machine elements.

Motion Adoption

Motion adoption by gears, rack and pinion is very common in machines like gear boxes, gear trains in machine tools like engine lathe. Big and huge gears can also be seen in rolling mills, sugar mills, etc. Miniature gears made of metal and nylon are used in machines like meters, watches, toys, household gadgets, etc. Gears are also used in construction of automation devices. The purpose of gears is to transfer torque from one shaft to another. The set of gears are also used to create a ratio of rotation between driving and driven elements. A typical example of this is a reduction gear box where rpm of driving shaft is reduced to, say one third for driven or output shaft. In gear box, worm and worm wheels are used. Worm and worm wheels are a kind of gears. In fact there are quite a number of designs of gear boxes to cater to the needs of a mechanism. In automation mechanisms, use of gears is not uncommon. Gears are also used to convert rotary motion into linear motion. Examples of this are a mechanical jack where a handle is rotated to lift the axil of a car.

Combination of Cam and Follower

Cam is a non-circular rotating portion of a shaft or a disc attached to it. The rotation of shaft may be continuous or for a certain degrees and then going back. Cam may be in the form of a disc with a profiled circumferential surface. Figure 5.3 shows a few cams with followers.

Referring to Fig. 5.3 A, this type of cam and follower can be seen fitted to sewing machines, and with the help of this

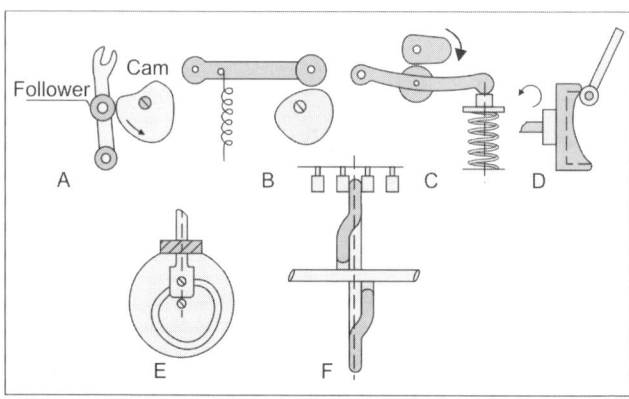

Fig. 5.3: Cams with followers

mechanism, thread is wound on the bobbin. While the cam is rotating, follower lever keeps on moving to and fro, shifting the thread left and right for getting it wound systematically. This means that one round of thread is just touching the previous round. Once one layer is completed, second layer automatically starts getting wound. 'B' is a round disc fitted eccentrically to the rotating shaft. Follower roller is touching circumferential surface of disc. Since the disc is fitted eccentrically, its rotation moves the follower up and down, down by spring tension and up by cam. Figure 5.3 C, is another example of a conical looking cam. The rotation of cam keeps on pushing and releasing rocker arm. In turn, rocker arm pushes valve stem down against compression spring. This arrangement can be seen in automobile engines. 'D' is a disc cam, but in this cam, face is the working surface. Up to a certain depth from circumference, face is given a definite shape that is thick and thin, thus forming a profile. The choice of such a cam depends on overall construction of drive system and space available for a particular movement of any machine element or unit of machine. 'E' is a continuously rotating cam. It is a plate having curved section for certain degrees, say 50° it rotates between equidistant rollers of 'a' under the table turret. As soon as curved portion passes between the rollers of turret, it gets turned softly, depending on the design of curve portion. Just before ending of curve portion, next gap between the rollers of turret is engaged, therefore, for the rest of 310° (360°–50°), turret remains stationary till curved portion arrives again. 'F' is again a face cam. A round disc has a precisely machined channel in which roller of follower is engaged. With this cam, axial movement of follower is positive for both to and fro movements. This type of cam may be found fitted in shaper, a machine tool.

Universal Joints

Universal joint is a combination of few machine elements which make it suitable to transfer reciprocating or rotating motion from one shaft to another while axis of two shafts may keep on changing angle. This means that sometimes axes of both the shafts are in line and at other moment, axis may be at an angle with each other. Universal joints are abundantly used in motor vehicles for power transmission from gear box to differential system of wheels. Figure 5.4 shows basic construction of a universal joint.

Universal joints are sometimes used in automation devices or arrangement.

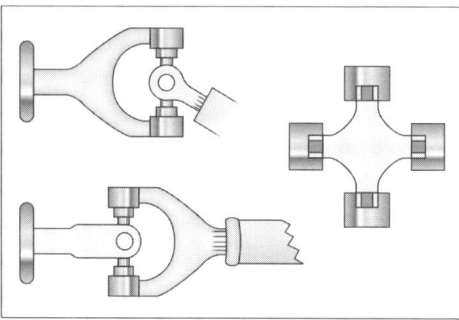

Fig. 5.4: Universal joints

5.3 PNEUMATIC, HYDRAULIC AND ELECTRIC ACTUATORS

Actuators are those devices which provide linear or rotary motion to other machine elements of machine. Actuators are basically pneumatic or hydraulic jacks. It consists of a cylinder, piston rod, inlet and outlet ports and seals. Figure 5.5 shows basic construction of a hydraulic jack.

Part (1) a is piston rod which is connected to machine element or mechanism which is to be actuated. Cylinder (4) is fixed at a suitable place with the help of a flexible joint (9). Piston rod passes through cylinder head (2) and rubber seals (3). Piston (5) is fitted with a pressure rubber seal ring. Fitting of piston and rubber seal ring is such that hydraulic oil cannot pass through running clearance between cylinder and rubber seal ring. When hydraulic oil pressure is applied through port (8), force generated due to oil pressure pushes the piston towards head end of cylinder together with piston rod. Flow of oil from pump to cylinder and cylinder to drain is controlled by a directional control valve in hydraulic circuitry. At a given moment, if oil enters a port then other port is connected to drain for oil to be pushed out by the face of piston.

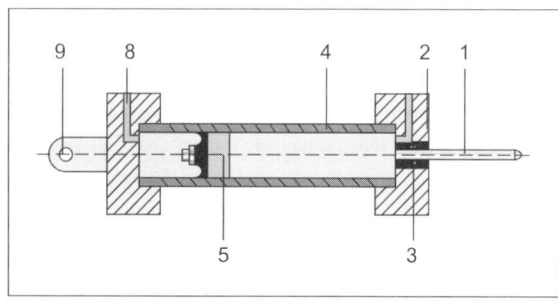

Fig. 5.5: Hydraulic jack

Hydraulic motors can also be called actuators because rotary and linear motions are provided by such motors.

Pneumatic jacks are almost similar to hydraulic jacks with the only basic difference that linear movement cannot be precise as air is compressible. Hydraulic jacks can be operated precisely for a desired length of stroke. Pneumatics are used where no much force is required and travel of actuator can be restricted by a suitable stopper. Another factor in favour of pneumatic actuator is that the place is free of oil wetness.

Electrical actuators are generally AC and DC solenoids. These are normally required to actuate valves, pushers, selector flaps and opening of some kind of door in the outlet channel of a machine. Electric motors may also be rotary actuators, providing continuous or intermittent rotary motion. Stepper motors are devices to provide 'stepless' or with steps, circular motion of precise requirement.

Precision Stepping Movement

In this paragraph, a typical example is described where precision stepping movement is an indispensable requirement. In extrusion of plastic film, variation in thickness is controlled in a very close tolerance, say 0.01–0.03 mm. In sheet extrusion, line thickness checking sensors are installed which continuously monitor the thickness of sheet. Signals from sensors are processed by microprocessors of attached computer. In turn, computer sends command to a stepper motor to rotate by, say exactly 11°. The next moment command may be for rotation by 8° in reverse direction. This rotation of stepper motor actuates servo system which actuates adjustment mechanism of extrusion die head to reduce or increase gap between die and plate to affect correction in the thickness of sheet. Another example of precision stepping movement is Electric Discharge Machining (EMD). In spark erosion machine (or EMD), cavity in steel block is formed by means of a copper, carbon or steel electrode. Gap between the surface of cavity and electrode surface is very important and has to be maintained for a specific distance, say 0.09 mm. This gap for proper electrical discharge bombardment is critical. This gap is electronically sensed and translated into required up or down movement of electrode. This is a continuous process and done by a stepper motor which drives a fine threaded shaft. Nut attached to EMD head or slide converts circular motion of screw to linear motion of head up or down according to requirement for maintaining a definite amount of gap.

Computerized Controls

In almost all production processes in small, medium and large industries, all types of controls are handled by computers which are loaded with suitable software. In case of adjustment of various process parameters which are interconnected are quickly controlled by computer. Manual control may not be possible because manual analysis of situation may take long time and then its implementation. Computer does the job in a very short time. Installed software is such that appropriate commands are given to various controls.

Robotics

Just visualize a situation where a vendor places a bottle of milk at the door step early morning. There may be ways to bring the bottle to kitchen. One, get up, leave bed, go to door, pick up the bottle and bring it and place on kitchen table. Second option, employ a domestic help to do the job and pay handsomely for early morning engagement. Thirdly, install a conveyor belt from door to kitchen with control switches. Well, first and second options may be practical, but the third option is not impossible but absurd.

In a house, garage, office or production place, there may be many small work which may be automated but would not be justified and almost impractical. Robot is a machine which is programmed to identify objects, find its way, handle the object as it should be or to replace it to another location. Robot is a self contained machine which may be designed to copy human movements for performing a job. It needs little space for moving around shop floor, in office, at home or even roads. There is 'near true' movie showing Robot (looking like a human being) running on the road with a ladies hand bag in one hand. A security officer on the round take that robot as a bad one, a thief, and he knocks the robot down just in front of the lady to whom the robot hands over the bag just before being knocked down. The security officer asks the lady if bag is hers. The lady says, "Obviously it is mine", have a breathing difficulty disorder, "He has just brought my bag which contains emergency medicine. He is just in time". Robots are used in assembly lines of automobiles for carrying out spot welding and other operations. There may be many other situations where use of expensive machine, Robot is justified. Another justifiable example may be use of robot by a bomb disposal squad. Robots are quite often used in research and development activities.

5.4 EXAMPLES OF AUTOMATION

Automation is a combination of a number of machine elements, mechanisms, control systems and source of driving power. Since automatic operation may be a combination of activities, some type of arrangement is required to synchronize the activities. This synchronization may be purely by mechanical means or through combination of mechanical, electrical, hydraulic, pneumatic and electronic. In modern automation system, use of microprocessor, computer, etc. is common. Automation may be categorized in the following groups:

- Receive and place
- Pick and place
- Static
- Performer

In 'receive and place' category of automation, components, material or input enter the automation device by itself. Automation device only carries it or just pushes the component to its due position. This means that item or object is placed on its correct position.

'Pick and place' is that automation where one of the functions of system is also to go to pick the item and then place it at the desired position.

'Static automation' is that arrangement where there is no moving part by virtue of any kind of drive system.

'Performer' automation does not necessarily handle any component or product but a number of operations are performed in a synchronized manner. This type of automation is abundantly found in automatic plants and machinery.

The author is providing herewith brief description of a number of automation systems. Care is taken to cover a variety of combination of movements of machine parts. Descriptions are supported by suitable typical figures.

Receive and Place Type

Figure 5.6 A shows a plastic component of about 28 mm diameter and having a neck of 8 mm diameter and 3.5 mm height. The crown diameter is 23 mm and height is 2.8 mm. The weight of component is about 2.4 grams. These components are to be fed to a cylindrical component which has a co-axial rod of about 3.8 mm diameter. These components came at feeding station to be loaded on a turret having equidistant seats. During stay period of turret, plastic component should get in the component having a co-axial rod. Refering to Fig. 5.6 B, plastic components are filled in hopper (1).

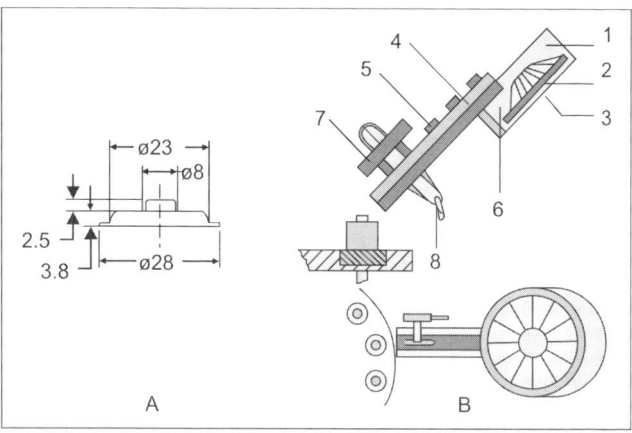

Fig. 5.6: Plastics component feeding system

It has a rotating disc which is attached to a shaft (3) that constantly rotates with a speed of about 15 rounds per minute. The rotating disc has a shallow conical shape having many shallow slots of about 29 mm width. While disc (2) is rotating, a number of components get into shallow slots of rotating disc. Components coming in front of feed channel get slipped into feed channel (4). In this way, channel remains filled up. At the outlet end of channel, there is a rocker arm with stopper pins. Dimensions of rocker and pins are so maintained that either of pins restrict movement of plastic component. As soon as a seat of turret with component reaches feeding station, rocker arm (7) releases one component while not allowing another component to slide down. Just before turret starts indexing, rocker arm pin restricts movement of plastic component. Rocker arm releases it only when next seat with component reaches in front of feeding channel. This process keeps on repeating itself.

Referring to Fig. 5.7, (1) is a magazine which is filled by shells manually or by a conveyor belt. Part (4) is an arm of machine which swings left and right by certain degrees, say 30°. On the top of arm platform (3), a small pick and carry mechanism is attached. It has a long tong (5) to prevent the shell falling down when arm moves to extreme left position. Pick and carry attachment is fitted with a holding arm (6). One end of holding arm has curvature to match shells to be held. The other end has a roller (8). The holding arm is pulled down by a tension spring (10). There is also a pin (9) which keeps the holding at such a position that gap (g) is less than diameter of shell and having a gripping force due to tension of spring. As soon as machine arm (4) swings to its extreme right position, roller (8) rides over a cam (7). Consequently, gap (g)

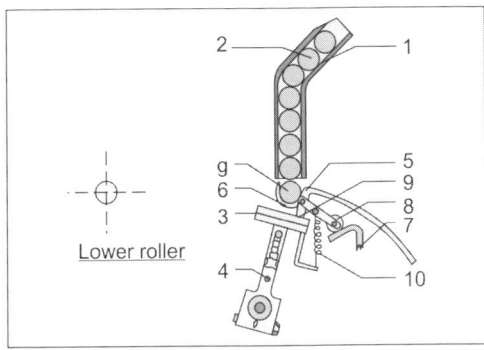

Fig. 5.7: Another 'receive and place' mechanism

increases and one shell gets dropped in the gap (g). As soon as the machine arm (4) starts moving towards its left side position, roller (8) gets released from cam (7) hence, full force of spring is transferred to (6) for firm grip of shell in the gap. When machine arm reaches its extreme left position, it stays there for a while. During this time, a pusher comes to push the shell on to machine spindle. After a little delay, machine arm swings back again to right to receive another shell from magazine. In this way, shells are automatically received by mechanism and placed to such a position that other machine element or mechanism 'take charge' of shell. In this way, this process keeps on going.

A third example of 'receive and place' mechanism is shown in Fig. 5.8. It is the mechanism which places the calot in front of extrusion die (7). Part (1) are calots which are guided in a feeding channel (2), (3). Feeder (4) is at such an extreme left position in feeder guide (5) and (6) that a calot falls between its sliding unit and spring loaded tong (8). In this position, tong (8) is in slightly open position against spring tension. As soon as feeder (4) moves towards right, tong grips the calot firmly enough that it does not fall. Feeder (4) moves right to reach its precise position. In this position, the center of calot coincides with the center of extrusion die and punch. Punch comes with high velocity and takes calot inside the die. While feeder is in its right side position, no calot can fall as lowest calot is held by upper surface of feeder slide. All the movements are mechanically synchronized.

Pick and Place

There are many situations in production processes where automation is done for picking up the component and placing them at a particular position. In this situation, component does not come in for carrying and placing, instead it has to be picked up (see Fig. 5.8).

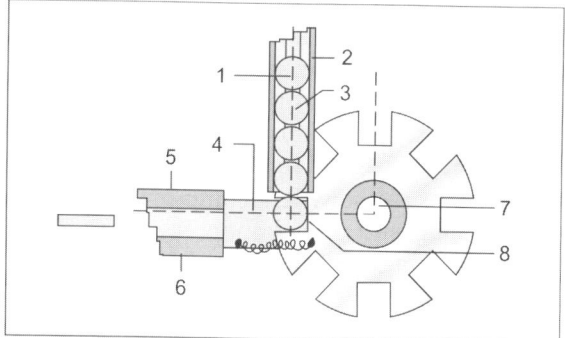

Fig. 5.8: Feeding system with orientation

Few examples of 'pick and place' automation are briefly explained with the help of figures.

Figure 5.9 shows a power press which has a series of dies and punches. Input to this press is shells of about 20 mm in diameter and 13 mm high. These shells are required to be converted to shells of 10 mm diameter. This is achieved by progressive draws in four stages. Part (1) is a channel which connects a feeder to station 'A' of press, that is, first draw die and punch. On next stroke of press, a shell present on die 'A' is drawn. Now it is to be transferred to position 'B'. On second stroke, 'B' is to be transferred to 'C' and another shell duly drawn from station 'A' to 'B'. In third stroke, 'A' is to be transferred to 'B'; 'B' to be transferred to 'C' and 'C' is to be transferred to 'D'. This is achieved by pair of transfer arms (in strip form). Motion of RAMs is mechanically synchronized with the movement of power press mini RAMs which are operated by a heavy and sturdy cam shaft. Movement of pair of arms is indicated by a movement depicting sketch. Referring to this sketch, arms grip all the components on four stations duly ejected by ejector

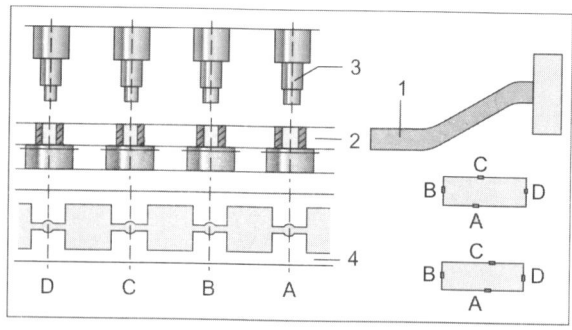

Fig. 5.9: Components transfer system

rods from below the dies. Arms move one pitch of dies, that means, central distance of one die from next die. Arms open out leaving the components on dies and dropping the last one on an ejection channel. Arms then travel back to original position and close into grip next set of components and repeating pick and place operation.

Next example is of a case where component is picked up from one turret of a machine and placed under the head of another turret of the same machine. Figure 5.10 shows basic arrangement of two turrets and transfer arm.

Figure 5.10 A is that of a component which is mounted on the head of turret (1). Each head of turret (1), when stays at station S1, has a component mounted. There is a transfer arm (4) fixed on the base of machine between two turrets. Vertical spindle of arm head (4) partially rotates clockwise and anti-clockwise with certain angle. On the top of spindle, an arm (2) is attached. At the other end of arm is fitted a gripper (3). While the turrets are in rotating mode, arm is also moving. When turret stops, just at that moment gripper (3) reaches and grips the glass stem of the component 'A'. Immediately after gripping, arm swings towards turret (5). While arm is swinging, gripper (3) also swings with component towards head of turret (5). It reaches turret (5) just after it stops moving. Now the component is just over a pusher which pushes glass stem (tube) on the head of turret's over head ring carrying heads with rubber grippers. Design of arm, its gripper and movement is quite complicated and unique. It is therefore explained in detail with the help of Fig. 5.11.

Figure 5.11 consists of two drawings 'A' and 'B'. 'A' is a line diagram depicting position of transfer arm (2) and gripper (3). OX is the position of arm and gripper when it has picked up component from turret (1) (Fig. 5.10). Arm then swings in anti-clockwise direction. Linkages between arm and gripper are so designed and

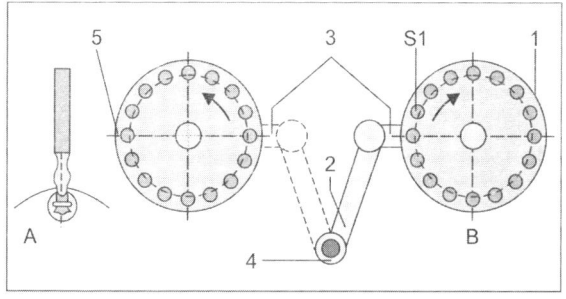

Fig. 5.10: Swinging arm transfer system

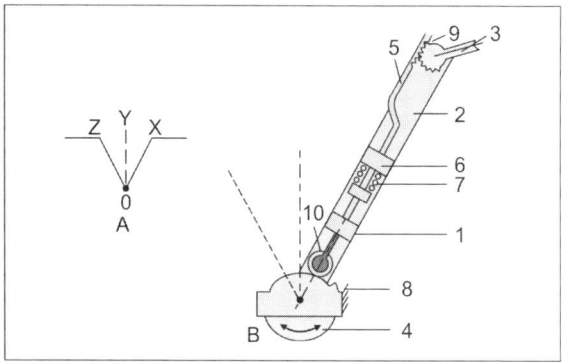

Fig. 5.11: Gripper actuation system

made that gripper also starts swinging in anti-clockwise direction with its swinging point at the end of arm OX. When arm reaches position OY, gripper has already swung to position almost parallel to arm. When arm reaches at position OZ, gripper has also swung further in anti-clockwise direction so that component is just above pusher rod of head of turret (5). Figure 5.11 B, shows the mechanism which provides motion to gripper while arm (2) swings. Gripper (3) has a rotating pin joint with a gear segment (9). A rack (5) is connected to gear segment. Rack (5) is rigidly fitted on an extension rod passing through holes in extended blocks (1) and (6) on swinging arm (2). At the end of extension rod, there is a roller (10) which is touching a fixed cam (8). This cam is so fixed with machine body that it is independent of arm head (4) movement. Consequently, roller (10) of extension rod acts as follower of cam and operates gripper. All construction elements of this mechanism are so adjusted that gripper (3) is pointing towards right when arm is towards extreme right. When arm is at its extreme left position, gripper is pointing towards left. Since the arm (2) is mechanically synchronized with movements of turrets, automatic picking of component (glass stem) by fingers (spring loaded or having vacuum system) and swinging to other turret keep on repeating itself with precision.

Another example of automation of 'pick and place' nature is where a glass tube is first oriented while coming down from a bulk hopper. Once oriented in vertical position, a vacuum picker holds vertical glass tube and carries it to gripper of turret head. The Basic construction is shown in Fig. 5.12.

A very interesting 'pick and place' automatic system is described here with the help of three figures. Figure 5.13 shows as to what is

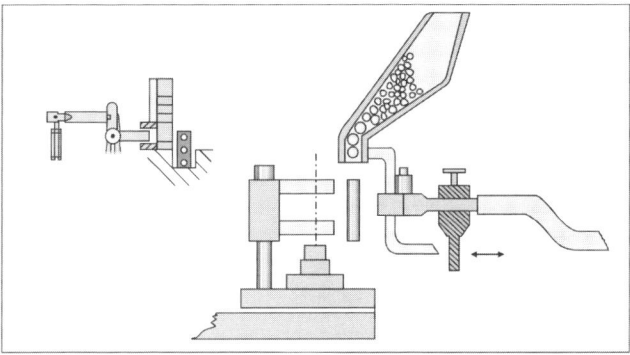

Fig. 5.12: Pick and place mechanism

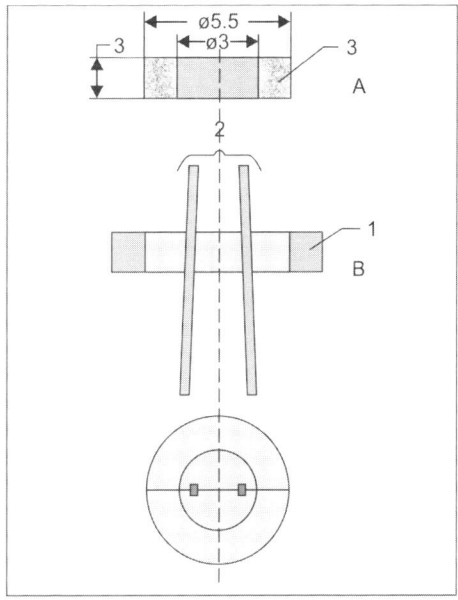

Fig. 5.13: Component feeding arrangement

required where (A) is a sintered glass bead. Approximate dimensions are given. (B) is the head of a turret. Fingers (1) are gripping two electrode wires. Some length of wire is above the finger (1) and some below the fingers. Indexing speed is fifteen indexing per minute. This means that each indexing takes four seconds. This time of four seconds is total time of stay of turret and then getting indexed. Out of four seconds, stay time is about three seconds and one second for indexing movement of turret. Stay time is the period in which sintered glass bead is to be fed.

Automatic pick and place system is such that glass bead gets dropped around upper electrodes (protruding wires) as soon as turret stops. System to pick such a small component and then to feed it around electrode is explained with the help of Figs 5.13, 5.14 and 5.15.

'Pick and place' system for feeding sintered glass bead consists of two sub-systems. First is shown in Fig. 5.14.

Referring to Fig. 5.14, part (1) is a bowl in which beads are filled to a certain level. Part (2) is a tube which has an opening (14) at the top. Bore of opening is so much that diameter of bead passes through opening. Gap between opening (14) and feeding pin (3) is such that any bead falling on the side of pin does not pass through

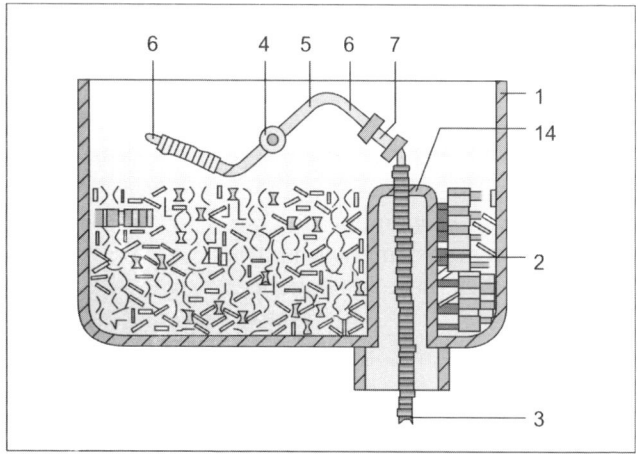

Fig. 5.14: Axial shift system by ratchet

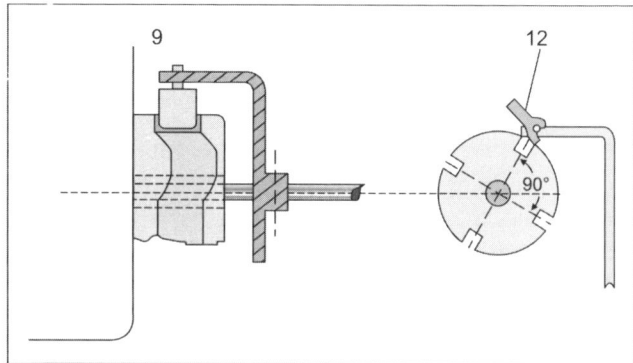

Fig. 5.14 A: Pick and feed system

the gap and actually fell down in the bowl due to taper shape of tube at the top. Part (4) is a spindle which passes through bowl walls. On the spindle is mounted a strip (5). At the end of this strip are mounted picking needles (6). Spindle (4) rotates in a peculiar manner. While rotating intermittently, carry pins pass through 'mass' of beads. Level of beads is so maintained that only pin (6) dip into beads and not strip (5). When spindle (4) rotates, needle is filled with beads which also move to such an extent that the end of feeding pin reaches near vertical feeding pin (3), only a few mm away from vertical pin (3) top end. All beads (7) slide down. Some fell around pin (3) and some in bowl. Spindle further rotates only after axially shifting to side so that pin (3) is NOT in striking range of vertical feeding pin (3). Further rotation of spindle (4) brings the other pin (6) in position. In this way, both pins (6) keep on feeding beads around vertical feeding pin. Part (12) is a ratchet operated by a cam underneath the table of machine where various cams are fitted. This ratchet rotates indexing wheel by 90°. There is a cam (9) on the diameter of which cam channel is machined. This cam is permanently fitted to bowl's outer wall. There is a follower bracket with roller engaged in track of cam (9). When spindle rotates, axial movement also takes place due to hump in cam track. Design and dimensions are so maintained that the following action takes place in sequence:

- Spindle (4) rotates.
- Pick up pin (6) enters beads in the bowl.
- Spindle rotates further till pick up pin's pointed end reaches near pointed end of vertical pin.
- Spindle starts to rotate further together with immediate axial shifting.
- Next pick up pin passes through beads, thus picking up few beads.

Referring to Fig. 5.14 A, one time stroke of ratchet during one complete indexing cycle of turret takes place. Each time spindle (4) rotates by 90°, hence moving ratchet wheel by 90°. This gives a little time to placing needle to stay aligned with vertical feed pin between two 90° timing of cam. Now comes the description of system which holds the vertical pin in place and yet places beads around electrodes. This is explained with the help of Fig. 5.15.

Part (1) a is vertical floating pin. Pin is called to be floating as it is not fixed anywhere. It is held in place vertically either by finger pair (5) or (4). These two fingers open and close alternatively. When finger (5) holds the pin, finger (4) opens. Then after a lapse of few

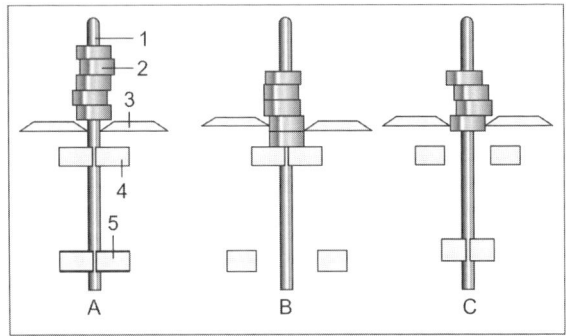

Fig. 5.15: Floating pin holding system

moments, finger (4) closes and finger (5) opens. Sharp edge finger pair (3) is meant to hold the beads from dropping and releasing only at a pre-adjusted sequence. It is worth noting that all the fingers (3), (4) and (5) get operated by cams underneath the machine table. All fingers have some mechanism for opening and closing. Mechanisms are connected to cam follower levers by means of connecting turn buckle rods. Sequence of operations is given below:

- All the three fingers are in closed position (see Fig. 5.15 A).
- Finger (4) is holding vertical floating pin in place and fingers (3) and (5) get open (see Fig. 5.15 B).
- Distance between lower face of finger (3) and upper face of finger (4) is a little more than one bead height, hence, one bead gets dropped and rests on upper face of finger (4).
- Fingers (5) and (3) close and then finger (4) opens. The bead which was resting on finger (4) drops down to rest on face of finger (5).
- Sequence repeats itself. Finger (5) opens, bead gets dropped around electrode wires protruding above turret head fingers. Timing of sequence is so set that finger (5) opens only when turret has just stopped indexing. (s ee Fig. 5.15).

In the following paragraph, another typical mechanism for the same work as explained above, is described with the help of Fig. 5.16 for picking and placing glass bead from feeding channel head to around protruded electrode in turret finger.

Referring to Fig. 5.16, Part (1) is a vibrating feeder in which beads are filled up to approximately specified height. Vibrator makes the beads to travel continuously to feeding channel (2) to keep it always filled up. Due to a stopper (3), beads do not flow out of

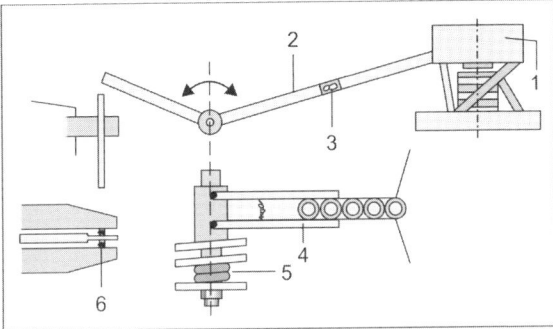

Fig. 5.16: Pick and drop system

channel, a swinging pair of gripper (4) is at picking station. At the position of swinging fulcrum, there is a mechanism which opens and closes gripper and swings from picking station to placing station. Operating mechanism is synchronized with turret movement with the help of under the table cam, follower lever and turn buckle link. Sequence of operation would be as follows:

- Gripper pair reaches picking station.
- Closes around a bead which is already there.
- Swings to a position where held bead is just over and axially aligned with protruded electrodes on the top of turret finger.
- Gripper opens up, bead gets dropped around electrodes.
- Gripper swings back again to picking station. In this way, sequence keeps on repeating itself. Swinging gripper swings through a friction drive which rotates (reciprocate cum rotation) somewhat more than required swing. There are stop screws under the gripper to precisely adjust stop position on both picking and placing stations.

5.5 PERFORMERS

Performer is the name given to those mechanisms which do not directly handle inputs or outputs of any production process. There are mechanisms which ultimately operate particular unit with or without sensing presence of input component at a particular station. There may be innumerable examples of such mechanism. In this chapter, two examples are given and briefly explained with the help of figures.

First example is of a machine which is used to fill and mark a miniature container with precise quantity of paste.

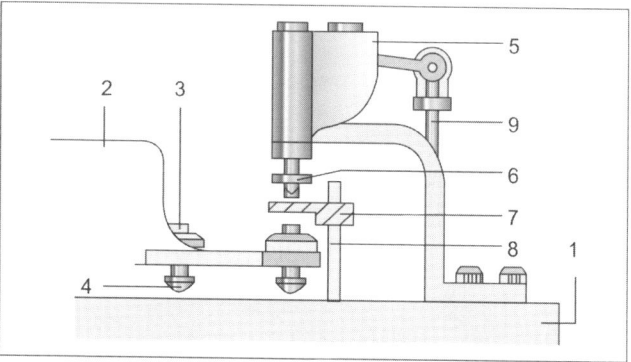

Fig. 5.17: Paste dosing system

Following is the list of parts of machine:

1. Table of machine
2. Turret with eight or ten heads
3. Component placement heads
4. Ejection pusher
5. Paste dosing and feeding unit
6. Paste feeding nozzle
7. Sensor plate
8. Sensor plate rod
9. Paste unit actuating turn buckle rod

The basic construction of paste filling system of a typical paste filling machine is shown in Fig. 5.17. Machine has a turret which also moves up and down. It has a number of heads in which miniature components are fed. Turret gets indexed only when it is in down position. As soon as indexing stops, turret moves up and stays there for 'no indexing' period of indexing system. This is the position of turret when component is fed in turret head at component feeding station. At paste feeding station, if component is present in turret head, lifts sensor plate which activates under the table mechanism to operate paste dosing and feeding mechanism. If by chance, there is no component in turret head, sensor plate will not be lifted and hence, mechanism under the table will not get operated. On third station, automatic marking unit will mark the container (small component) by marking ink or stamping. Both being rotary units (not described), it needs to be briefly explained as to how sensor plate controls operation of dosing and feeding unit. It is explained with the help of Fig. 5.18.

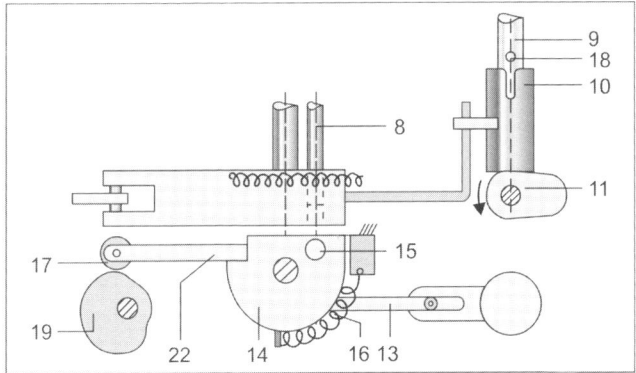

Fig. 5.18: Component presence sensor system cum actuator

Part (9) is a paste actuating push rod which has a follower tube (10) at the lower end. (11) is a cam which lifts follower tube to the extent of its throw. Part (8) is a sensor plate rod which is inside a hole (15) of a plate (14). This plate is located on a pinion so that it can revolve. On the circumference of plate (14), there is a spring which keeps plate pushing against pin (15). There is a lever (22) whose one end is fitted to plate (14) and the other having a follower roller (17). Lever (22) is actuated by cam (19) fitted on the main cam shaft extension under the table. All the movements are synchronized with the movement of turret. Operational sequences are given below.

When there is no component inside turret head:

• Turret reaches at 'no indexing' position and rises.

• Since there is no component in the head, sensor plate is not lifted.

• Rod (8) remains in the hole of plate (14).

• Cam (19) is at a position with 'no throw' just below follower roller, but follower roller does not move to touch cam.

• Plate (14) remains stationary against pin (18).

• Cam (11) actuates follower tube (10). This tube is below the pin fitted in pusher rod (9). No vertical movement of pusher rod takes place.

In case, there is a component in the head then sequence of action of mechanism would be as follows:

• Sensor plate rod (8) would be out of hole of plate (14).

• Less radius portion of cam (19) would be there and follower (17) with lever (22) will swing due to spring (16) compression.

- Consequently, lever (13) would rotate tube (10) to such an extent that its slot is no more under the pin (18).
- Hence, paste unit would operate and paste would be fed inside the mini container/shell.

In this way, machine would keep on working. Paste unit delivering metered quantity when mini container is present in the head and unit would not operate, if there is no component in the head. There may be other type of construction of mechanism as well to achieve same function of the system.

There may be a number of operations which may be carried out on separate machines. For example, operations carried out on a brass barrel are as below:

- Lining
- Bulging, both ends
- Thread rolling, 1st end
- Thread rolling, other end

All the four machines are connected to each other. Output of first machine is the input of second machine and so on. This is achieved by various automation devices.

Hope, the reader has gained pretty good idea of automation. Now some cases of tools with combined automation are briefly described with the help of figures.

5.6 ROLL FEEDING TO TOOLS

Referring to Fig. 5.19, where tool (1) is for punching out blanks from roll of brass sheet (in rolls). The functioning of whole arrangement is such that there is no need of an operator to do watching.

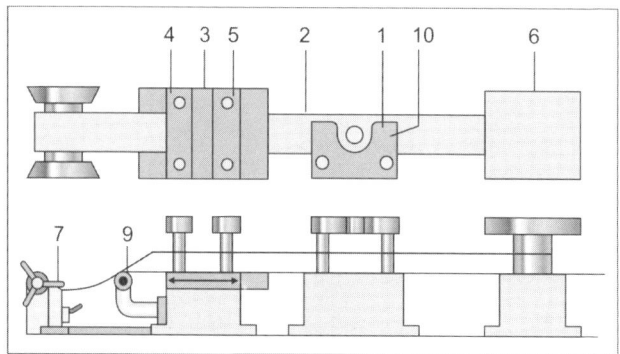

Fig. 5.19: Auto feeding system for sheet metal strip

The operator has to load another roll when running roll leaves the rotating stand. Part (2) is the roll strip (word strip is generally used on shop floor for pieces of sheet metal L meter long and W mm wide). Part (3) is the feeding unit which precisely feeds the desired length of roll strip to tool. Generally, stroke of feeding is diameter of blank plus 2–5 mm, depending upon diameter of blank and its thickness. Tie between two blanked holes may be between 2–5 mm (see Fig. 5.20). Feeding unit consists of two vertically reciprocating bridges (4) and (5). Up and down movement of "bridges" of small value, say 10 mm. Motion of these two bridges is synchronized with movements of power press RAM. Bridges (5) have two motions, up and down, forward and backward. Sequences of operation would be as follows:

- Both the bridges are brought to up position.
- Press RAM is in up position.
- Operator feeds the roll strip through space between stripper plate and cutting rings so much that leading edge of roll strip is beyond cutting edge of ring, say 3 mm.
- Now the lever or button of feeder is activated, thus, feeder unit and RAM movements are synchronized.
- Bridge (4) moves down and grips coil strip in position.
- Bridge (5) travels towards bridge (4) and grips coil strip on the lower railing of bridge (5), which is integral part of bridge and moves together with top cross member of bridge (5).
- Press RAM moves down, cuts the blank and pushes it down through blanking die and under the press table.
- As soon as cutting punch moves up, coil strip gets detached. Due to stripper plate (10), stationary bridge (4) releases its grip on coil strip and after a fraction of a second (time period), bridge (5) moves towards die together with coil strip. Amount of

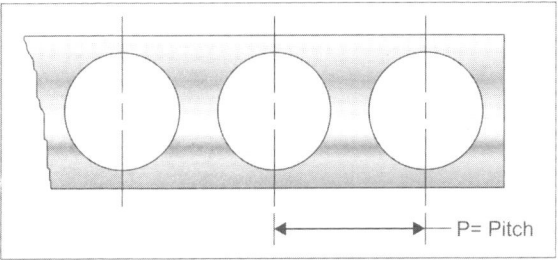

Fig. 5.20: Pitch of blanking on strip

movement is equal to set pitch of blanking, meaning diameter of blank plus desired die width.

• Again blanking takes place and bridge (4) grips coil strip and bridge (5) releases coil strip and moves towards bridge (5).

• In this way, automatic feeding of coil strip keeps on taking place.

After the die, there is another unit (6) which keeps on chopping the scrap into small length for the purpose of convenience in removal of scrap from press site.

Also note that on the left hand side of press, there is roll holding stand that keeps on releasing coil strip of desired length so that there is always a loop, consequently, coil strip is always available for bridge (5) to pull it without excessive force. There is a sensor which controls loop formation. Coil holder automatically keeps on releasing coil strip by partially rotating in unwinding direction.

Feeding units are generally standardized. Surface on which coil strip slides has a definite height from base of unit. Feeding unit is generally mounted, on bolster plate of press on which press tool is also mounted hence, top of cutting ring should be in level with coil strip sliding surface of feeding unit. This consideration is taken care of during designing of tools. In case due to other factors, if it is not possible then feeding unit would have to be modified for height adjustment to suit die height.

Another example of automatic feeding of coil strip for a cut and cup tool is illustrated by means of Fig. 5.21.

In Fig. 5.21, part (1) is the bolster plate of power press on which cut and cup tool (2) is set. Part (3) is the stripper plate of lower member of tool. Part (4) is cutting punch cum drawing die fitted in

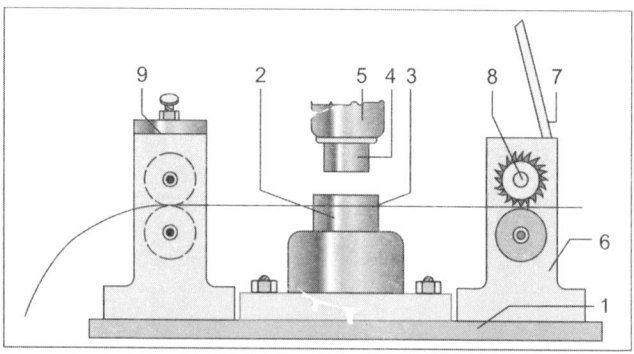

Fig. 5.21: Another synchronized feeding system

RAM. Part (5) is the RAM of power press. Part (6) is the roll feed unit fitted on bolster plate. It carries a pair of rollers, which are spring loaded and connected with each other with the help of spur gears. As shown, rotation of upper roller is in anti-clockwise direction and that of lower roller clockwise, consequently, scrap roll strip is pulled towards right side. Indexing rotation of rollers is achieved by ratchet mechanism which is connected with adjustable eccentric on crankshaft extension spindle. Eccentricity is adjustable to achieve desired indexing length of coil strip (See Fig. 5.22). A chopping unit may also be installed after right side feeding roller.

Another example of a Hitch indexing system is briefly explained with the help of Fig. 5.23.

In Fig. 5.23, scrap roll strip (1) is shown getting indexed towards left. There is a Hitch feeder which grips only one edge of roll strip scrap. Other edge is guided by guide plate (2). Part (3) is a Hitch feeder which reciprocates according to pitch set at eccentric

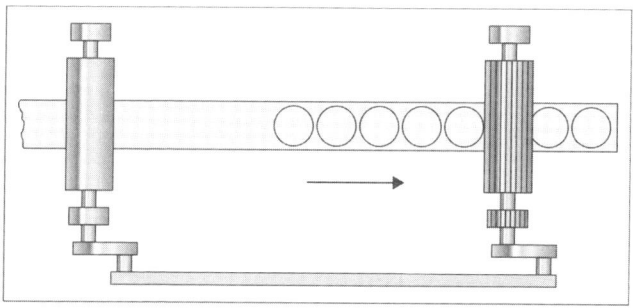

Fig. 5.22: Rollers feeding system

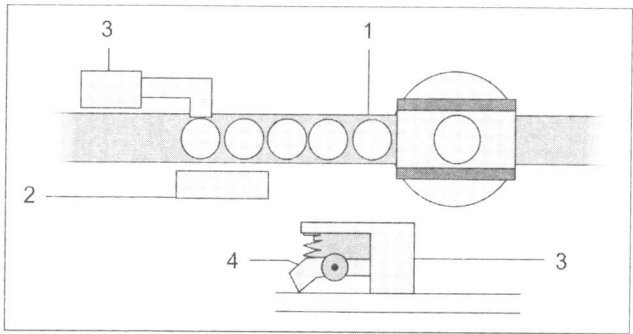

Fig. 5.23: Hitch feeder

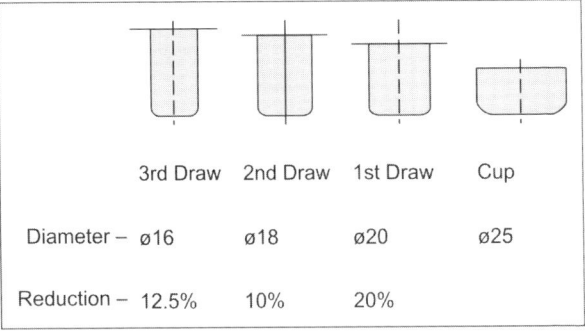

	3rd Draw	2nd Draw	1st Draw	Cup
Diameter –	ø16	ø18	ø20	ø25
Reduction –	12.5%	10%	20%	

Fig. 5.24: Prograssive draws

mechanism attached to crankshaft end. This feeder is used for light work on small presses. It is effective for aluminum and brass because scroll (4) of sharp edge can penetrate a little when pulling action starts. In handling steel roll strip, it might not be very effective. Sometimes there may be a slip in grip, consequently, improper indexing.

Example: Automation picks and places action in a multi-station light duty press where progressive operation of draws, redrawn and trimming take place. The functioning of this type of automation is explained with the help of Fig. 5.25.

5.7 HIGH SPEED LIGHT WEIGHT PRESS

Figure 5.25 shows a high speed light weight press. Such type of presses are generally used for progressive draw operations such as shown in Fig. 5.24.

Figure 5.25 A shows constructional concept of press. Part (1) is the frame/body of press. Part (2) are groove cams mounted on cam shaft (3) which is driven by a driver box (4). Small square RAMs (5) are so attached to groove cams that RAMs move up and down with the rotation of cams. Part (6) are individual die blocks in which die inserts of appropriate bore and radii are fitted. These are replaceable inserts. Part (7) is a transfer arm which carries gripping fingers. As soon as punches (attached to RAMs) go up, transfer arm grips component and moves by one pitch of cavities/dies. Then punch comes down and draws take place. As soon as punches go up, transfer arm moves back to its first position, ready for repeating grip and transfer operation.

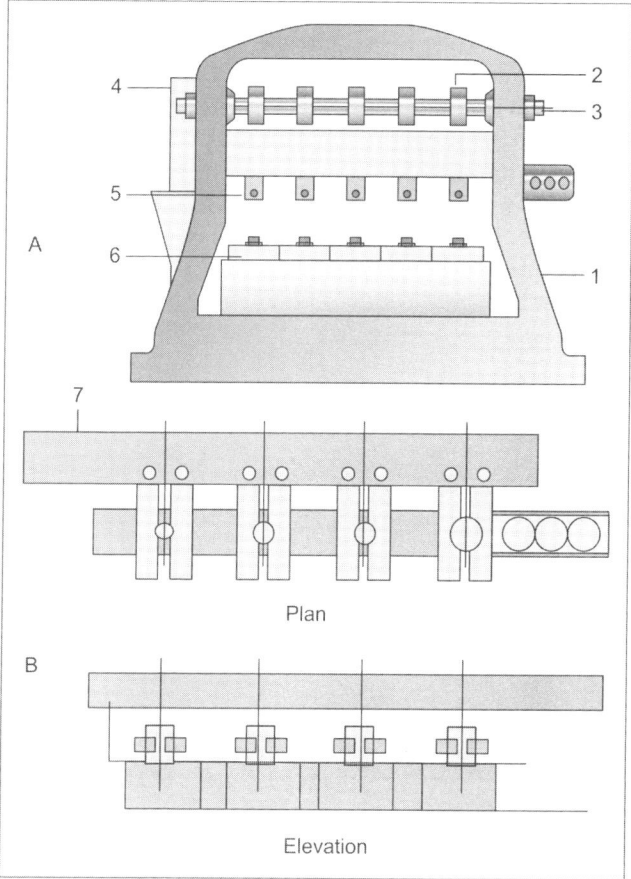

Fig. 5.25: High speed, light duty press

5.8 ROBOTICS

Robot is a machine which is programmed to carry out certain operations such as pick and place. These are generally used on production lines to perform light and heavy work. Designing and manufacturing of a robot is done keeping in view the nature of work.

Robots are also used in assembly work of electronics assembly. Other sturdy robots are used on car assembly lines. Robots are also used for picking big size steel panels of say, railway engines, from conveyor belts to heavy power press for operation such as piercing, stamping, lancing, etc.

Figure 5.26 shows a schematic set up for transferring a sheet steel plate (2) from a conveyor belt (1). There is a robot (4) whose pick up head (3) can be moved in all the three X, Y and Z axes. Robot is so programmed that its magnetic head (3) picks up the plate and carries it to place it over the die (6) in a hydraulic press (5).

Arm comes out of the die, stays out while tool closes to perform operation. As soon as tool opens, robot arm comes again and picks up pressed plate (say, a side panel of railway engine) and carries it to place it on another conveyor. Sequence of operations of power press conveyor belts and robots are synchronized by means of PLC (Programmable Logic Control).

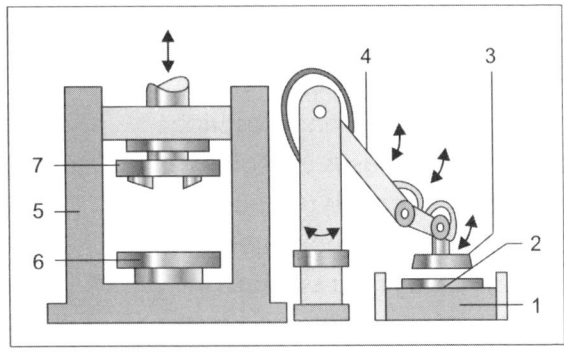

Fig. 5.26: Robotics pick and place system

6

Metrology

This chapter consists of the following aspects:

6.1 List of measuring aids and description
- Calculation of least count
- Method of taking measurements
- Photographs of micrometer and vernier caliper
- Types of gauges and description

6.2 Description and uses of dial test indicator

6.3 Description of profile projectors

6.4 Surface finish and measurements

6.5 Definition of lays, Ra symbol and value
- Profilometer construction and description

6.6 Magnifiers and microscopes

6.1 LIST OF MEASURING AIDS AND DESCRIPTION

Metrology is a science which deals with the measurements of dimensions, angles, weight, volume, pressure, vacuum, temperature, speed, etc. In sheet metal press tool making, it is of utmost importance to have fairly good idea of accurate measurement of length, width, height, depth, diameter, bore and angles by a variety of measuring instruments. It is also important to know correct method of handling a measuring instrument and doing accurate observation. Gauges and gauging system also comes under preview of metrology. For example, radius of a draw die cannot be normally measured directly. Value of a radius can be compared with a **radius gauge** which has a number of leaves having various value of radii marked on it. Figure 6.1 gives an idea of external and internal comparative measurement by a Gauge.

Figure 6.1 shows a draw punch with a radius R_p and draw die with an internal radius R_d. Value of radius R_p of draw punch can be accurately measured by means of a profile projector, but internal

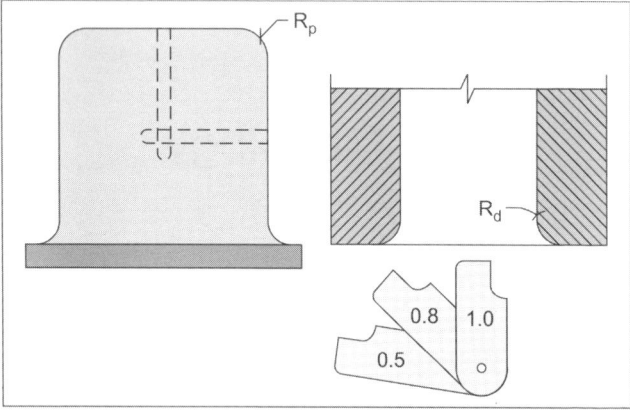

Fig. 6.1: External and internal radii checking

radius R_d of draw die cannot be measured by a profile projector. The only normal way to measure is through use of a radius Gauge. It may be noted that it is not an exact measurement of radius value. It is a visual comparison with a radius gauge. After above example, it is in the fitness of things that a die and tool maker should know about various measuring instruments, equipments, electronic Gauges, etc. Following is the list of various **measuring instruments** with brief description, which are normally used for measurements or aids to measurements:

- Steel rule
- Vernier caliper
- Digital caliper
- Outside micrometer
- Feeler gauge
- Slip gauge
- Profile projector
- Illuminated magnifier
- Wall thickness measuring micrometer
- Slide caliper
- Dial caliper
- Inside micrometer
- Depth micrometer
- Dial indicator
- Sin bar gauge
- Height gauge
- Hand held microscope

Steel Rule

It is used for rough measurement of a dimension. Suppose a tool and die maker has to pick up a flat steel piece from which a finished plate of 110 mm × 90 mm is to be made. Then by approximate measurement by a steel rule, die maker may pick up a piece measuring approximately 120 mm × 95 mm or a bigger plate, what ever is available nearest higher to finished size. Figure 6.2 shows a steel rule.

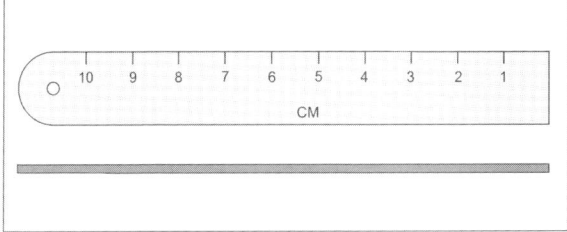

Fig. 6.2: Steel rule

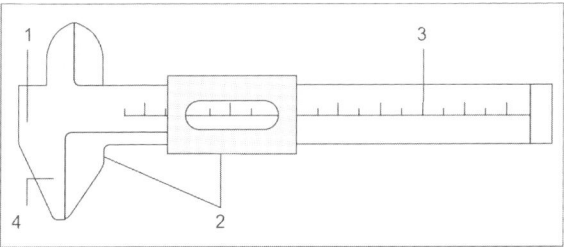

Fig. 6.3: Slide calipers

Such steel rulers are normally made from stainless steel or spring steel duly chromium plated.

Slide Caliper

It provides a better accuracy of measurement of length as compared to steel rule. It is more convenient to measure by this instrument.

Referring to Fig. 6.3, Part (1) is the main body of slide caliper and part (2) is the sliding arm which carries a window cage where a line is marked to read on the main scale line (3). First of all, jaws (4) are opened enough to bring the object in between them. Jaws are then closed over the object. In such a slide caliper, measurement is that where mark on slide is nearest to marked distance on the main scale.

Vernier Caliper (See Photo 6.1)

Refering to Figs 6.4 and 6.5, vernier caliper is much superior measuring instrument as compared to slide caliper. Its measurement reading accuracy is 0.02–0.01 mm, depending on scale and vernier graduations. There is an adjustment slide (5) mounted on the main scale body which is also connected with sliding jaws body by means of a screw (8) and there is a round serrated nut at a location in adjustment slide. There are two locking thumb screws (4) to lock sliding jaws body over the main scale body and (5) to lock adjustment slide over the main scale body.

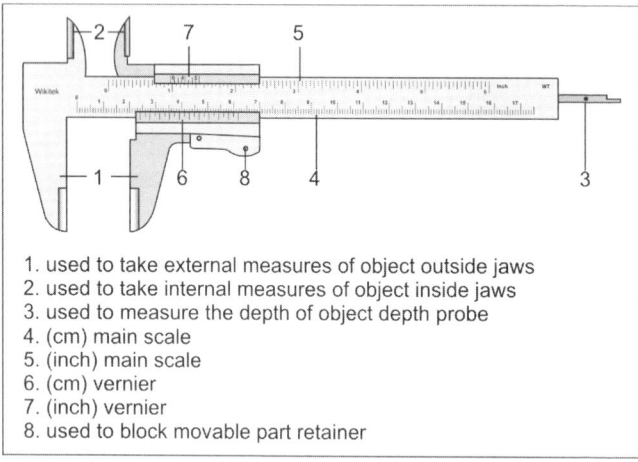

1. used to take external measures of object outside jaws
2. used to take internal measures of object inside jaws
3. used to measure the depth of object depth probe
4. (cm) main scale
5. (inch) main scale
6. (cm) vernier
7. (inch) vernier
8. used to block movable part retainer

Photo 6.1: Vernier calipers
Source: Wikipedia, Free Encyclopedia
Dated: October 7, 2013

The main scale body and sliding body have two jaws each—one set to measure diameter or outside dimension of any job or workpiece.

Figure 6.6 shows use of vernier caliper for measuring diameter and counter bore. For carrying out the measurement, first of all, loosen both the thumb screws so that jaw body and adjusting body can be freely slided over the main scale body. Bring the diameter of workpiece in between the jaws. Tighten thumb screw of adjustment slide which becomes fixed to the main scale body. By

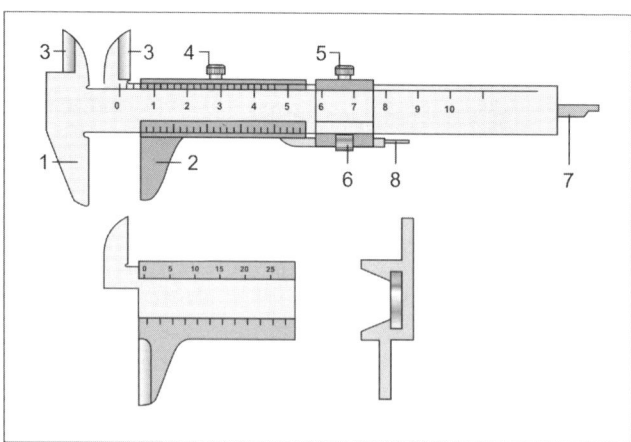

Fig. 6.4: Parts of verier calipers (Enlarged view of sliding jaw)

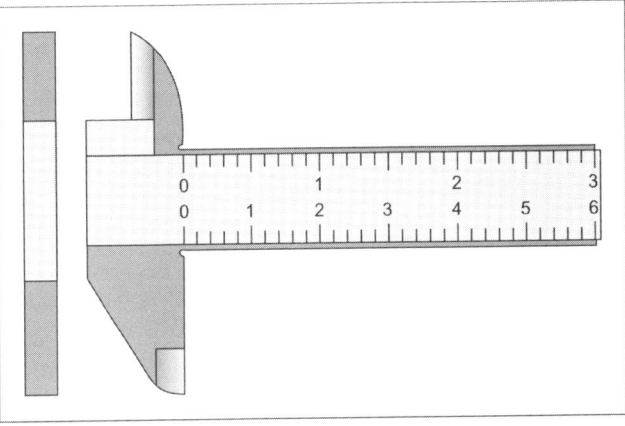

Fig. 6.5: Parts of vernier calipers (Enlarged of main jaw (body)

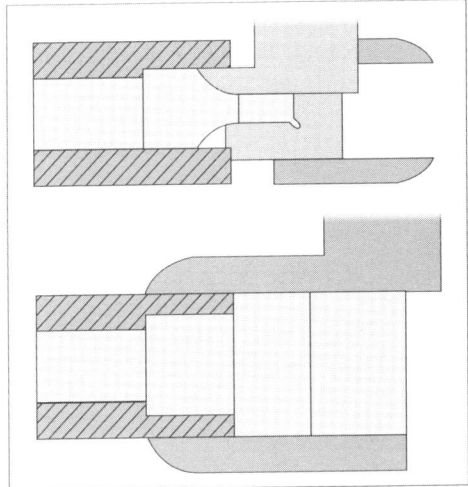

Fig. 6.6: Inside and outside measurement

rotating adjusting nuts (6), jaws may be further closed or opened minutely, depending on the rotation of adjusting nuts.

Now comes the touch feeling of a person which should be sensitive enough to feel proper touch of jaws on workpieces. Too tight or too loose adjustment may give an erroneous reading. Same method of measurement applies to measurement of internal dimension.

Let there be three persons measuring the same diameter with the same instrument and the result is as follows:

- First person ——————————31.26 mm
- Second person ——————————31.24 mm
- Third person ——————————31.28 mm

Hence, the difference in measured value is 0.02 mm. This difference is quite possible when dimension is measured by vernier caliper.

Dial Caliper

A vernier caliper fitted with a dial is called dial caliper.

Amount of movement of sliding jaws can be read on dial rather than looking for matching of the main scale graduation line on sliding jaw. Figure 6.7 shows the shape of a typical dial caliper.

In dial caliper, generally adjustment slider is not provided as it is done in vernier caliper.

It is still a more convenient and accurate to directly read the reading on a window **LCD screen** (liquid crystal diode). There is a thumb lock lever which may be released when desired reading is obtained. Sliding jaw with mounted digital display window can be slided with pressed thumb lock.

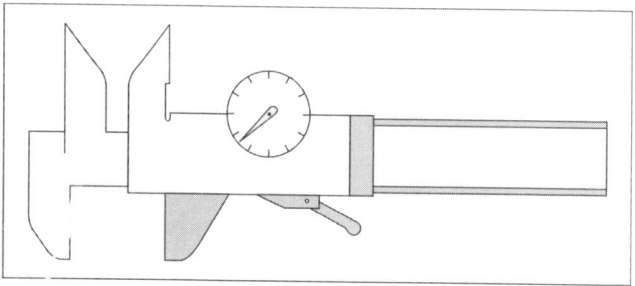

Fig. 6.7: Typical dial calipers

Reading a Vernier Caliper

The method of determining actual reading is explained with the help of enlarged view of scales in Fig. 6.8. It shows enlarged view of vernier caliper's main and vernier scales. Enlargement is about double the size. There is a main scale which is marked on the body of fixed jaw. Five divisions of one centimeter each are shown and each division is further divided into ten, marked to show fifty millimeters. Length of vernier is 49 mm which is divided into ten divisions, each having further five divisions. This means a total of 50 divisions.

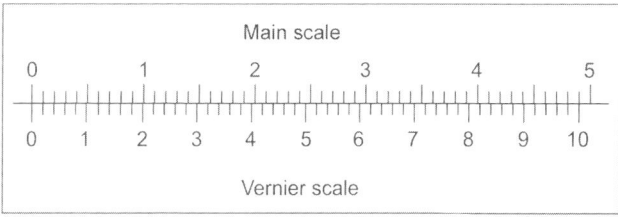

Fig. 6.8: Reading a vernier caliper

If vernier scale zero marked line is shifted to marked line of first division on the main scale then the reading would be exactly one millimeter. In case sliding jaw (vernier scale) is shifted so that marked line of first division on vernier scale coincides with first marked line on the main scale then the reading would be 1/50 mm which equals 0.02 mm which is the smallest measurement or the least count. If line of 2nd division of vernier scale is matched with 2nd division line of the main scale then the reading would be 2/50 mm which equals 0.04 mm, so the formula for determining actual measurement reading is as under:

Number of full divisions crossed by zero line of vernier scale + Number matching division of vernier scale × least count

Example

Open jaws reading =

Full divisions on main scale + Value of bigger division of vernier scale + value of smaller division coinciding main scale division line

$$= 54 \text{ mm} + 0.5 \text{ mm} + 0.08$$
$$= 54.58 \text{ mm}$$

Least Count Determination Method

Referring to Fig. 6.9, a vernier scale is constructed by taking 49 main scale divisions and dividing them into 50 divisions, i.e. 49 mm divided into 50 parts, therefore,

$$1 \text{ VSD (vernier scale division)} = 49/50 \text{ mm}$$
$$= 0.98 \text{ mm}$$
$$1 \text{ MSD (main scale civision)} = 1 \text{ mm}$$

Substituting in formula

$$\text{LC (least count)} = 1 \text{ MSD–VSD}$$

Least count = 1 main scale division – 1 vernier scale division

$$= 1.00 \text{ mm} - 0.98 \text{ mm}$$
$$= 0.02 \text{ mm}$$

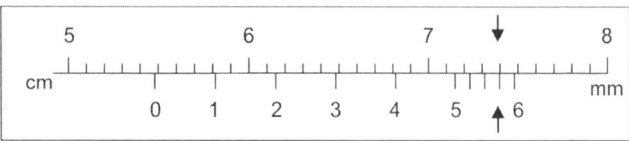

Fig. 6.9: Indication for least count calculation

British Scale

The upper edge of vernier calipers main scale generally has British measurement unit, inch. Normally one inch has ten equal divisions and this each 1/10th division is further divided into four divisions as shown in Fig. 9.6a. Hence, one smallest division has a value of 1/40th of an inch.

Vernier caliper divides it (1/40) into further 25 divisions, hence, 1/40 inch × 1/25

= 1/1000 inch or 0.001 inch

Let us take an example of an actual measurement (Fig. 6.10).

0.450 Plus reading of vernier. 10th line of vernier is coinciding with the main scale line. Hence, value would be, value of one

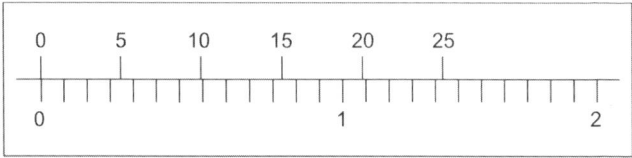

Fig. 6.9 A: British scale on calipers

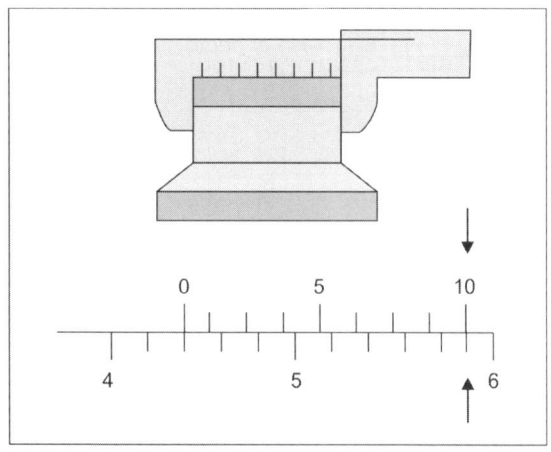

Fig. 6.10: Example of actual measurement

vernier scale division which is 0.001 inch multiplied by coinciding division, that is 10.

Therefore, 0.001 × 10 = 0.01

Hence, total reading

0.450 + 0.01 = 0.460 inch

Figure 6.11 shows the use of depth probe which is provided in a vernier caliper.

Part (1) is a workpiece which has a slot. Depth of this is to be measured. To do so, first of all, slide the probe a little out, place end surface of the main scale body on the workpiece surface (3). With the protruding depth probe, end a little inside the slot.

Now push the sliding jaws so that depth probe end touches the inside surface (base) of slot. Now lock the sliding jaws, take the vernier calipers out and take the reading. Vernier caliper are available in various length measuring capacities, such as 15, 30, 50 and even 100 centimeters.

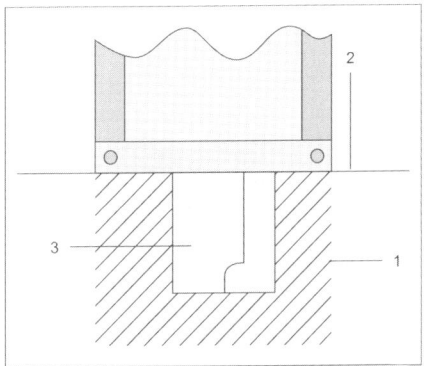

Fig. 6.11: Use of depth probe

Outside Micrometer

(See Photo 6.2 and 6.3)

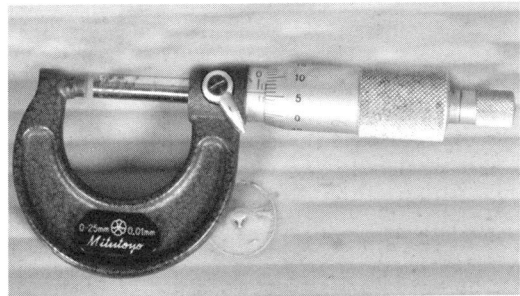

Photo 6.2: Out side micrometer

Photo 6.3: Parts of outside Micrometer

Tube Wall Micrometer

It is a normal micrometer with the exception that face of anvil is not straight. It has a curvature to facilitate measurement of wall thickness of a tube or a sheet metal cap. Figure 6.12 shows anvil and measuring position.

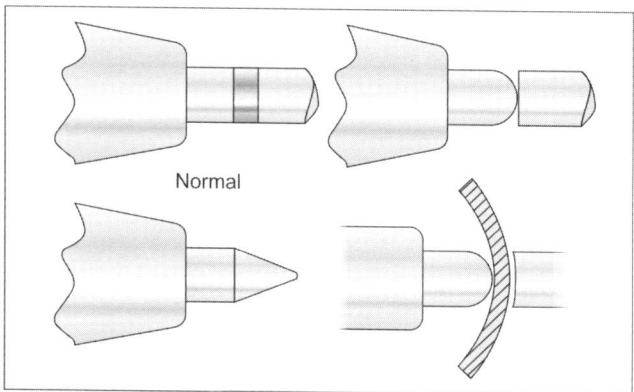

Fig. 6.12: Tube wall micrometer

External Micrometer

External micrometer is a measuring instrument by which a length of an article can be accurately measured. Repetitive accuracy is much better as compared to vernier caliper. Referring to Figs 6.13 and 6.14, an external micrometer consists of a frame (1) on which there is fixed anvil (2). Face of this anvil is made of tungsten carbide. Face is highly plane, lapped and polished. This ensures accuracy in measurement and long performance life. On the other end of U shaped frame, a sleeve (4) is fixed. A spindle (3) is assembled inside the sleeve with precise fine thread, generally with a pitch of 0.5 mm. Threads assembly inside sleeve is such that practically there is no backlash.

Spindle (3) is rigidly fixed with thimble (5). On rotating thimble, spindle moves axially towards or away from anvil (2). Sleeve is axially marked with a straight line (6). This line is marked with division. Each division on top side of the line is one millimeter and marks below the line divides one millimeter division into half, that means, 0.5 mm.

Thimble is marked with 50 divisions at the top position. One complete rotation of thimble (say, from zero to back zero) moves the thimble exactly 0.5 mm. Since spindle is rigidly fixed to thimble therefore, it also axially moves by 0.5 mm. If thimble is screwed

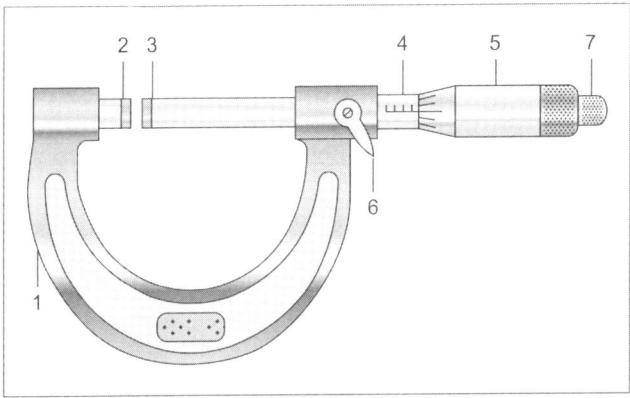

Fig. 6.13: Parts of outside micrometer

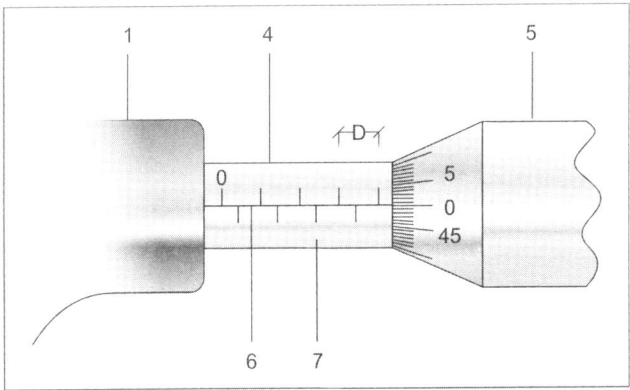

Fig. 6.14: Parts of outside micrometer (scale)

in completely then faces of anvil and spindle sit over each other precisely and at this point, zero line of thimble should coincide with zero line of marking on sleeve.

To ensure that spindle every time closes with equal force, a ratchet mechanism (7) is provided. When ratchet is turned clockwise (seeing towards ratchet), it starts slipping as soon as spindle face meet resistance due to presence of any workpiece or directly at anvil if there is no workpiece.

Let the diameter of a ground pin is being measured and the position of thimble is as shown in figure. Note the thimble edge has crossed 4 mm division line but not 4.5 mm line. The 49th division marking on thimble is coinciding with axial line on sleeve. Hence, the reading would be as follows:

$$= 4 + 0.49$$

$$= 4.49 \text{ mm}$$

Sometimes, it is found that on closing spindle face on anvil face, zero line of thimble does not exactly match axial line of spindle. Most likely it may be due to some micro dust particles in between the faces of anvil and spindle. Proper matching of two zero lines may be achieved by careful cleaning. In case error persists then sleeve may be slightly rotated inside the frame by applying force with the help of a C-spanner which is normally provided with micrometer (See Fig. 6.15)

Micrometers are generally available in a variety of measuring ranges such as 0–25, 25–50, 50–70 and so on. In micrometer, ranging from 25–50 mm or 50–70 mm, checking of zero error is done by means of standard checking gauge as shown in Fig. 6.16.

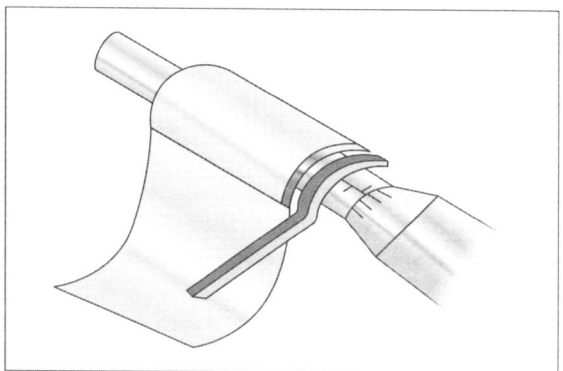

Fig. 6.15: Zero error adjustment

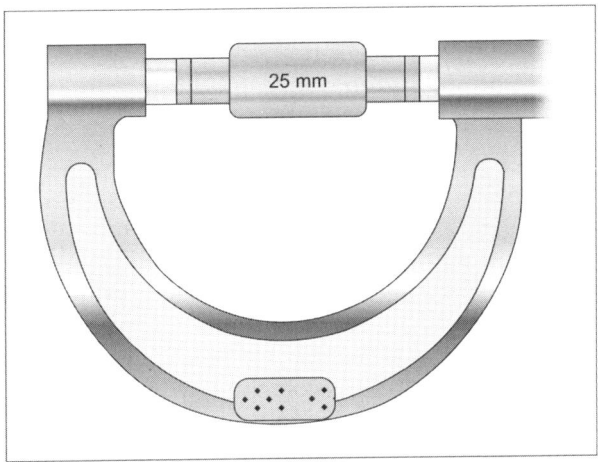

Fig. 6.16: Error checking with standard gauge

An important precaution is that micrometer should not be allowed to get warm or too cold because it will cause an error in measurement.

Internal Micrometer

Internal micrometer is used to measure internal dimension such as bore, width of a slot, etc. The basic construction of internal micrometer is shown in Fig. 6.17.

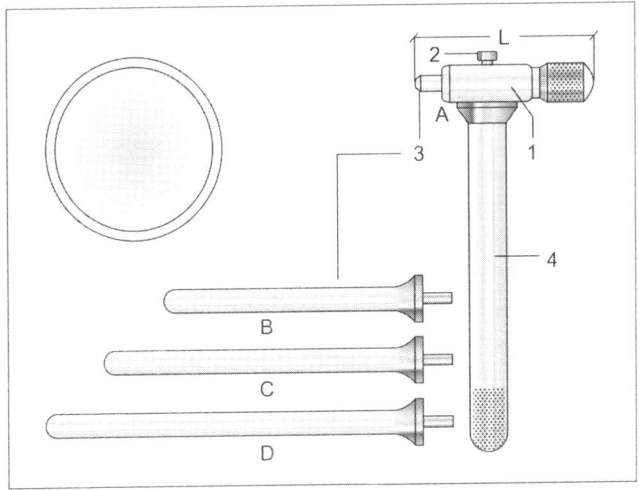

Fig. 6.17: Internal micrometer

Referring to Fig. 6.17, part (1) is the body of inside micrometer. L is the minimum bore size which may be measured due to dimensions of instrument. Let it be 25 mm. Adjustment in dimension L is generally 13 mm and it is done by unscrewing thimble.

Range increasing inserts (3) may be put in place as per need. For example, measuring a bore of 45 mm would need a longer insert 3 B. Still bigger bores may need 3 D insert.

Measurement reading techniques would be the same as in outside micrometer.

Depth Micrometer

In Fig. 6.18, measurement of slot depth is done by vernier calipers which has its own accuracy limit. In case depth is to be measured very accurately then depth micrometer is used. Figure 6.18 shows a situation where vernier caliper depth probe cannot be used properly. Use of depth micrometer is expected to give reliable result.

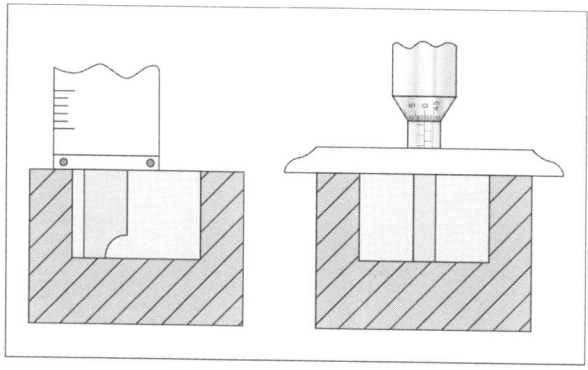

Fig. 6.18: Depth micrometer

Inside and Outside Caliper

Figures 6.19 shows various inside and outside caliper. These are aids to measure dimensions. Figures 6.20 and 6.21 show need of using caliper.

Feeler Gauge

Feeler gauge is used to measure the gap between two tools or machine part. It is a set of three to six leaves of different thickness. Leaves are made of spring steel sheets lapped to accurate and uniform thickness. Figure 6.22 shows a feeler gauge having six leafs.

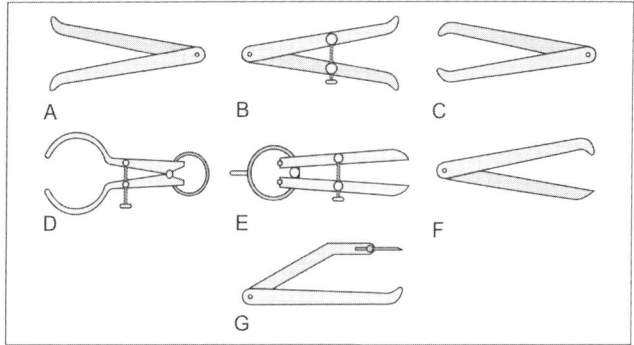

Fig. 6.19: Inside and outside calipers

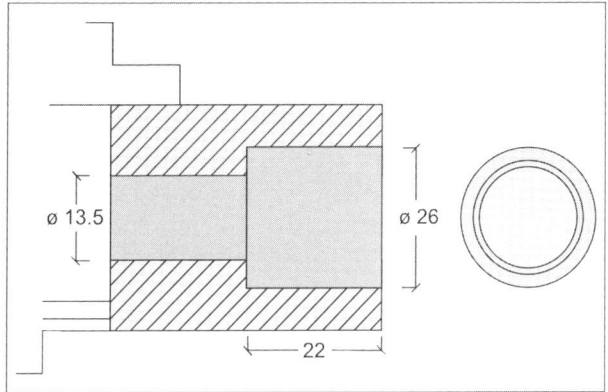

Fig. 6.20: Showing need of using calipers

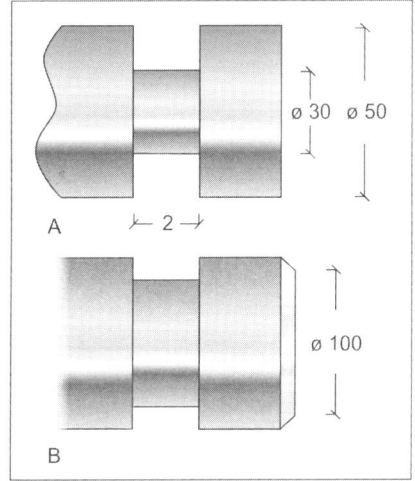

Fig. 6.21: Showing need of using calipers (thickness)

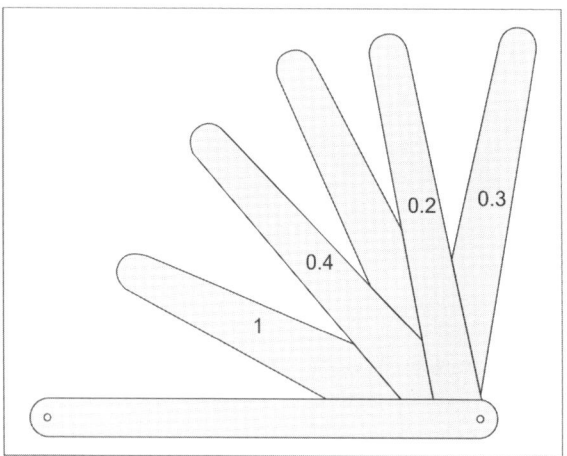

Fig. 6.22: Feeler Gauge

Figure 6.23 A shows the part of a tool having gaps G1 and G2 provided between a base plate and top cover plate. This gap is meant for passing sheet metal strip of 0.35 mm thick. Desired value of G1 and G2 is 0.50 mm, so to measure gaps G1 and G2, feeler gauge leaves of 0.2 and 0.3 mm thickness would be combined and tried to be inserted in the gap. If combination leaves get in easily and freely, it would indicate that gap is more than 0.5 mm. Then a combination of higher thickness would be tried, so by trial and error, approximate value of gap G1 and G2 may be obtained.

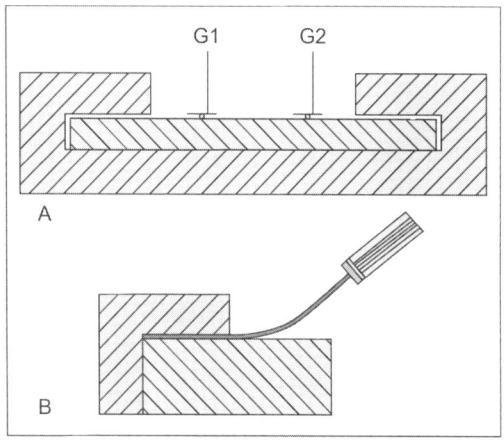

Fig. 6.23: Use of feeler gauge is shown

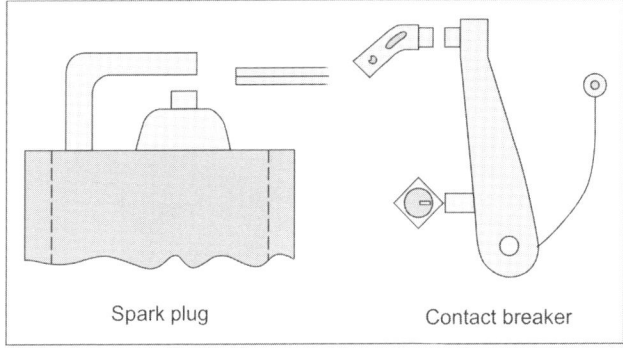

Spark plug   Contact breaker

Fig. 6.24: Illustration of use of feeler gauge

Use of feeler gauge is also made to check gap between two electrodes of an automobile spark plug or contact breaker point. This is illustrated in Fig. 6.24.

Slip Gauge

Slip gauges are the set of small blocks (say 35 × 25 × different thickness) made of hardened and tempered steels. Surfaces are highly plane and lapped. Blocks of selected thickness may be placed one over another to achieve desired thickness. Placement of one block over the other is done after carefully cleaning the mating surfaces. One block slides over the other rather than just putting over the other. Thickness of blocks are so made that almost any size may be made under prescribed range of particular set.

Sine Bar Gauge

These are generally used to accurately measure the angle in the situations where a protractor cannot be used. Sine bar gauges are also used to set the tool or machine parts with accurate angle formation between them.

6.2 DESCRIPTION AND USES OF DIAL TEST INDICATORS

Dial Indicator

It is an instrument to check the variation in a dimension or alignments. Minimum variation which can normally be indicated is 0.001 mm. Internal mechanism in dial gauge converts axial movement of touch probe into circular motion which is indicated by means of a needle which rotates over a division marked dial. Figure 6.25 shows basic construction of a dial indicator.

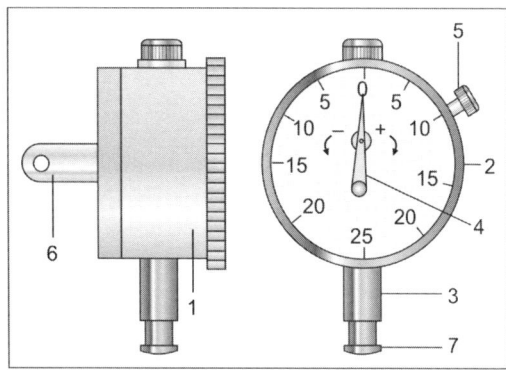

Fig. 6.25: Basic construction of a dial indicator

Part (1) is the body of dial indicator on which a dial ring (2) (also called Benzel) is mounted. Dial is an integral part of ring. Dial can be rotated by rotating the ring. Part (5) is a lock screw for locking the ring at any desired place. Part (3) is the spindle which is internally connected with indicator needle. It is already mentioned. When point (7) is brought to touch and press a little, needle may deflect clockwise which is marked + ve on dial. While touching a point on a surface, it is advisable that dial gauge is so placed that at least quarter of a full round of needle should take place. This is to ensure that touch point will always remain in touch with surface under testing. Now to facilitate reading of deflection, dial may be rotated to bring zero mark under the dial needle (arrow). While sliding the surface under touch point, needle (just like 'hands' of a watch) may either deflect or not when workpiece under test is slided.

Dial indicators are generally mounted on a dial stand which has a heavy base with straight and lapped surface. Stand base may be placed on a surface plate or any machined plane surface.

Dial stands may have the following features:
• Magnetic base
• Stand post tilt adjustment
• Universal holding joints

Figure 6.26 shows dial stand with tilt adjustment facility. Screwing or unscrewing of thumb screw tilts the mounting rods (pillars). Tilting system is spring load, therefore only screwing and unscrewing makes the mounting rod to tilt or become erect.

Figure 6.27 shows the testing of parallelism of die set plates.

For proper functioning of a tool, it is necessary that upper surface and lower surface of die set should be parallel, so before machining

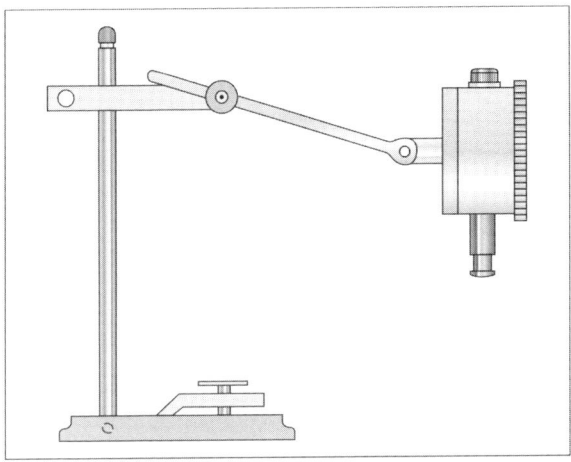

Fig. 6.26: Dial indicator with stand

the die set for fixing tool members, it is tested for parallelism of upper and lower surface of die set. Procedure for testing parallelism is explained with the help of Fig. 6.27.

Part (1) is the cast iron or granite surface plate which is big enough to accommodate die set and dial indicator stand with sufficient space available for sliding the base of stand as required. Part (2) is the lower plate of die set on which die set pillars are rigidly mounted. Part (3) is the linear ball bearing on which sleeves of upper plate move up and down with precision sliding clearance.

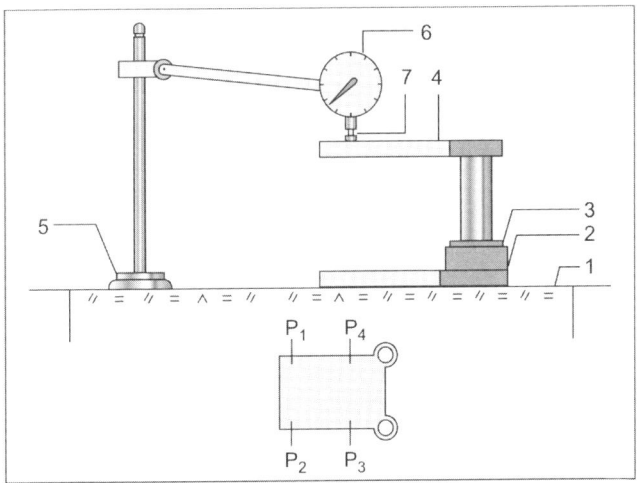

Fig. 6.27: Testing of parallelism

Parallelism of die set plates should be under specified accuracy range while the upper die set plate is at the top position, in the middle or at the lowest position. Generally, parallelism is checked for top and bottom positions only.

Now the die set is placed over 'surface plate' (1) with upper plate at top position (It is done by putting a spacer between two plates of die set). Dial indicator spindle probe is brought over surface (4) and dial indicator is so adjusted that its needle moves about quarter of a round in clockwise direction. Then dial ring is rotated to bring zero mark under the needle. Now dial indicator is ready to perform testing.

The position of dial spindle point (7) is shifted to point P_1 by sliding the base of dial stand. After noting down the deflection indicated by dial needle, spindle point is shifted to other point such as P_2, P_3 and P_4 and deflection readings are noted down. In a particular case, typically deflection reading is found to be as below:

$P_1 - 0$ $\qquad\qquad\qquad$ $P_2 - 0$

$P_3 - + 0.03$ $\qquad\qquad$ $P_4 - + 0.03$

This means that points P_3 and P_4 are 0.03 mm higher than P_1 and P_2. Above readings indicate that upper plate is slightly tilted towards front of die set (away from pillars). It may be OK, if permissible deflection is 0.05 mm for a plate dimension of approximately 200 mm × 140 mm.

Another example of dial indicator use in a typical turning process of die set is explained with the help of Figs 6.28 and 6.29.

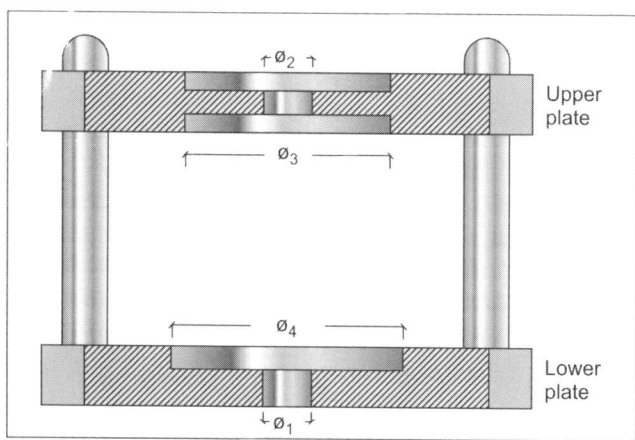

Fig. 6.28: Example of use of dial indicator

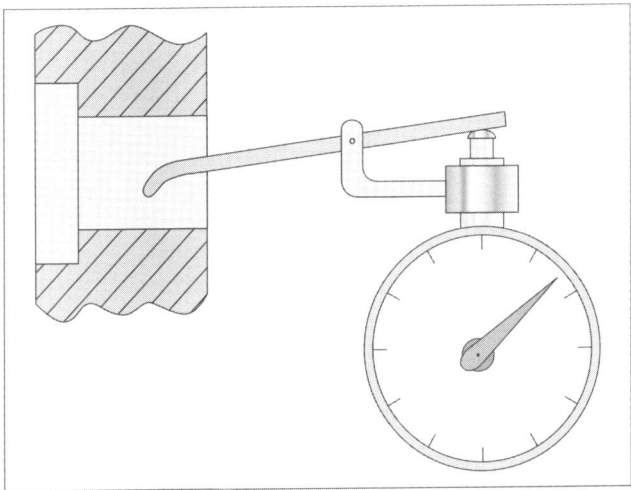

Fig. 6.29: Another example of use of dial indicator

Referring to Fig. 6.28, diameter ϕ_1 of lower plate and ϕ_2 of upper plate/block are first turned in one setting either by loading it on surface plate of lathe or holding it on a 4 jaws chuck. Now the plate is unloaded and adjusted again with ϕ_1 towards face plate. ϕ_2 is made to 'run' true by delicate adjustment of workpiece on face plate of lathe while checking true running of ϕ_2 by means of a dial indicator as shown in Fig. 6.29.

Adjustment is carried out till there is no deflection in needle and workpiece is firmly tightened on face plate. ϕ_3 is now machined.

In a die set, it is essential that bore ϕ_4 which is to be turned should be concentric with bore ϕ_3 of upper plate/block. After placing upper plate on pillar, lower block is loaded on the plate on lathe and ϕ_3 is made true running. Since ϕ_3 is true to ϕ_1, therefore, ϕ_1 is also made true. Upper plate/block is taken out of pillar and ϕ_4 is then turned. Consequently, ϕ_1 and ϕ_4 would be true, this is what is required.

There are dial gauges provided with a lever to retract spindle/plunger to produce a gap between plunger pointer and the surface on which it was resting. This is to facilitate the `shifting of workpiece to bring another point under plunger pointer. Lever is then slowly released to take reading (see Fig. 6.30).

In this particular model of dial indicator, a small dial with a needle may be seen on the main dial. It indicates the number of rounds which the main needle has taken. This facilitates taking of reading.

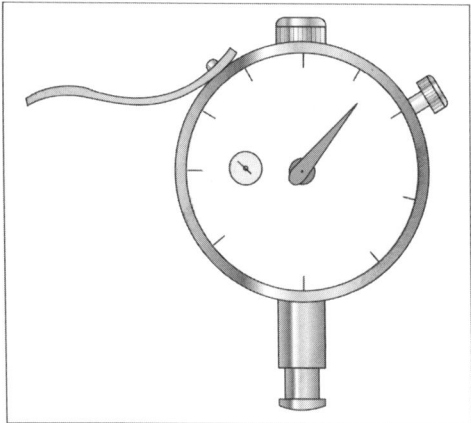

Fig. 6.30: Lever type dial indicator

Sometime, there are narrow bores or slots which are to be tested for trueness. Normal size probe cannot enter into narrow bore, so another type of dial indicator is available. Shape of such a typical dial indicator is shown in Fig. 6.31.

Arm of various lengths may be fixed according to need. Diameter 'd' of spherical end varies from 1.5–3 mm, this dial indicator actually senses angular displacement and not linear. With the help of mechanism inside, angular displacement is converted into linear displacement.

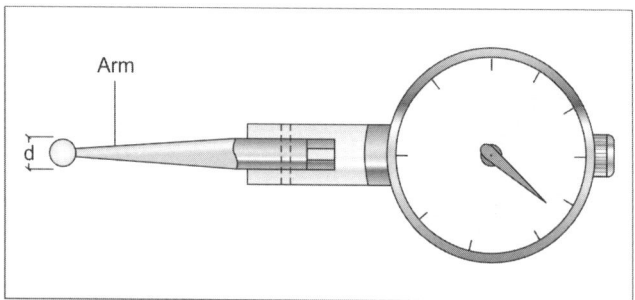

Fig. 6.31: Typical dial indicator

Height Gauge

Height gauge is a height measuring instrument used in the making of sheet metal press tools and other workshop requirements (see Fig. 6.32).

Part (1) is the base of height gauge which is heavy enough to provide stability to height gauge. Part (2) is the vertical scale which

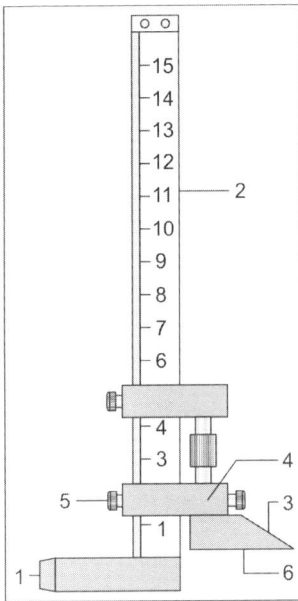

Fig. 6.32: Height gauge

is generally 300 mm in height. Vertical scale is exactly at the right angle to working face of base. Part (3) is a measuring blade. If height gauge is placed over a surface plate and measuring blade is lowered so that face (6) of blade touches surface plate in this condition, zero mark of sliding scale (4) coincides with zero mark of vertical scale (2).

Figure 6.33 shows a heavy steel part of a die under making. It is desired to check dimension X. Part and height gauge is placed on the surface plate. Sharp edge of blade is lowered a little away from part so that face of blade does not rest on part. Now the sharp edge is delicately brought to touch vertical face of part. Gradually,

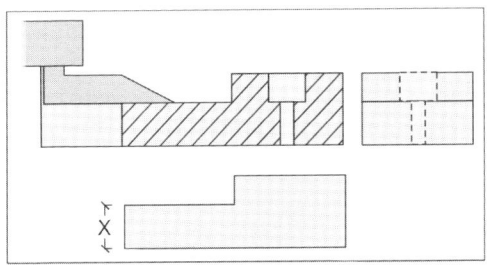

Fig. 6.33: Heavy part under checking

blade is raised by rotating adjustment screw of slide block. A stage would come when blade can be just delicately pushed in over the part surface. Now the reading is taken on vernier scale and it is the value of X.

It is further desired that a line is drawn on the face with the same setting of X. For doing this, general practice is that the face of steel part is smeared with copper sulphate solution which will produce a very thin layer of copper over the surface. Now height gauge is shifted (by sliding) over surface plate in such a manner that sharp corner of blade touches the surface on which line is to be marked. While keeping the corner touched with a slight pressure, it is drawn over the surface by holding the height gauge at the base and moving to mark the lines. Copper layer is scratched to show a distinct line.

6.3 DESCRIPTION OF PROFILE PROJECTORS

Profile Projector

In sheet metal press tool making, sometimes it becomes necessary to check dimensions and profile of machined part such as a piercing die or a stamping punch of very small size which is difficult to measure by a conventional measuring instrument. Once a tool is completed, it is tried out. Trial components taken out are checked for accuracy in measurement to ensure if it conforms to components drawings. Enlarged view of two typical components is shown in Fig. 6.34.

Figure 6.34 'A' shows a piercing die duly machined to give profile as per drawing dimensions. Few dimensions are given in the figure to give an idea of size. Since the profile is complex, it is very tedious to measure by conventional measuring instruments, so a profile projector is needed where light passes through profiled opening of die insert, say a round disc. Similar is the case with brass sheet component as shown in Fig. 6.34 'B'.

Components shown in Fig. 6.34 'C' is a typical metallic clip of a pen cap. Word 'prince' is stamped with a depth of say, 0.08 mm. Dimensions and alphabets cannot be measured accurately by means of conventional measuring instruments. An illuminating profile projector is needed for this purpose.

Figure 6.35 shows two threaded components of sheet metal.

Sheet metal components of big diameter is shown in 'A' and small diameter in 'B'. For seeing the profile of 'A' component, a projector of bigger size would be needed so that component 'A'

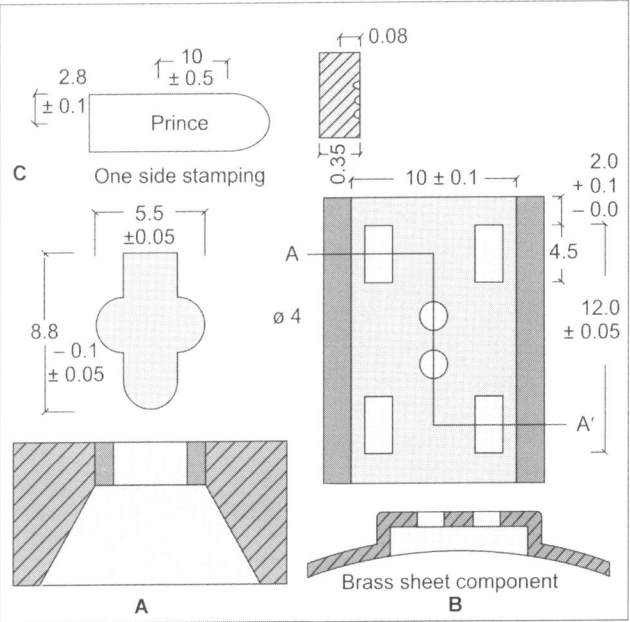

Fig. 6.34: Piercing die, component

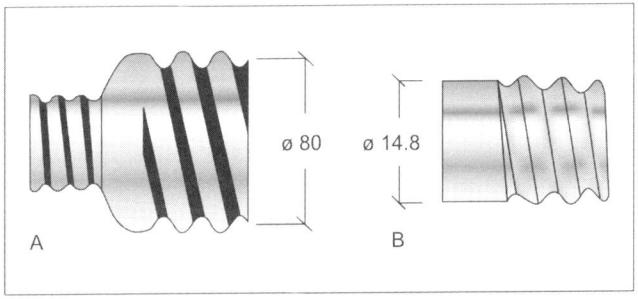

Fig. 6.35: Enlarged view of threaded component

can be accommodated. For small component, a smaller model of profile projector may be used.

In the following paragraph, a few types of profile projectors are briefly described so that a tool maker or tool designer has better understanding for selection and use of profile projector.

Profile projectors may be classified in the following manner (see Diagram 6.1).

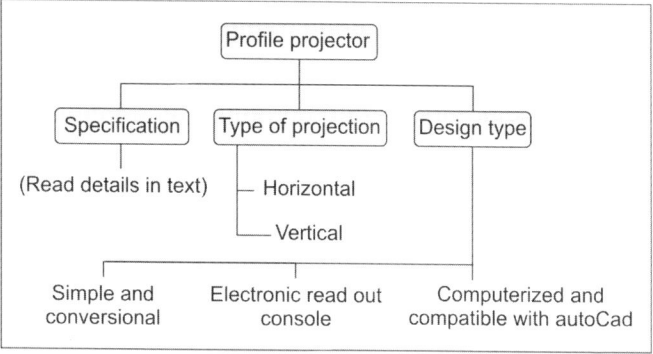

Diagram 6.1: Classification of profile projectors

The basic shape/design of a horizontal projection type of optical profile projector is shown in Fig. 6.36.

Part (1) is the body of profile projector, (2) is vertical member of the body with a bright light source in horizontal position. High intensity light is generally produced by halogen-tungsten lamp of high wattage, say 1000 watts. A small air circulating fan is provided inside part (2) to keep the temperature of lamp in permissible limits. Part (3) is a vice with a micrometer screw to precisely shift the object which is loaded on it. In many cases, there is a provision of lifting and lowering the vice. Beam of light coming out of illuminator (6) horizontally reaches receiving lens (4). Light spot

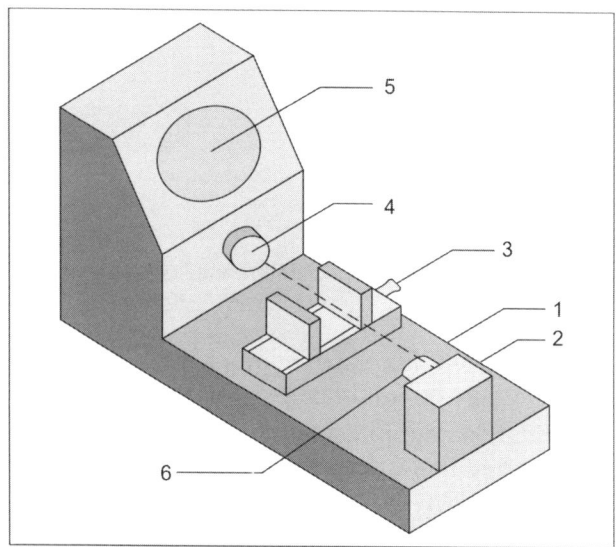

Fig. 6.36: Horizontal optical profile projector

after entering lens (4) gets enlarged by means of a designed optical 'circulatory' consisting of lens, prisms, mirrors, etc. The height of light source and lens is so made that a job held in vice comes in the path of light ray. Profile of portion of any object brought in the path of light is projected on the screen (5) of projector. How many times the enlargement of actual size depends on power of lens (4). It may be 10X, 20X, 50X or 100X. In many models of projector, there is facility of changing magnification lens (4). In simpler model, there may be only one fixed magnification lens. Figure 6.37 shows the image of thread of a brass sheet cap of miniature lamp.

Profile of thread may be compared by placing a shadow graph on projector screen. A shadow graph is a scale drawing of standard thread drawn on tracing paper.

Similarly, profile of an eyelet of 0.24 mm thick brass sheet may also be checked (see Fig. 6.38).

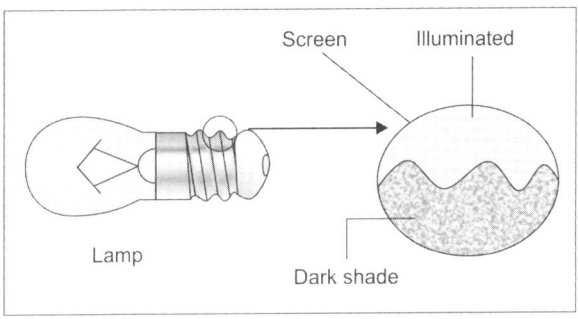

Fig. 6.37: Image of thread profile

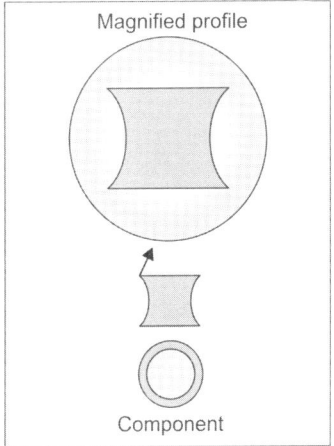

Fig. 6.38: Profile of an eyelet

Dimensions may be either compared by means of a shadow graph or a vernier caliper or a scale shadow graph may be used.

The basic shape/design of a vertical profile projector is given in Fig. 6.39 which shows a floor stay model. It is a much bigger model as compared to one shown in Fig. 6.36. It is normally placed in small room type of place where lights may be put off or dimmed while using this optical profile projector. Part (1) is the body/stand in which optical 'circuitry' is there. Part (2) is the screen of rectangular shape where image is formed. Part (3) is changeable magnification lens. Part (4) is a stage or glass fitted platform where a flat sheet component may be placed. For example, a small gear of a clock train may be examined as shown in Fig. 6.40.

As usual, profile may be compared by means of shadow graph drawn 50 times the original drawing dimensions.

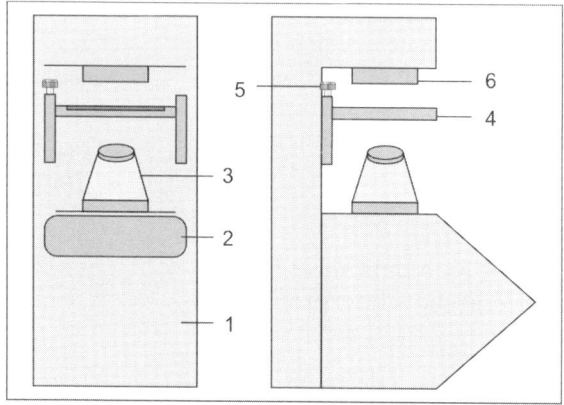

Fig. 6.39: Vertical profile projector

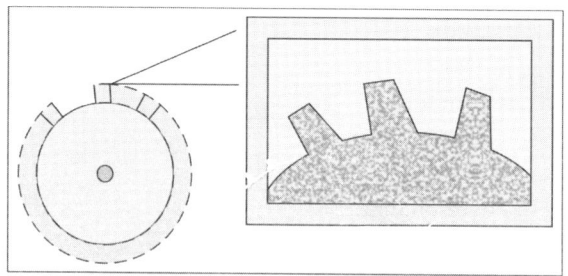

Fig. 6.40: Large profile of portion of small component

Modern Profile Projector

Modern profile projectors are designed and made to provide facilities which are user friendly and accurate such as stage table which can be moved in all the three X, Y and Z axes with micrometer movement reading.

Sizes of profile projectors vary in dimensions. Portion of component as big as shown in Fig. 6.40 can also be projected. Further improvement in many models is that it is equipped with electronic read out system. This means that various dimensions including angles and radii can be directly read out on screen of console by operating various controls provided.

There are most modern profile projectors which are equipped with facility of interfacing with computer to make use of AutoCAD. Printout of profiles may also be taken.

Suppose if it is desired to see enlarged view of embossing of a coin then how could it be seen. There are special profile projectors which are equipped with spot light by means of fiber optics, end of which may be manipulated for distance from object and angle of incidence light to give a sharp image. In a microscope, same is done. Only difference in a microscope and profile projector is that in microscope, we see by bringing our eye near the eyepiece, whereas in profile projector, image is projected on the screen of profile projector.

Specification

A typical specification data sheet is provided below (in Table 6.1) to give an idea as to show how profile project manufacturers specify various models.

S.No	Description	Model no.	Remark
	Table. 6.1: Specifications data sheet		
1.	Table dimension	300 × 180 mm	
2.	Glass dimension	180 × 125 mm	
3.	X measurement travel	140 mm	
4.	Y measurement transverse	100 mm	
5.	Z focusing travel	80 mm	
6.	X Y read out resolution	0.001 mm	
7.	Axial measuring accuracy	3 + L /100 mm	
8.	Screen diameter	310 mm	
9.	Angle read out range	0–360°	
10.	Angle read out resolution	0.001°	
11.	Maximum contour distortion	0.08 %	
12.	Magnification lenses	10 X, 20 X, 50X	

Measuring Aids

Measuring aids consist of inside caliper, outside caliper, a pair of divider caliper, odd leg caliper, etc. These aids are generally used for transferring measurement to a micrometer or vernier caliper. Figure 6.19 shows few of above mentioned measuring aids.

There are situations during machining or inspection when vernier caliper or inside micrometer cannot be used. Such situation is shown with the help of Fig. 6.20.

Workpiece shown in Fig. 6.20 is under turning operation. Dimensions shown are nominal. Bore 13.5 cannot be measured by inside jaws of vernier caliper as its length is about 15 mm in a 20 cm vernier caliper, so if initially rough measurement is to be taken then caliper 'A' may be used. Caliper points are brought inside the bore with points expanded so much that it slightly touches the bore on both the sides at its maximum. Caliper is then brought out and the distance between the two points is normally measured by vernier caliper. When the bore is reaching near completion then accurate measurement is required. To make an accurate measurement, caliper 'B' is used (see Fig. 6.19). This caliper consist of an adjusting thumb nut with fine threads, hence, opening of caliper can be gradually adjusted. It is brought inside the bore and adjusted to such an extent that the point of caliper just touches the wall of bore on both the sides. Now here comes the importance of 'FEEL' of a measuring person. An experienced person is capable of 'registering' the touch feeling in his/her mind. He takes out the caliper and measures the opening of point either by vernier caliper or micrometer. While doing so, he maintains the same feeling of touch to get as much accurate reading as possible. In practice, it is found that a good feeling of touch can repeatedly reproduce reading within plus or minus 0.01 mm.

Caliper 'C' and 'D' are used for aiding measurement of diameter of a portion of part which may not be directly measured by a vernier caliper. Such a situation is shown by Fig. 6.21.

Figure 6.21 'A' shows that the diameter 30 of an under cut of 2 mm wide is difficult to measure by vernier caliper, therefore an outside caliper may be used. Further, diameter of workpiece 'B' is so large that jaws of vernier caliper cannot reach up to diameter of job if a 20 cm vernier caliper is used.

Divider 'E' (Fig. 6.19) is a sharp point caliper used either to take distance between two marked points on a plate or scribe a circle on a plane surface. It also has spring loaded arms and thumb nut for precision opening and closing of sharp point caliper.

'F' is a caliper which has a slight inside bend on one arm. Other has a sharp point. It is used to scratch a line on plane surface of a plate, parallel to edge. Bent end of arm is placed on edge face and drawn with sharp point pressed on surface. Consequently, a line parallel to edge is scratched. 'G' is also used for the same purpose as 'F', but for bigger jobs.

6.4 SURFACE FINISH AND MEASUREMENT

In our day-to-day observations, we come across many objects and surfaces. On touching the surface, we have different feeling. When finger is moved on a glass pan, there is a feeling of smoothness. Movements on cardboard offers a feeling of roughness. Surfaces turned, shaped or milled cast iron surface give entirely different feeling of roughness or smoothness. These two terms are antonyms.

If a sharp pointed needle held in hand and moved across a surface, a rough idea of surface finish (rough or smooth) may be formed. Roughness of a surface is due to presence of micro size 'hills' and 'valleys' on the surface. Normally, these 'hills' and 'valleys' are not clearly visible by naked eyes. These irregularities (hills and valleys) may be seen to some extent by a magnifying glass, say of X50 magnification.

If it is desired to see the roughness still more magnified then high power microscope may be required. An approximate idea of 'hills' and 'valleys' formation may be gathered from Fig. 6.41.

Part (1), (2), (3) and (4) are the peaks of 'hills' and (5) and (6) are the 'valleys'. A surface having less distance between (say, in micrometer) peaks and valleys would be smoother as compared to the surface having more distance.

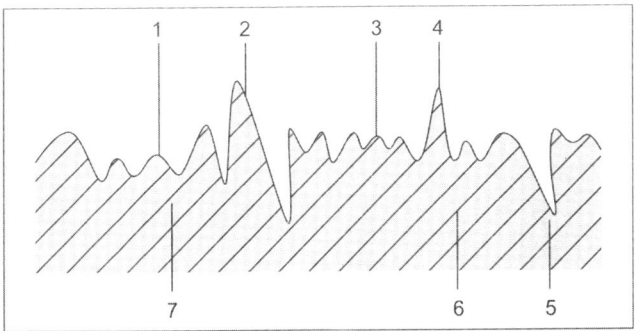

Fig. 6.41: 'Hills and valleys' shown enlarged

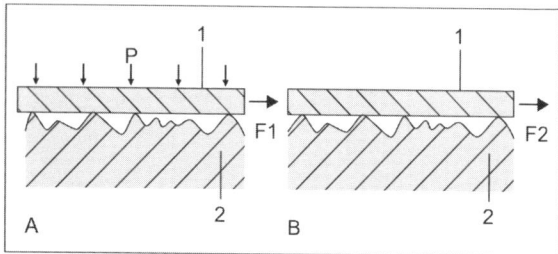

Fig. 6.42: Enlarged view of surface roughnes causing resistance to movement of aluminum strip

Let us now discuss the significance of surface finish in sheet metal tools.

Figure 6.42 'A' shows an aluminum strip placed on a steel surface (2) with a certain pressure P on almost full contact area. If strip is pulled with a force F1, there would be a biting effect on strip surface as sharp peaks would penetrate in aluminum. When strip is taken out by a force F, then it would be found that there may be long scratches on aluminum strip surface. What are these scratches? Minute material is dislodged from strip and stuck to hills of surface roughness. If the process is continued, valleys would be filled up with aluminum and ultimately seizing (or cold welding) of aluminum with steel surface might take place.

Figure 6.42 'B' shows steel surface where peaks of hills are flattened to some extent by rubbing with an emery paper attached or wrapped on a file or by an emery stick. This means that surface roughness is improved to some extent. Consequently, peaks do not remain sharp, they become flattened to some extent.

This change in surface finish or texture reduces scratches on sheet metal and pulling force F2 becomes less as compared to F1. Example of aluminum strip is taken as it is softer than brass. Brass is also affected but to a lesser degree.

Now another example of a cut and cup tool is taken (see Fig. 6.43).

Before draw starts taking place, blank (1) is held by draw die (2) face and pressure pad (3). Holding of blank is with quite a force. When drawing takes place, sheet material slips over radius R of draw die. Passing of sheet material over radius 'R' is with a great pressure over the surface of radius 'R'. Here comes the effect of surface finish of radius of draw die. Figure 6.43 'B' shows enlarged view of radius of draw die. 'C' shows enlarged view of peaks of hills and valleys. This offers a great resistance to the flow of sheet material so much so that sometimes, draw punch ruptures the base

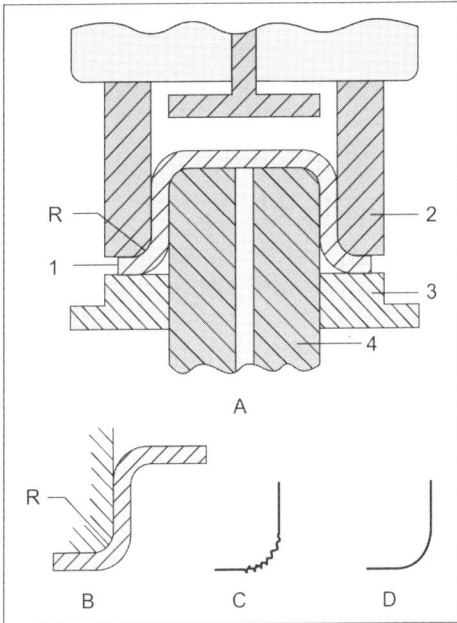

Fig. 6.43: Example of shell cup draw

of partially drawn cup. 'D' shows a polished radius where peaks of hills are removed by lapping and polishing.

In small size, say 100 mm, draw dies, drawing radius is formed by manipulating turning tool point. This manipulation is done by turner, who moves cross slide and carriage in a pattern that a radius is turned. This radius is full of 'hills and valleys'. It is minimized or eliminated to a great extent by progressively lapping with rough, fine and ultra fine abrasive sticks. Since steel is in its unhardened state, great care is taken during lapping. Excessive pressure may further make the surface bad. Once lapping is completed, '00' grade polishing emery paper is used to finally polish the radius and it is done very delicately.

Once draw die is hardened by heat treatment, diameter is polished by emery paper to remove somewhat blackened layer. This is required to prepare blanking punch cum draw die for grinding of diameter to final size by removing grinding allowance.

Bore of draw die is polished by hand or high speed rotating polishing wheels. Now radius is polished with hard wooden stick and diamond polishing paste of fine to ultra fine grade. Final polishing is done with a clean soft cloth while draw die is rotated at high rpm. This gives a shiny smooth surface which helps in drawing operation.

Large diameter draw dies, say of 350 mm bore are normally turned on a programmable lathe (CNC). Generation of draw die radius is quite accurate and smooth. Draw die bore is also quite smooth.

Nowadays, modern tool rooms are equipped with computerized numerical control (CNC) machines. Accurate radii with a tolerance of + or − 0.015 mm and profiles may be turned very accurately. Even in conventional tool room, a turner, a machinist or a grinder operator has to be given instructions as to what type of surface finish is required on a particular spot/position of tool part which is to be machined as per provided drawing. Drawing so provided carries indication of finish and it is done by referring to finish number of a **sample plate** which is kept in tool room to be seen by turner or machinist. Figure 6.44 gives an idea as to how a sample plate looks like.

This plate is generally made of stainless steel or steel duly electroplated. It has four or five blocks on one face of plate. Block '1' has roughest surface, '2' better, '3' still better and '4' very fine and smooth surface. So far the matter of surface finish discussed depended on visual judgment of a turner, machinist, and supervisor or quality assurance persons. This method is good enough as far as small tool rooms are concerned and meeting in-house requirements of sheet metal press tools.

When it comes to produce standard components of tools for customers or for market on a regular basis then high standards of surface finish is required to be maintained either as per customer's specification or general market demand. So here comes the necessity of scientific method of measurements of surface finish to ensure its quality and/or convey the surface finish parameters to other party, which might be necessary from commercial point of view as well.

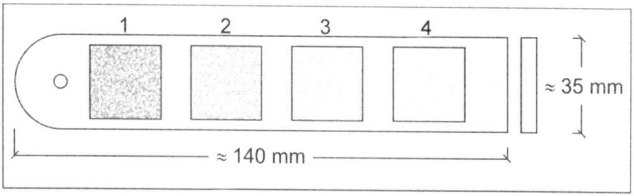

Fig. 6.44: Finish sample plate

6.5 DEFINITION OF LAYS, RA SYMBOL AND VALUE

Finish or Surface Texture Standards and Measurements

Definition of Ra, a measure of roughess, is explained with the help of Fig. 6.45. So far **surface** finish or **texture** is discussed in qualitative terms that are description of finish or comparison. Few decades back, importance of surface finish of machine parts and tool parts was felt and sustained work on machining process development.

Surface finish measurement system has now brought remarkable mechanical and optical equipments to very accurately measure surface texture and to even plot graphs and automatic analysis is carried out by computer with which sensing probes (sensing head) are attached.

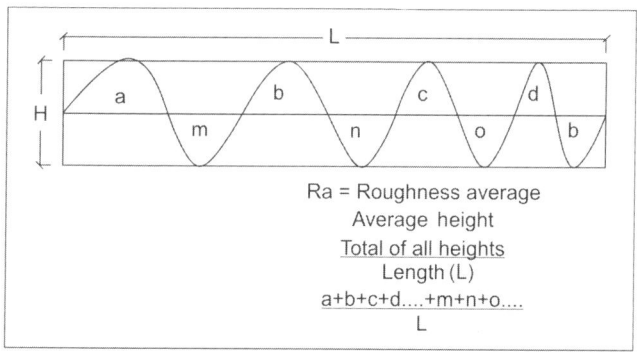

Ra = Roughness average
Average height
$$\frac{\text{Total of all heights}}{\text{Length (L)}}$$
$$\frac{a+b+c+d....+m+n+o....}{L}$$

Fig. 6.45: Definition of Ra

Machined surface normally has the following components/ aspects to be considered:

- Lays
- Surface roughness
- Waviness

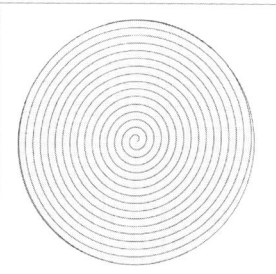

Fig. 6.46: Circular 'Lays'

i. Lays

Lays are the pattern of lines/impression produced on a surface by a particular machining process. While facing a workpiece on lathe, would produce continuous circular lines throughout the face which may be represented as shown in Fig. 6.46. This is called circular Lays. If a surface is straight machined by a shaper, plainer or milling, a pattern full of straight lines is generated as shown in Fig. 6.47.

Lays of a surface would be quite different if, for example, a steel plate surface is machined by a cup end mill cutter as shown in Fig. 6.48. Grinding, spark erosion, honing and lapping would produce different lays.

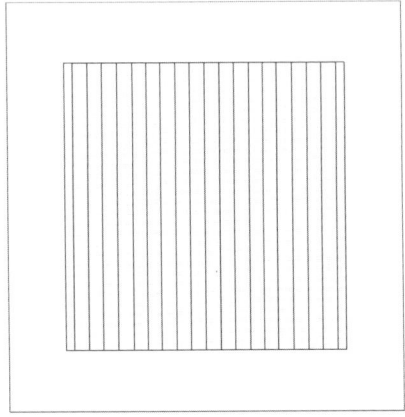

Fig. 6.47: Straight line 'Lays'

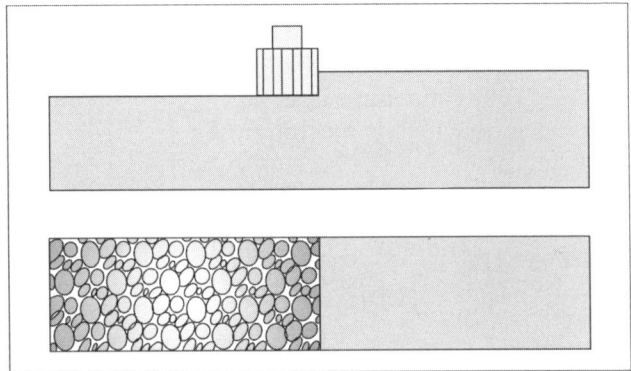

Fig. 6.48: Circular spots 'Lays'

ii. Surface Roughness

Surface roughness is the amount of heights of 'hills' and depth of valleys from a mean reference line and distance between peak to peak or valley to valley. It is further explained with the help of Fig. 6.49.

Generally, surface roughness is denoted by alphabets Ra, which means roughness average.

Roughness profile shown in Fig. 6.49, is shown many thousand times magnified on YY' axis and a few hundred times on XX' axis. Let L be the length covering shown hills and valleys.

Average height = Sum of heights of peaks and valleys/length L

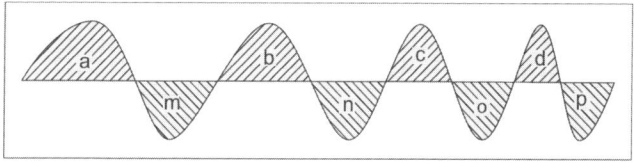

Fig. 6.49: Idea of surface roughness

iii. Waviness

Waviness is the pattern of surface roughness which changes momentarily at periodic intervals. Such a happening has been depicted with the help of Fig. 6.50 which shows surface roughness with a wave form. Part (1) is a roughness pattern which is almost continuous throughout the length with the only difference that at periodic intervals, 'X' roughness pattern gets raised momentarily at (2). This keeps on repeating. Such a form of surface pattern is called 'waviness'.

There may be many causes of occurrence of waviness such as travel of machine tool or table movement is not smooth. It has periodic 'jumpy' action. Other reason may be that some heavy power press or other machine is in operation nearby, causing vibration in machine which is machining the surface.

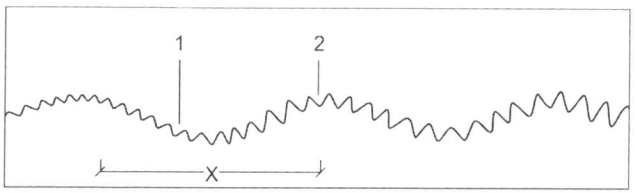

Fig. 6.50: Example of waviness depicted

Measurements

There are two basic categories of surface texture/roughness measurement, which are as follows:

• Contact measurement

• Non-contact measurement

In contact measurement, a stylus is drawn across or along lays. Instrument is called 'profilometer'. The basic design of a profilometer is shown in Fig. 6.51, where part (5) is the reciprocating arm of a **profilometer**. The Basic construction of a typical surface tester is shown in Fig. 6.52 which shows a typical portable profilometer where (1) is the body. Front of profilometer carries an LCD screen and operation control buttons. Part (2) is an LCD screen, (4) is a reciprocating arm, (5) is pick up carrier rod, (6) is a pick up which carries a diamond point tracer.

When reciprocating arm reciprocates, diamond point moves ups and downs according to hills and valleys of surface. Displacements (ups and downs and transverse motion) of diamond point tracer are 'processed' in mini logic control of instruments and data such as Ra (roughness average) is displayed. This instrument has a facility to be connected to a computer loaded with appropriate software to generate surface roughness pattern, graph, data, etc.

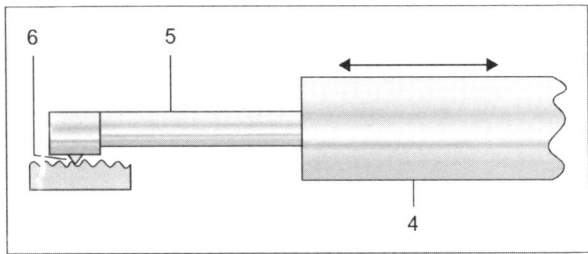

Fig. 6.51: Basic design of profilometer

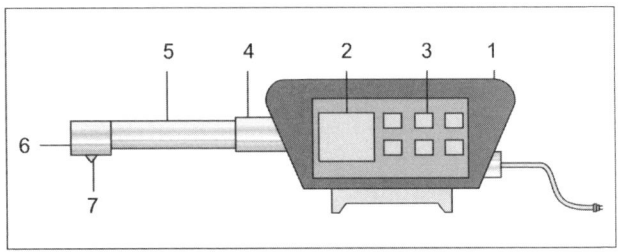

Fig. 6.52: Typical portable profilometer

Printout of graph together with data may also be taken out. The disadvantage with contact measurement system is that absolute depth of valley cannot be reached if it is sharp beyond sharpness of diamond point of pick up head. Such a difficulty is highlighted with the help of Fig. 6.53.

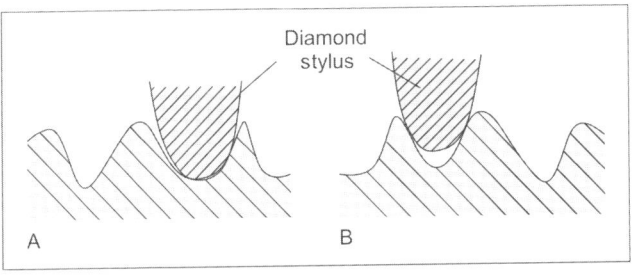

Fig. 6.53: Sitting of stylus point

In Fig. 6.53 'A', radius of diamond stylus is less than radius and width of 'valley', therefore, stylus point is completely sitting on the deepest surface of valley. In Fig. 6.53 'B', stylus point radius is bigger and wider than valley of roughness, consequently, it is 'hanging between the walls of valley' and there is a gap between stylus point and lowest point in the 'valley'.

For more accurate measurement of roughness, non-contact measurement is used. Non-contact measurement system consists of the following techniques:

- Interferometry
- Focus variation
- Electrical capacitance
- Photogrametry
- Confocal microscopy
- Structured light
- Electron microscopy

General Ra values for various machining operations are given below:

- Turning ——————————25–1.5 µm (micrometer)
- Shaping ——————————— 12–1.6 µm
- Milling ——————————— 6.5–0.8 µm
- Grinding ——————————— 1.6–0.1 µm
- Honing ——————————— 0.8–0.1 µm
- Lapping ——————————— 0.4–0.005 µm
- Super finish ——————————— 0.2–0.05 µm

Development of all the above techniques could be made possible by the use of electronics and optical systems.

In large mass production plants, such as automobile engine cylinder blocks, surface finish checking is automated by the use of robots that brings in and out sensor head of roughness measuring equipments from one cylinder to another.

For better visual concept, the following website may be visited on World Wide Web (www) on internet:

http://www.obsnap.com

(as visited on May 16, 2013, 8 AM IST)

More videos may be seen on website

http://www.youtube.com

On home page of this website, write down surface finish and then click search button. A page would open containing website of many parties. On clicking selected party, website would open and a short video may be watched. (as visited on May 16, 2013, 8.30 AM IST)

Gauges

In production of sheet metal components, maintenance of critical dimensions is a tough job. In assembly of components, there may be few sizes of two components which have to be precisely assembled. Figure 6.54 shows example of two sheet metal components which are to be assembled together with precise fit. Bore of component 'A' is 22 mm with a tolerance of +0.02, –0.02. This means that $\varnothing_1$ may be

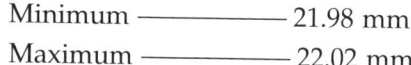

Minimum —————— 21.98 mm

Maximum —————— 22.02 mm

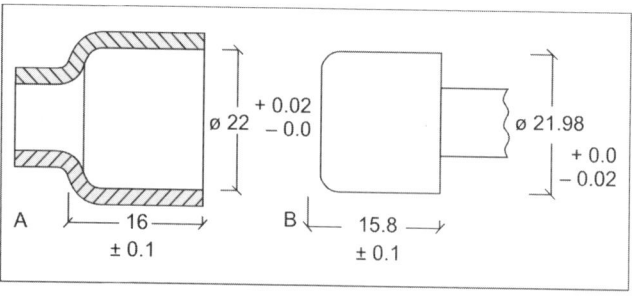

Fig. 6.54: Plug gauged component

Hence, there is a tolerance of 0.02 mm on positive side. Similarly, tolerance in diameter of components is 0.02 mm on negative side.

It is very difficult to maintain this tolerance due to many variable factors such as

- Variation in sheet thickness
- Variation in blank under holding pressure
- Wearing of radius of draw die
- Wearing of bore of draw die
- Dis-balancing in under holding pressure due to worn out pressure pins
- Variations in viscosity of lubricating medium for sheets

Since there are quite a chances of variation in dimensions of bore of 'A' or 'B', it is necessary that operator of power press should check the diameter or bore from time to time while production is going on. Checking of these dimensions by vernier caliper or micrometer is not a practical proportion, so a checking device is provided to operator to check and it is called a gauge. Figure 6.55 shows a gauge to check diameter 21.98, + 0.00, – 0.02 of component 'B'. Diameter of component should not be less than 21.96 (21.98 – 0.02) mm, so the NO GO bore Ø3 in the gauge should be made 21.96. Hence, a component having correct size of 21.96 would not pass through NO GO hole and it should happen. It may be noted that component is never forced through the gauge. Delicate attempt may be made to pass the component. In case component passes through gauge hole then it means that diameter Ø2 is less than 21.96 mm, hence,

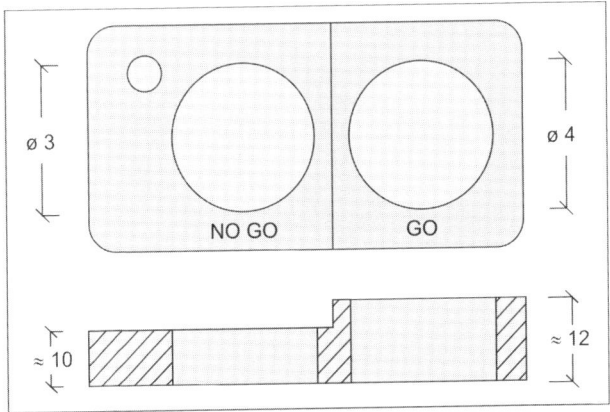

Fig. 6.55: GO, NO GO ring gauge

components may be rejected. Large size of Ø2 is 21.98. It has to pass through gauge bore Ø4. For 21.98 to pass through, Ø4 should be made 21.995. Any component not passing through Ø4 is failed as it has become bigger than specified size 21.98 Ø mm, that is why, gauge plate is marked with GO and Ø3 as NO GO. Sometimes, step is made on the face of gauge to differentiate between GO and NO GO side. On a corner, a hole is provided to hang the gauge at safe and convenient place.

Now coming to component 'A' of Fig. 6.54. In this component, bore $Ø_1$ is to be checked which is specified as 22 + 0.02, – 0.00 mm. This means the component may have bore

$$\text{Minimum} \quad\text{————}\quad 22.00 \text{ mm}$$
$$\text{Maximum} \quad\text{————}\quad 22.02 \text{ mm}$$

During production, it has to be ensured that minimum size does not go below 22.00 and maximum does not exceed 22.02 mm, so to control/check this, a GO–NO GO plug gauge is used. Figure 6.56 shows a typical plug gauge.

GO plug should go inside bore Ø, 22.00. This ensures that Ø1 of components is not smaller than specified size. If it does not go then component bore is below 22.00 mm, hence, may be rejected. NO GO end of gauge should not enter the bore Ø1. If it enters then it shows that bore Ø1 is bigger than specified 22.02 mm.

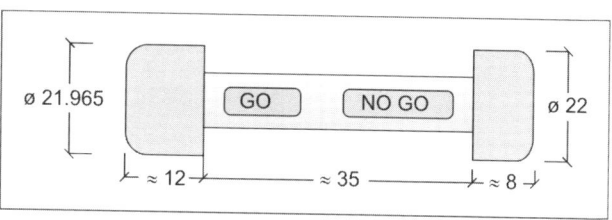

Fig. 6.56: GO, NO GO plung gauge

There may be other type of gauges such as profile gauge, location gauge. These would be briefly described in the following paragraph.

i. Profile Gauge

A profile gauge is to check the profile of a sheet metal component. It is not a GO-NO GO gauge. It is just a comparator. By placing the profile gauge over the component, a comparison can be done. Figure 6.57 illustrates this point.

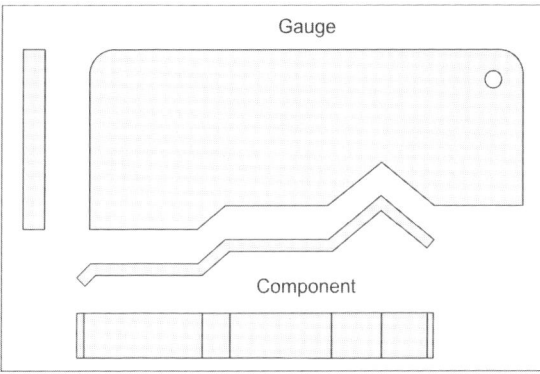

Fig. 6.57: Profile gauge

ii. Location Gauge

Sheet metal components may have a number of features. Some may be dimensionally related to other features. Such a situation is highlighted with the help of Fig. 6.58.

Figure 6.58 shows a component in which center of hole is located at a distance of 8 mm from inside wall. There is a tolerance of + 0.2, – 0.00 in these dimensions. Furthermore, there are two slots. End of slots towards hole is located at a distance of 18 mm from inside wall of component. This dimension also carries a tolerance of + 0.2, – 0.1. During piercing and slotting operation, there is a chance on shifting of location of hole and slots due to worn out location

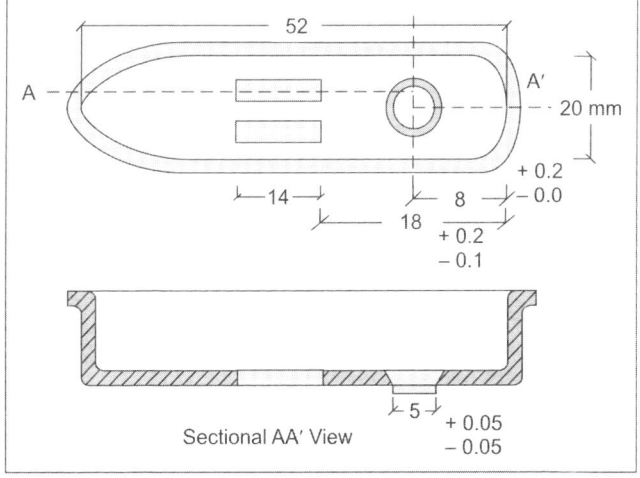

Fig. 6.58: Component with location of features

plate or reduction in size 52 mm in drawing operation. This would cause component to become loose in the 'nest' (location plate) of piercing cum slotting tool. As a result of this, there may be undesirable variation in size 8 + 0.2, − 0.0, so it is necessary that operator of press should have a location gauge to check the correctness of component from time to time. A location gauge may be as per design shown in Fig. 6.59.

On comparing sizes of gauge with corresponding component sizes, it would be noticed that suitable clearances are provided to take care of specified tolerance in sizes of component. If the component conforms to specified sizes then it would sit properly over the gauge, meaning that pin of the gauge would enter hole in component and raised rectangular pins would enter component slots.

In case component does not sit properly on to the gauge then it is an indication that component hole or slots are out of location. Then the operator may take necessary action as per instructions of supervisor. Instruction may be to stop production for carrying out corrective measures.

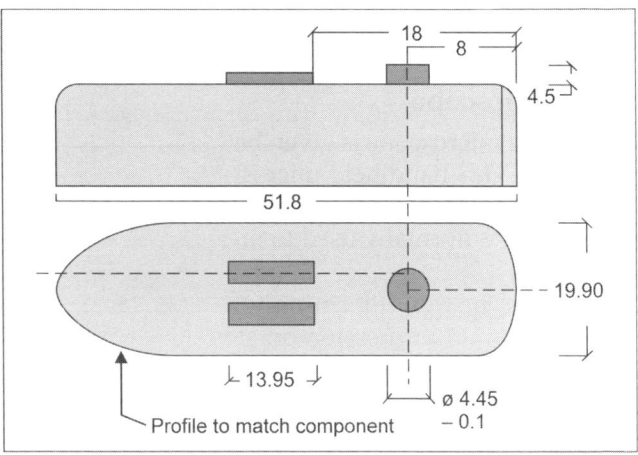

Fig. 6.59: Location gauge

6.6 MAGNIFIERS AND MICROSCOPES

Illuminator Magnifier

This instrument is very useful for tool maker for clearly seeing the details of a small part of a tool already made or under making. Before citing a few examples, photograph of an illuminator magnifier is given below.

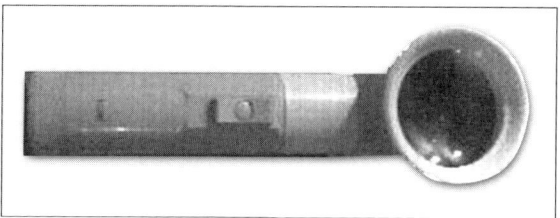

Photo 6.4: Illuminator magnifier

For observation, magnifier is placed over the object with lens towards eyes. Light is put ON by operating the switch button. Object gets illuminated and magnified. It makes easier for observer to see details, such as thread profile of a fine screw or raised miniature marking on an embossing or stamping insert of a tool. Sometimes, it may be needed to know the size of an impression. This is possible by the use of a transparent disc having scale marking. This disc may be snapped in the round opening of magnifier. On seeing from lens, magnified thread pitch or distance between two markings on embossing/stamping insert may be read.

Generally, two or three discs are provided with different scale, linear and angular markings.

Hand Held Microscope

Photo 6.5 of this microscope is given below to convey an idea of its shape and size. This hand held microscope has a magnification of 50 X and a visible scale, in vision with a least count of 0.05 mm. This microscope is normally used to measure a minute impression on material.

For example, in the determination of hardness in Vicker's pyramid number, an impression on steel has to be measured for the sides of generally square impression due to penetration of indenter. After recording the measured size, Vicker's pyramid number can be calculated. It needs a little practice to use this type of handy microscope.

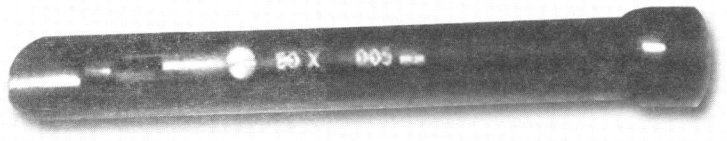

Photo 6.5: Hand held microscope

7

Sheet Metal Components

This chapter consists of the following aspects:
7.1 Varieties of components
7.2 Manufacturing objectives
7.3 Description of few components
7.4 Thread rolled components
7.5 Blanking and fine blanking

7.1 VARIETIES OF COMPONENTS

The design and making of sheet metal press tools very much depend on the design objective for component. In this chapter, attempt is made to explain and illustrate as to how a particular component can be manufactured. There may be only one or a number of operations required to complete the component. There may be examples of hundreds and thousands of sheet metal components used in a variety of end products which include fields of mechanical, electrical, electronics or combination of these and other fields such as optics. Some components are picked up at random and are explained with the help of figures and photos. Manufacturing features of each would be discussed which influence the design and making of press tools for such components.

Large varieties of sheet metal component may be categorized in the following groups:

- Big size components
- Small size components
- Development stage
- Bulk production
- Medium size components
- Miniature size components
- Batch production
- Dimensional accuracy

Referring to Diagram 7.1, Components may be of steel, stainless steel, titanium, brass, phosphor, bronze, aluminum or copper. Depending on the thickness of sheet, weight of a particular component may vary. Moreover, approximately overall length, width, height, diameter, etc. may largely vary.

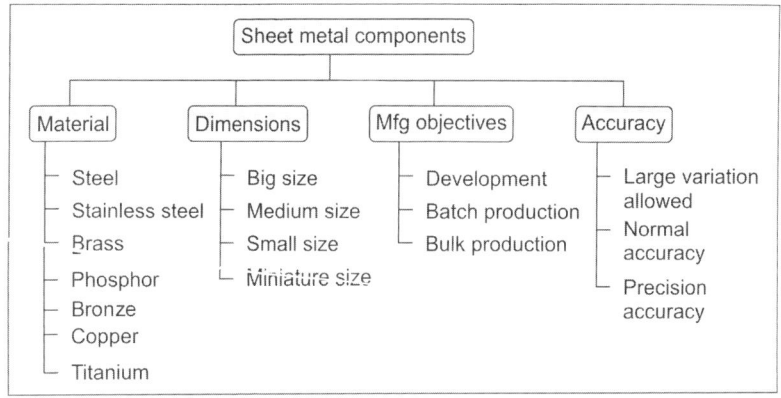

Diagram 7.1: Sheet metal component aspects

Example: The steel cover of a tractor may measure 1 M × 1.5 M × 0.5 M and weigh around 3–5 kg. A gear of clock or flat pins of USB (universal bus) of computer may hardly be less than one gram. With this example, it may be appreciated that there is going to be a vast difference in designing and making press tools. It is quite likely that tool designer for big size components and miniature size components may be different individuals. Designer of a big size component tool may find it difficult to design a tool for miniature component and vice versa.

A sheet of steel, for example, may have different thickness and hardness. Sheet required for manufacturing tractor engine bonnet/cover may be 1–1.8 mm thick. The hardness may vary from 180–210 BHN (Brinell hardness number) and/or tensile strength of 276–1882 MPa. A washbasin of kitchen (see Photo 7.1) may be of say, 0.8–1 mm thick stainless sheet of spring hard quality.

A tool designer has to take into consideration the sheet hardness, tensile strength as the spring back after draw or bending operation depends on hardness and grain direction. Angles and radii in tools are so kept that final dimensions of component is achieved.

Components such as tractor engine cover, railway engine side cover, electrical equipments box cover and many sheet metal components are used in ships and military tanks, etc. may be considered as big size components.

Medium size components may include cover plate of air cooled generator set, cooking utensils, slot angles, spoke plats and front brake plats of two wheeler, irregular housing for electrical control panels and a large number of such items.

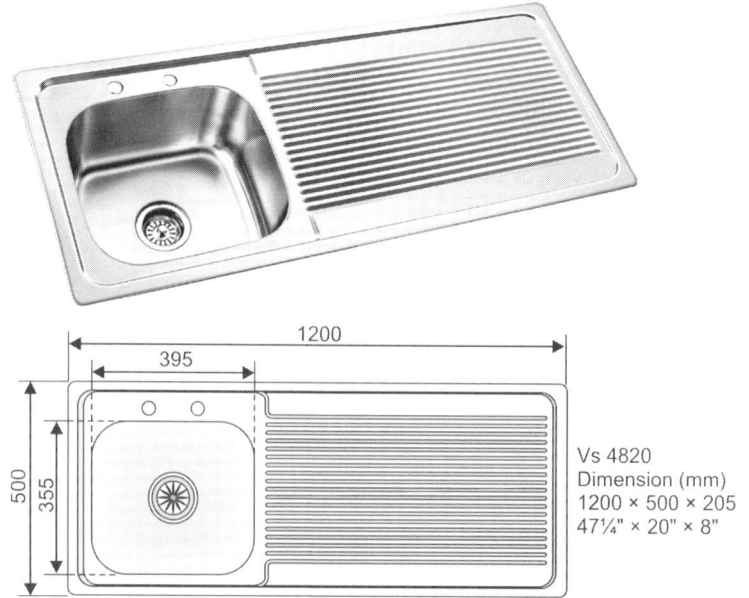

Vs 4820
Dimension (mm)
1200 × 500 × 205
47¼" × 20" × 8"

Photo 7.1: Kitchen sink

Small size components category may consist of components such as shown in Photo 7.2.

Small size components may be of any material such as brass, aluminum, steel, titanium. Weight may vary from few grams to few hundred grams.

Miniature size components may be those as used in electrical switches, relays, components used in electronics such as flat pins,

Photo 7.2: Sheet metal components

hollow pins of connectors and edge connectors, clock gears and many more parts as used in instrumentation and control system. These components are generally manufactured from half hard spring hard brass, phosphor, bronze, tungsten, etc. Such components may weigh from say, 0.5–5 grams.

Press tools required for manufacturing miniature components may be multi-impression follow on and/or progressive tools.

Weight of tools from few to hundred kilograms are against tools weighing in tons for big size components.

7.2 MANUFACTURING OBJECTIVES

Many a times, product development team prefer to have prototype components to ascertain its functional suitability and to carry out an opinion survey to find out as to what opinion customer may have about visual effect of the components assembled on a product.

To give an example, switch body of a metallic flashlight is displayed during opinion survey. Shape of switch is shown here with the help of Fig. 7.1.

Most of the viewers expressed their opinion that switch does not look attractive, consequently, this design is discarded and prototype of another design is prepared. It is shown in Fig. 7.2.

This design is also subjected to opinion survey. Most of the viewers expressed their opinion that this design looks attractive.

For manufacturing prototype, simple draw punch and die is made out of mild or carbon steel plate piece as available. No die set is used. Similarly, bending, piercing and sliding tools are made. In this way, tool making with development objective is fulfilled.

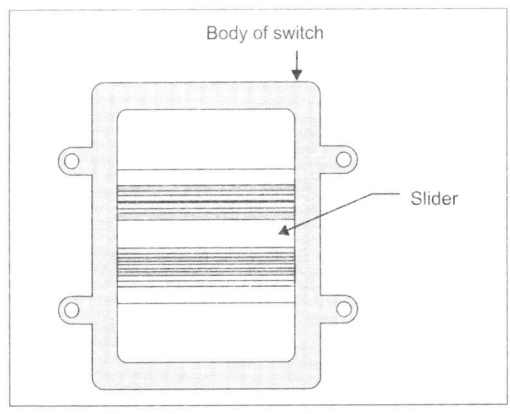

Fig. 7.1: Component, switch body

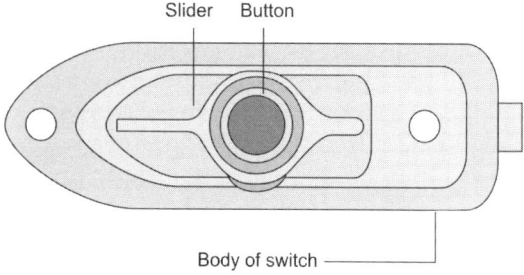

Slider Button

Body of switch

Fig. 7.2: Complete switch design

7.3 DESCRIPTION OF FEW COMPONENTS

Sometimes, only few thousands of components per month are needed which may be produced by single impression tools, but the tools are properly made in die sets wherever it is needed. Tools are run on hand press or power press depending on size of tools and tonnage required.

In case components are required to be continuously produced to maintain a regular supply in bulk quantities then generally multi-impression, progressive or follow on tools of good quality are used. Examples of such **components** are shown in **Fig. 7.3**.

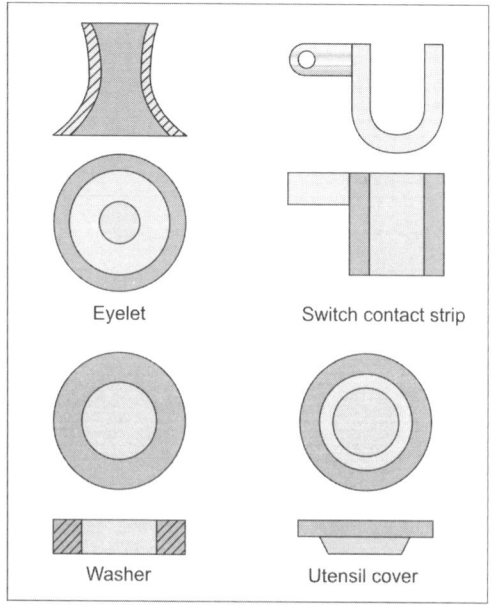

Eyelet   Switch contact strip

Washer   Utensil cover

Fig. 7.3: Examples of components

While a tool is designed, accuracies required in components are taken into account by designer. Accuracy of a sheet metal component may be related to various dimensions, flatness and clean and straight edges.

Dimensional accuracy means as to how much tolerance is provided for a particular size in components drawing. Figure 7.4 shows example of a component showing large tolerance.

Figure 7.5 is the example of components where closer tolerance is specified as compared to components shown in Fig. 7.4.

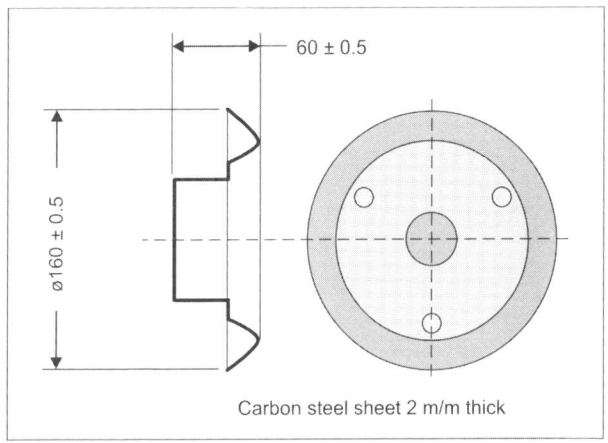

Fig. 7.4: Steel sheet components

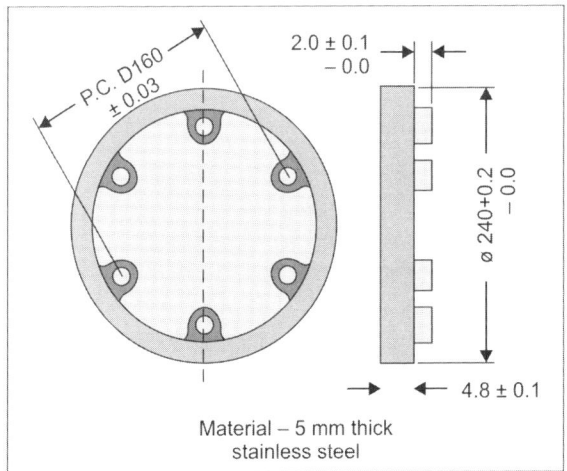

Fig. 7.5: Precision sheet metal component

In this case, sizes of blanking punch, die and piercing punches and dies have to be machined to achieve desired accuracy. Furthermore, proper cutting clearance is to be provided. Pitches of holes have to be 60° within two minutes. It is better to specify cordial distance between the center of holes, because the position of assembly of brake disc on hub of front wheel of two wheeler may be non-selective.

7.4 THREAD ROLLED COMPONENTS

In numerous sheet metal components, very close and precise tolerance in some dimensions are needed. Significance of this is that components should get assembled very precisely. Few examples of such components are shown in Fig. 7.6.

Two brass sheet components 'A' and 'B' are shown. Component 'A' has to screw in and out over component 'B'. It is difficult to maintain final dimensions as the same depends on the following:

- Dimensional accuracy in the sizes, diameter and bore of drawn cups
- Accuracies and surface finish in thread rolling rollers, upper and lower
- Matching and accuracy in rotation of rollers
- Accuracy in the wall thickness of cups
- Consistency in the hardness of cup walls
- Speed of rolling
- Time period of rolling

Another example of precisely accurate component is shown in Fig. 7.7.

In this case, tool design has to ensure that pitch 16 has to be maintained within given tolerance, so the sequence of operation and location system has to be decided. Tool may be progressive so that one component in each stroke is obtained.

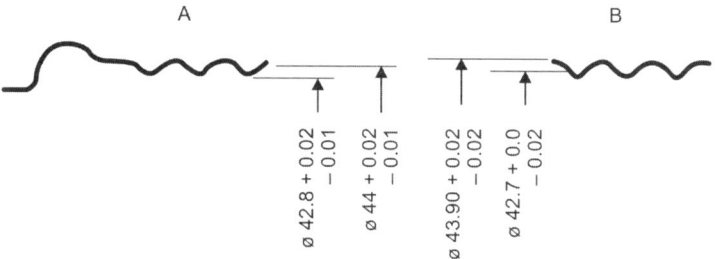

Fig. 7.6: Precision thread fitting

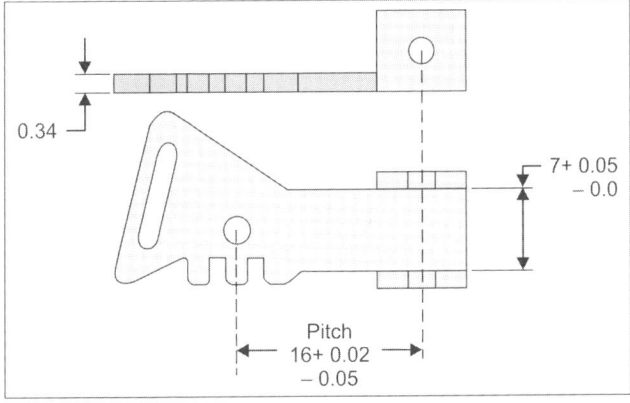

Fig. 7.7: Another example of precision component

7.5 BLANKING AND FINE BLANKING

If the commonly blanked edge of a slug/blank is seen magnified then generally it looks like the one shown in Fig. 7.8.

Fine Blanked Component

A finely blanked edge would look like the one shown in Fig. 7.9. To achieve fine blanking, an entirely different tool has to be designed and made. Further, a double acting hydraulic press is needed.

Fine blanking is explained with the help of Fig. 7.10.

Referring to Fig. 7.10, part (1) is aluminum strip of 4 mm thickness. It is held with a pressure between faces of pressure ring (2) and blanking die (4). Part (3) is a blanking punch and (5) is a pressure insert which exerts force on the area/profile which is to be blanked. Once the blanking action is over, punch (3), blank and

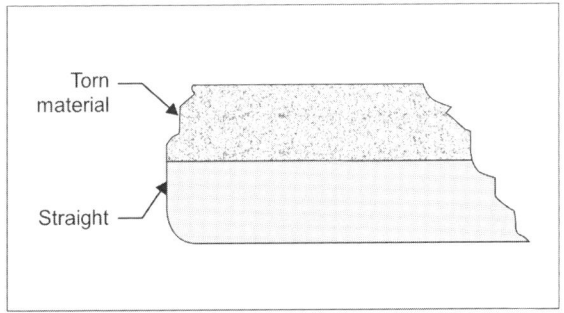

Fig. 7.8: Magnified edge condition of normally blanked part

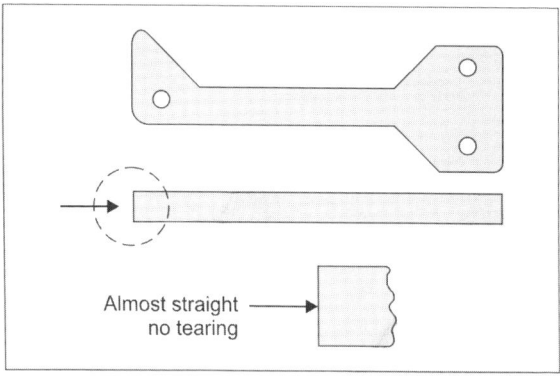

Fig. 7.9: Finely blanked component

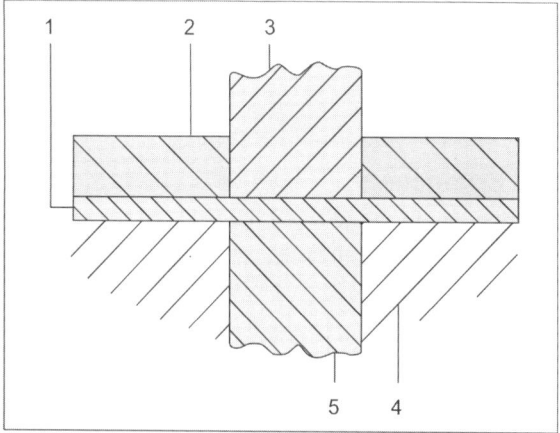

Fig. 7.10: For explaining fine blanking

pressure insert move up together while keeping the blank pressed between faces of blanking punch and pressure insert. This means that component (blank) is ejected out from the top of the die when blank has crossed sheet/strip and punch (3) is still moved up further to free the component which may be removed by some mechanical means or air jet.

With this technique, walls of blank (components) are almost flawlessly smooth, without any sign of tearing, hence, tool designer has to design a tool accordingly.

8

Sheet Metal Press Working Operations

This chapter consists of the following aspects:
- List of various operations and description
- Single and multi-impression operations such as blanking
- Scrap percentage consideration
- Spring back in bending
- Tool design for spring back compensation
- Wrinkle formation in cup drawing
- Avoidance of wrinkle formation in cup drawing
- Reduction consideration in cup drawing
- Cause of stress at wall brim of drawn cup
- Description of various operations such as forming, trimming, beading, embossing
- Types of threads and thread rolling
- Thread rolling machine
- Thread rollers

LIST OF VARIOUS OPERATIONS AND DESCRIPTION

There are innumerable sheet metal components of various shapes, dimensions and materials. Such components are produced by a variety of operations and their sequence. Once basic principles are clear, a designer and maker would be in a better position to design and make tools.

Press working of sheet metal means that sheets are given a variety of shapes and sizes with the help of suitably designed tools and correctly selected presses which may be hand or power operated. Hand presses may be manually operated, pneumatically or hydraulically. Small capacity presses are used for producing tiny components. These are also used for assembly work such as riveting. Heavy mechanical and hydraulic presses are used to perform various operations of press working of sheet metals. Some commonly used operations are listed below:

- Cutting and shearing
- Blanking
- Trimming
- Bending
- Stamping
- Forming

- Drawing
- Deep drawing
- Piercing
- Beading
- Notching
- Thread rolling

Photo 8.1 shows a number of sheet metal components for readers of this book to visualize as to what operations and its **sequences** are required to produce shown components. Sequence of operation for a particular component can be well visualized once the concept of various sheet metal working operations are understood. Therefore, list of operations is briefly described in the following paragraphs:

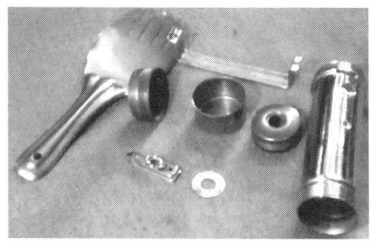

Photo 8.1: Sheet metal components

Cutting and Shearing

On most of the occasions, it is necessary to cut small strips from stock of large strips or rolls.

In Fig. 8.1, a typical brass sheet of 60 cm wide, 95 cm long and 0.35 mm thick is shown. Grain direction, due to rolling, is shown by small arrows parallel to edge of sheet. Strips of 5 cm width are required to be cut. It is worth noting that cut strips would have

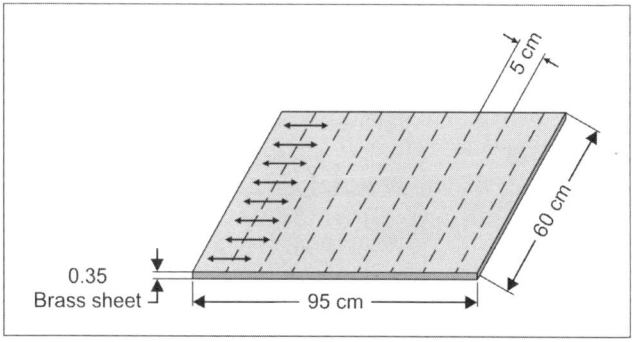

Fig. 8.1: Cutting of strips from sheet or large strips

grain direction at right angle to long edge of strips. For cutting one or two pieces, pair of scissors can be used, but it is not a feasible proposition for production of large number of strips. For this purpose, shearing machines (guillotines) are used.

Figure 8.2 shows a guillotine in action. Figure 8.2 A shows basic concept of a guillotine. Part (1) is a big cast iron bed of the machine. Big sheet is placed over it. Part (2) is upper moving shear blade which has an inclined cutting edge having an inclination of few centimeters. (3) is lower shear blade which is suitably fixed to machine bed (1). While upper shear is in up position, there is a gap between two shears. Sheet to be cut is pushed forward through the gap to touch a stopper. It is so adjusted that required width of cut strip is obtained. Figure 8.2 'B' shows shearing action in progress. Pressure pad (6) presses the sheet before shearing action starts. Shearing blade (1) travels down and shears the sheet to strip (3), while lower fixed shear supports the sheet.

Inclination in cutting edge of upper moving blade is provided so that shearing takes place gradually (see Fig. 8.3). This is to avoid sudden shock load on shearing blades and machine system. Guillotines are designed and built for a maximum thickness of sheet metal and maximum and minimum width.

Force required to cut strips from a sheet may be calculated by the following formula:

Force = Cutting area × **Shear strength** of material

F = A × fs   where F = Force required in tons

A = Area, sq cm

fs = N/Sq cm

Area A is the product of sheet thickness and effective shearing length while inclined shearing blade is shearing the sheet progressively.

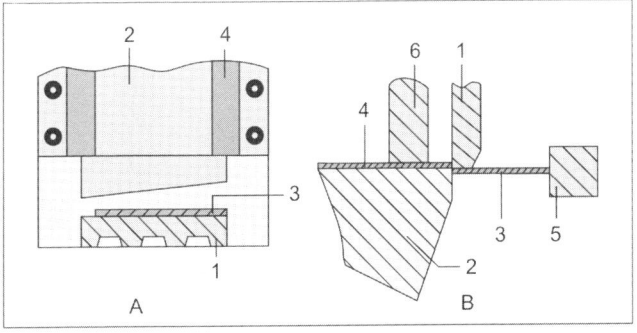

Fig. 8.2: Guillotine

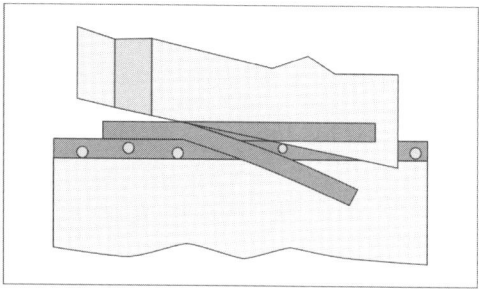

Fig. 8.3: Shearing with inclined blade

If there is no inclination in upper shearing blade then cutting area would be

$$A = \text{Sheet thickness, } t \times \text{width of sheet, } w$$

This would cause sudden and jerky load on the machine. Many a times, it is necessary that grain direction should be along the length of strip. In that case, sheets are placed on the guillotine table in such a way that grain direction is parallel to shearing blades. Consequently, cut strips would have grain direction along the length of strips.

Blanking

Figure 8.4 shows a phosphor-bronze component and its blank which is 6.80 mm wide having a tolerance of ± 0.10 mm. Strips of 6.8 mm width can be cut on guillotine but a close tolerance of ± 0.10 cannot be ensured, therefore, it is necessary to produce this strip by a blanking tool.

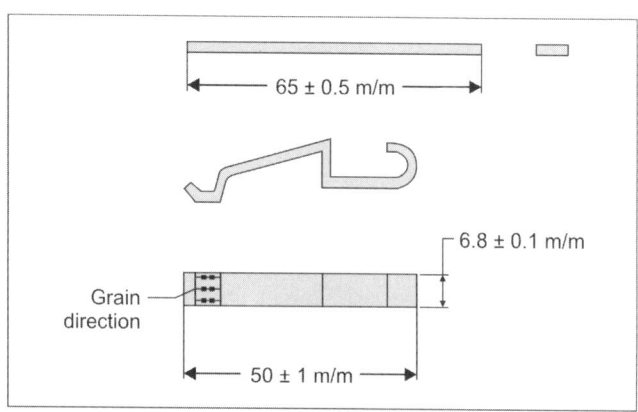

Fig. 8.4: Bent component and its blank

Basic design and construction of a typical tool is shown in Fig. 8.5 where part (1) is pillar type die set. It is fitted with stem (2), blanking punch (3), blanking die (4), stopper (5) and guide plate (6). Distance between guide plates is maintained so that strip of specified dimension and tolerance can move freely with minimum possible shift along width. Blanking punch (3) and die (4) are well matched with a cutting clearance of about 0.05% of sheet thickness for phosphor–bronze. Cutting clearance finely varies with sheet material and thickness. Cutting clearance for steel strip of 0.36 mm thickness would be 0.06–0.08 mm, depending upon shear strength of strip material.

Blanking principle is explained in the following paragraph with the help of Fig. 8.6. It shows a blanking punch passing through strip and then in blanking die. A perfect straight cut edge is shown. This does not happen. What actually happens can be seen in Fig. 8.6 B. Blank is sheared out of strip. Shearing action is in a slant

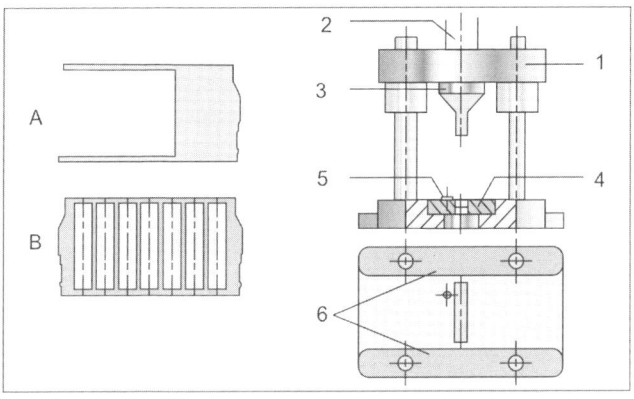

Fig. 8.5: Blanking tool

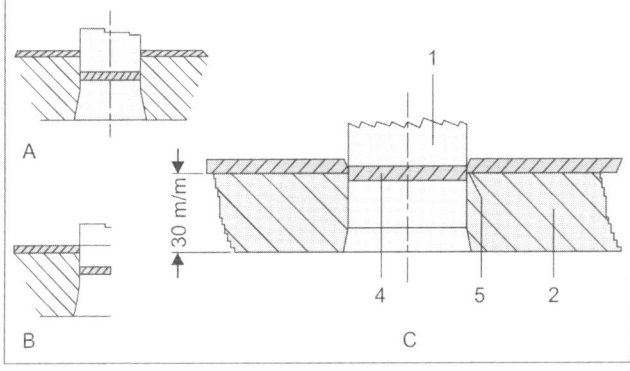

Fig. 8.6: Blanking principle

manner as shown. Force required to punch out a blank is calculated for a typical zinc blank.

Formula for calculating blanking force is

$$F = \text{Peripheral length} \times \text{thickness} \times fs$$
$$= 1.2 \times 6 \times 0.32 \times fs$$
$$= 2.34 \times fs$$

where fs is shear strength in MPa
$$= N \text{ (theoretical)}$$

Shear strength, fs for zinc ranges from 214 to 325 MPa, therefore, theoretical value of force 'F' required to blank out one piece can be worked out. Actual shear strength of zinc sheet very much depends on rolling conditions.

In practice, blanking tools with five or seven impressions are generally used, so five or seven blanks are obtained in one stroke of power press. For a five impression blanking tool, theoretical force required would be 5F which equals to approximately 90 tons. In practice, a press is selected with a tonnage at least two times the theoretical force required. Hence, a power press of 200 tons is required to blank five blanks of zinc, as shown in Fig. 8.7. Power press of standard tonnage is selected. Use of power press of still higher rating is preferred because wear and tear in power press would be much less as total blanking load is much less as compared to power press tonnage.

Speed of blanking for one impression blanking tool may be defined as number of strokes per minute. This generally depends

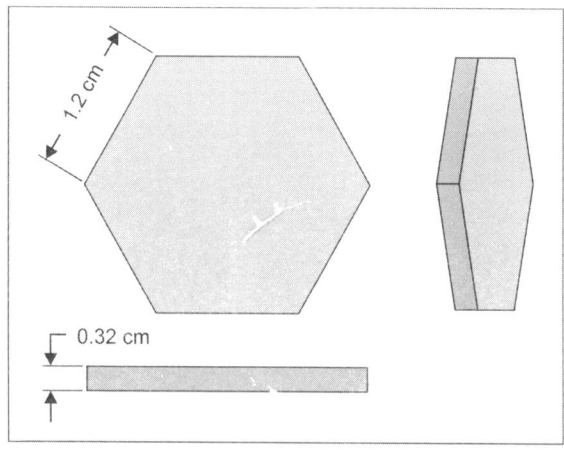

Fig. 8.7: Zinc calot

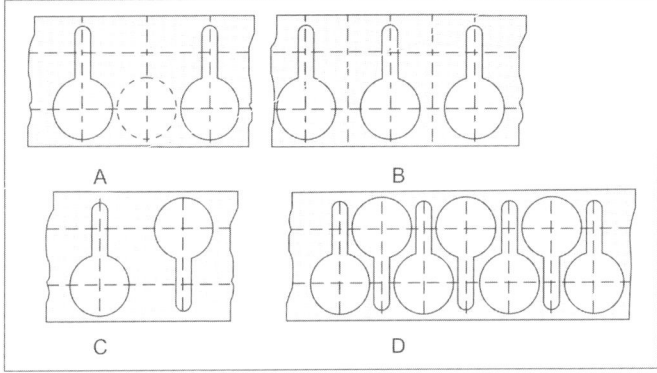

Fig. 8.8: Layout of multi-impression blanking

upon the range of strokes per minute of power press. It may range from 30 to 58 strokes per minute or even more. Sturdiness of tool and sheet feeding mechanism may also be factors on which selection of speed depends.

Multi-impression tools are designed in such a way that generation of scrap (left over) is minimum. Figure 8.8 illustrates layout of multi-impression blanking tool for a phosphor–bronze blank.

Sheet blank as per dimension given in figure–A tie of 1.2 mm may be considered for the purpose of calculating percentage of scrap. Width of strip would be

Length of blank + 2 × tie

$$W = 22 + 2 \times 1.2$$
$$= 24.4$$

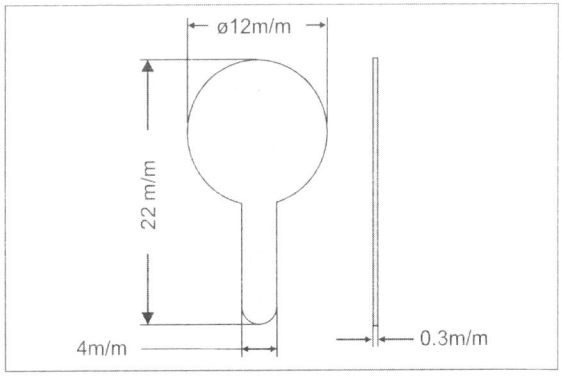

Fig. 8.9: Area calculation of blank

As per layout shown in Fig. 8.8 'A', pitch of dies (openings) would be 26.4 mm, so in a strip of one meter length, number of blanks obtainable would be

No. of blanks = 37.8, say 37

Further, as per layout shown in Fig. 8.8 'B', pitch of two dies would be 18.4 mm, so in a strip of one meter length, number of blanks obtainable would be

No. of blanks = 53.3, say 54

Approximate area of one blank is calculated in the following manner: Referring to Fig. 8.10.

$$A1 = 113 \text{ sq mm}$$
$$A2 = 32 \text{ sq mm}$$
$$A3 = 6.28 \text{ sq mm}$$
$$\text{Total Area} = 151 \text{ sq mm}$$

Total Area of 37 blanks = 55.87 sq cm

Area of 1 meter long and 24.4 mm wide strip = 244 sq cm

$$\text{Area of scrap} = 244 - 55.87$$
$$= 188.13 \text{ sq cm}$$

Percentage of scrap = 77 %

Area of 54 blanks = 81.5 sq cm

Area of scrap = 162 sq cm

Percentage of scrap = 66%

From above calculation, it is clear that layout of blanking die (opening where punch precisely gets in) as shown in Fig. 8.8 'C' may be preferable because it generates less scrap as compared to layout shown in Fig. 8.8 'B'.

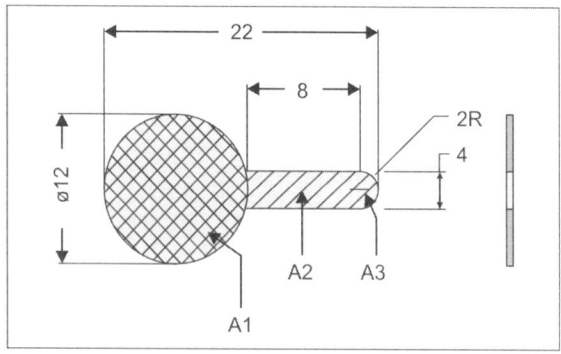

Fig. 8.10: L shape component

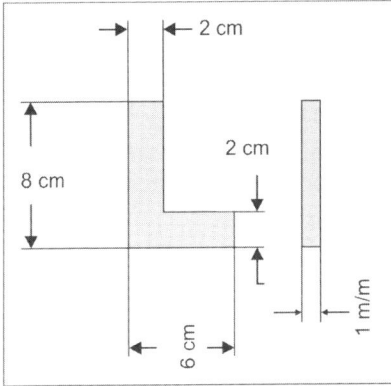

Fig. 8.11: Two possible layouts

Another typical component of mild steel sheet is shown in Fig. 8.11.

Two possibilities of layout are considered, as shown in Fig. 8.12. In case of Fig. 8.12 'A', number of blanks which can be obtained in one meter length of strip work out to be 15 and for Fig. 8.12 B, number of blanks work out to be 30.

Area of one blank = 8 × 2 + 4 × 2

= 24 sq cm

Area of 15 blanks = 360 sq cm

Area of one meter long and 10.6 cm wide strip = 1060 sq cm

Area of one meter long and 8.8 cm wide strip = 880 sq cm

Hence,

Scrap area for layout as per Fig. 8.12 'A' = 880 – 360

= 520 sq cm

Percentage of scrap = 59 %

Scrap area for layout as per Fig. 8.12 'B' = 1060 – 720

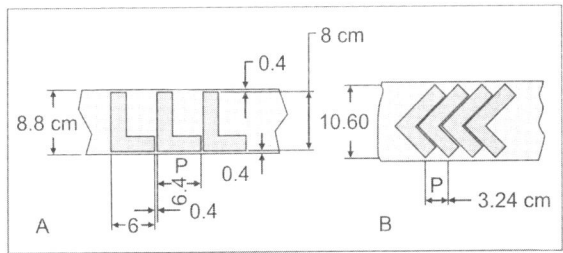

Fig. 8.12: Trimming of components

= 340 Sq Cm

Percentage of scrap = 32 %

Above two examples demonstrate that percentage of scrap can be minimized by considering various possibilities of layout and calculating percentage of scrap. A layout with minimum generation of scrap may not be practically possible to adopt. This may be due to tool making limitations, so, for such blanking process, it is considered where blanks are pushed through die. There may be typical possibilities where blanks are not completely blanked out of strip or sheet metal in the form of roll. Blanks are kept attached to strip for further operation in progressive tool design. Blanking operation is often combined with other operation such as cupping. Hence, a tool which carries out both the operations is called 'cut and cup' tool.

Trimming

In press working of sheet metal, trimming is an operation by which excess sheet is removed by circular cutting or flat cutting like a blanking operation. Examples of both the operations are explained with the help of Fig. 8.13.

Figure 8.13 'A' shows an aluminum cup which is produced by a 'cut and cup' tool. Brim of such cups are not perfectly straight. It needs to be straightened by trimming and to achieve specified height of such cups which are not perfectly straight. In this typical example, approximate height of untrimmed cup is 32 mm. Required height of cup is 29 mm. Figure 8.13 (B) shows a rotating circular trimming tool which cuts out a scrap ring to give a cup of 29 mm uniform height.

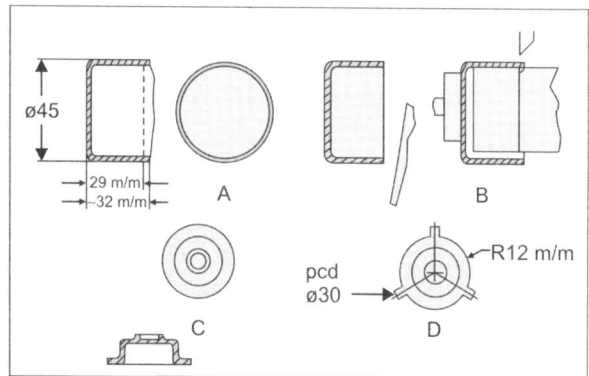

Fig. 8.13: Hole and slug diameters

Figure 8.13 'C' shows another shallow cup component which has a flange of about 35 mm in diameter, which is not exactly circular. Fig. 8.13 'D' shows the shape of desired flange which is about 5 mm less than roughly circular flange as shown in Fig. 8.13 'C'. Final shape of brim of component (as shown in Fig. 8.13 'D') is achieved by a punch and die of suitably matched dimensions and design. This operation with the help of tool is called trimming.

As a general rule, die dimension produces blank dimension and punch dimension produces 'hole' dimension. This general rule is taken into account if desired dimensions have very close tolerance. Let the diameter of a blanking punch is 30.00 mm and the bore of die is 30.03 mm. Then the hole generated by the tool would be 30.00 mm and the diameter of blanked disc would be 30.03 mm (see Fig. 8.14).

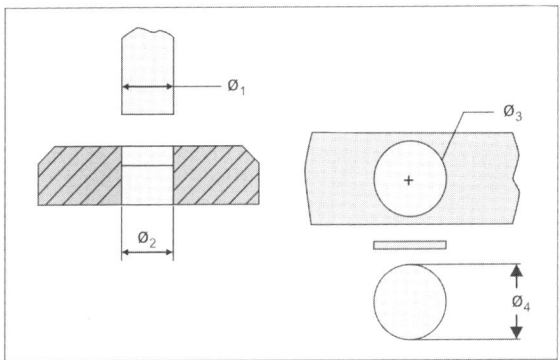

Fig. 8.14: Few components before and after blanking

Bending

Bending is an operation by which portion of a strip or a component is made to have a different plain. It may be at a certain angle with respect to original plain. Figure 8.15 shows a few components before and after bending.

Bending is a simple operation as long as bending angles or dimensions are not critical. Bending assumes a form of delicate operation if bending angles or dimensions become critical. Figure 8.16 is the drawing of brass strip before and after application of bending operation. Bending radii r_1, r_2, r_3 and r_4 are important. In Fig. 8.16 'B', inner bending radius is 0.5 mm and in Fig. 8.16 'C', inner bending radius is 2 mm. Angle of bend in both the cases is different. Bending is carried out by a tool shown in Fig. 8.17.

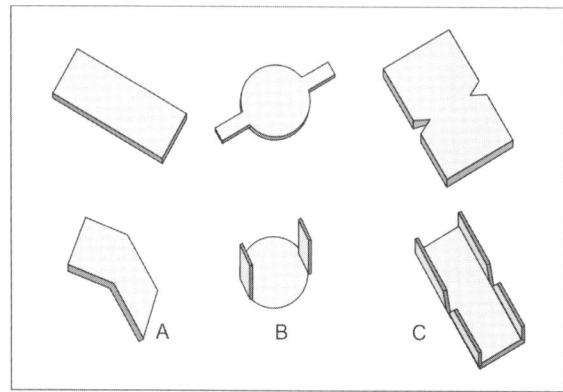

Fig. 8.15: Component before and after bending

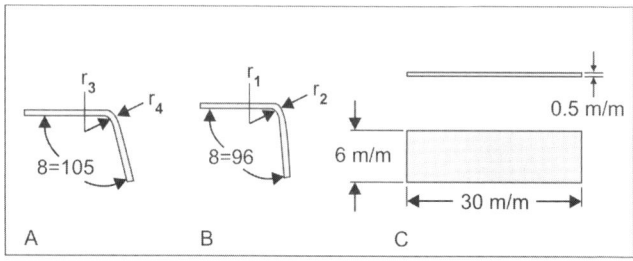

Fig. 8.16: Bending tool

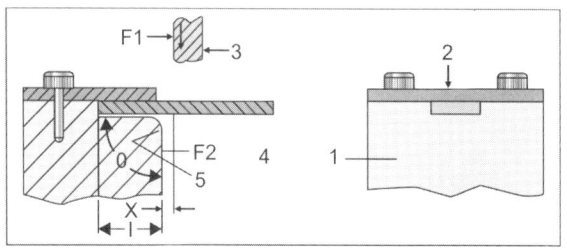

Fig. 8.17: Spring back in bending explained

Referring to Fig. 8.17, part (1) is a die block which has a slot 19.5 mm long, 0.6 mm deep, (2) is a cover plate. Part (3) is a bending punch which is X mm away from die face. In this typical example, X is kept as 0.55 mm. Part (5) is a bending radius. When punch (3) travels down, it pushes the strip down which gets into the gap between punch and die faces F1 and F2. Face F2 of die block is at right angle to slot face. After completion of bending operation, component is taken out of slot. On examining the component, it is

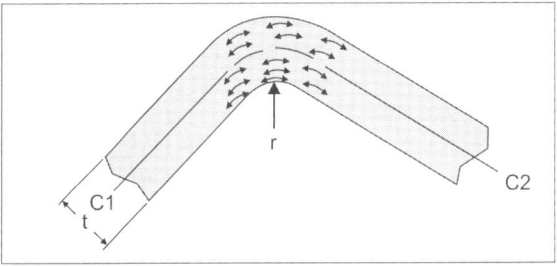

Fig. 8.18: Bending tool with spring back compensation

found that angle of bend is more than 90°. This is because of 'spring back' characteristic of brass strip. Why does spring back take place? It is explained with the help of Fig. 8.18.

Figure 8.18 shows a magnified portion of bent strip. In the middle of thickness of strip, an imaginary central line is drawn by dashes. On bending of strip, grains above central line tend to stretch. Maximum stretching of grains is near the surface of strip. Grains below the central line tend to compress. Maximum compression is near the opposite surface of strip. This stretching and compression of grains cause the tendency in the strip to 'go back' so that internal stretching and compression forces are reduced. Thus, bent portion goes back a little after bending punch is pulled up. This phenomena is called 'spring back'.

Amount of spring back depends on the following factors:
• Thickness of strip
• Width of strip
• Hardness of strip material
• Grain direction
• Radius of bend
• Amount of extra bending
• Gap between punch and die working faces

Sheet of brass and phosphor–bronze are produced with varying hardness. For example, an annealed phosphor–bronze sheet may have a hardness of 100 BHN (Brinell hardness number) and the one having spring hardness of 119 BHN. Brass sheets may be 'quarter hard', 'half hard', 'spring hard' and extra hard'. Strip of quarter hard brass will spring back less after bending as compared to spring hard strip.

Amount of spring back for a particular strip basically depends on the ratio of bending radius and sheet thickness. It may be considered before designing and making a bending tool. Due to practical limitations, calculated spring back may not be exactly achieved. This may be, for example, due to the fact that strip actually does not have that Brinel hardness number which is taken for calculation. Bending radius may not be exactly what is taken for calculation. Grain direction in strip is not that which is assumed for the purpose of calculation. Gap between bending punch and die may be more than assumed for calculation. Slot in which strip is placed may be too high for strip to be too loose.

All the above reasoning does not mean that calculations for spring back is of no use. No, it is useful to calculate spring back before finishing the tool. Calculation of spring back provides designer of tool with the nearest bending angles to be taken into consideration. Some fine adjustments may be done afterwards by actual bending trials. The following paragraph provides some information regarding calculation of bending spring back.

Parameters,
- θ1= Angle of bend in component
- θ2= Angle of bend in the die
- θ3= Spring back, (θ2 minus θ1)
- R1= Bending radius (internal)
- T = Thickness of strip
- M = Material of strip
- BHN = Brinell hardness number
- Grain direction

The basic design of bending tool illustrated so far cannot take care of spring back if angle of bend in the component is 90° or more. A modified design of bending tool is shown in Fig. 8.19 which can be adjusted for achieving desired angle of bend in component.

In case bend strip is in V shape then basic design of bending die may be the one as shown in Fig. 8.20.

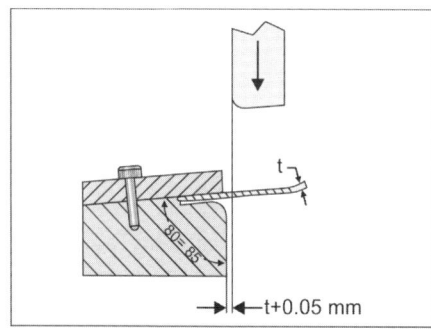

Fig. 8.19: V shape bending tool

Some strips have got a number of bends as shown in Fig. 8.21.

Material of strip as shown in Fig. 8.21 is phosphor–bronze. All the bending would be done at a time by tool of basic design as shown in Fig. 8.22.

Most propably, one or two times correction in upper and lower tool may be necessary to achieve correct angles in component.

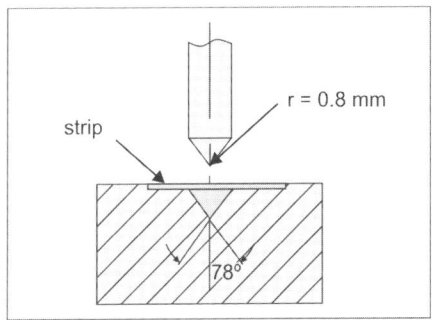

Fig. 8.20: Component with many bends

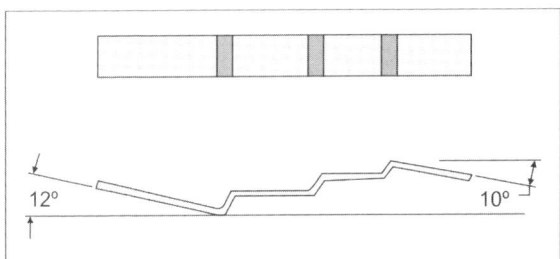

Fig. 8.21: Multiple bends cum stamping tool

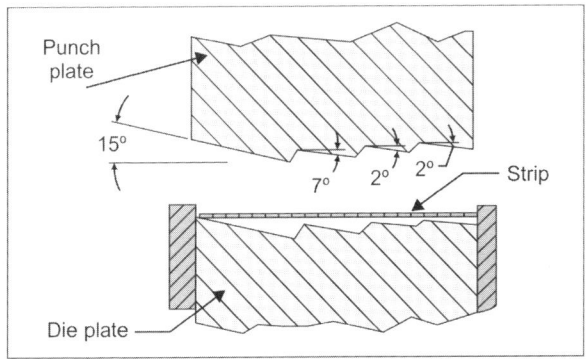

Fig. 8.22: Examples of stamping

Stamping

Stamping is an operation in press working of sheet metal by which an impression is created over the surface of a sheet metal part. Stampings are generally of two types:

Fine and shallow

Rough or fine and deep

In case of fine and shallow, only one sided impression is created. Examples are fine marking on fountain pen metal clip, bottom of a pressure cooker, on metallic body of a pin stapler, etc. Fig. 8.23 shows example of both types of stampings.

In Fig. 8.23 'A', a punch is shown which has raised letters. Height of such fine letters is kept about one fourth the thickness of sheet metal. In Fig. 8.23 'A', it is shown as 0.15 mm for a strip thickness of 0.6mm. The other side of strip is completely flat.

In Fig. 8.23 'B', punch is shown which has raised ring. Die has a sunken ring with increased dimensions to accommodate sheet thickness. On stamping impact, a raised ring is formed on sheet metal strip. This is illustrated in Fig. 8.24.

Instead of raised ring, there may be any monogram or letter or numbers. In case of a coin, a blank of coin material is pressed/stamped by two dies on a double acting press, hence, material flows in sunken details of die. Consequently, raised details develop on both sides of coin. This operation is called coining.

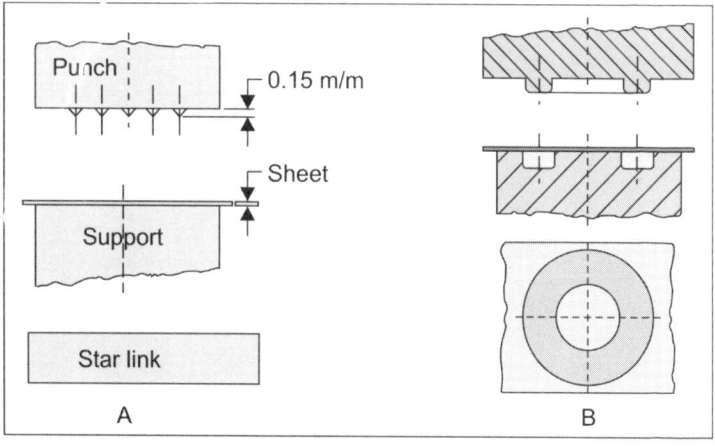

Fig. 8.23: Raised ring formation

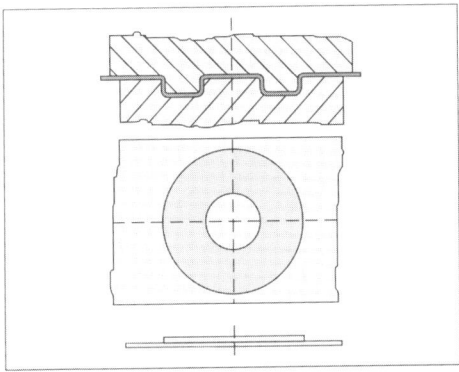

Fig. 8.24: Typical cup, shell components

Drawing

Drawing is an operation in press working of sheet metal by which flat sheet metal is partly or fully converted to the shape of a shell. It may be circular, rectangular or square. Figure 8.25 shows few typical cup, shell or container.

Referring to Fig. 8.25, 'B' is a shallow round shell having no flange. 'C' is a shallow rectangle container having flange, 'D' is a dome drawn in a sheet metal, 'E' shows a strip or coil of sheet metal to which drawn shells are attached by means of three curved ribs. Ribs are lanced at station before drawing to hold a blank for drawing. This is to ease drawing operation.

Referring to Fig. 8.26, 'B' is a brass sheet blank of say, 100 mm in diameter and 0.48 mm thick. 'd' is die of 76 mm bore. 'p' is a draw

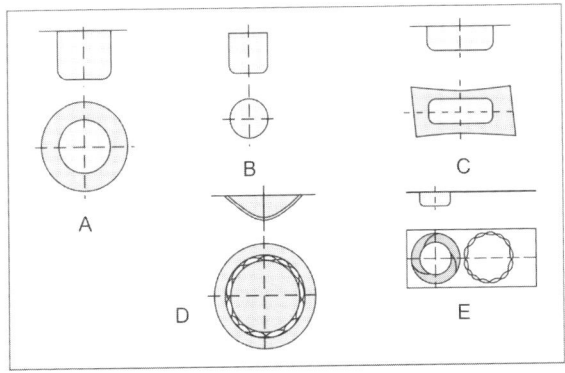

Fig. 8.25: Draw operation and Lancing

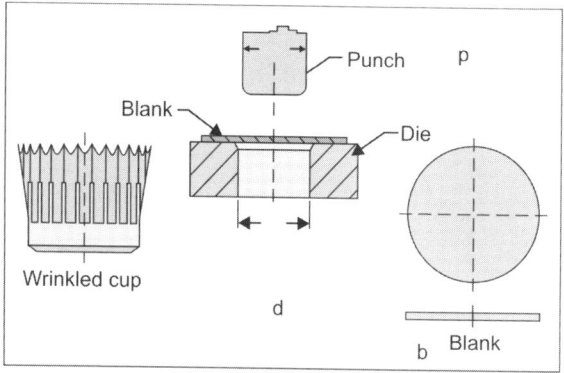

Fig. 8.26: Fractured component due to wrinkle formation

punch of approximately 72 mm diameter. Blank is placed over the die face and draw punch is brought down. It keeps on moving down till blank is completely cramped and gets into the die. Cramped cup (cup with wrinkles) is taken out which looks like as shown in figure. Formation of wrinkles is due to the fact that circumference of blank is forced to get reduced from $\pi \times 100$ to $\pi \times 76$ mm. In this typical example, circumference of blank cannot get reduced, therefore, it takes form of wrinkles so that it virtually gets reduced.

Another situation may be that diameter $\varnothing_2$ of draw punch is equal to bore of die minus twice thickness of blank minus 0.05 mm. If draw is attempted then resulting component would be something like shown in Fig. 8.27.

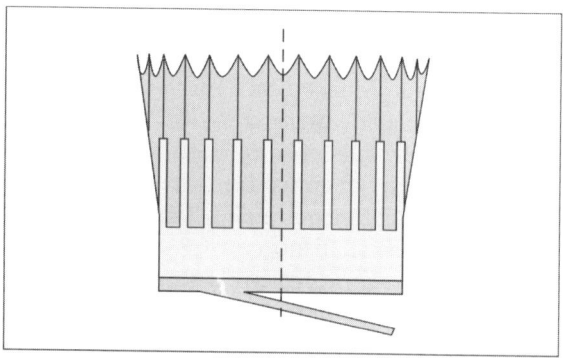

Fig. 8.27: Arrangement for under holding pressure on strip/blank

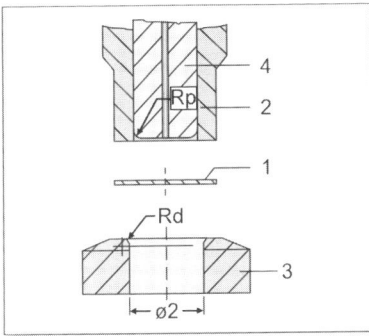

Fig. 8.28: Marked blank and cup

For some distance, say about 3 mm, walls of cup will be straight and almost with no wrinkles. On further movement of punch, wrinkles would form and therefore blank cannot enter further. If punch is forced down then bottom of wrinkled cup would snap.

This problem is overcome by keeping the blank pressed while drawing is taking place. This is further explained with the help of Fig. 8.28.

Referring to Fig. 8.28, (1) is sheet metal blank and (2) is a pressure pad which presses blank over the face of die (3), before the draw punch face touches the blank. Pressure between the faces of pressure pad and die face is kept so much that formation of wrinkles is avoided while punch is pushing the blank into the die. Draw punch and draw die have radii Rp and Rd respectively. Values of Rd significantly influences the draw quality of cup. For a particular blank, thickness and quality of material are two factors to decide value of Rd. As a general rule, value of Rd may lie between 2.5–3.5 t where t is the thickness of sheet.

Figure 8.29 'A' shows a circular blank which is marked by scratching or etching ink. Marked face of blank is placed towards the face of draw die. After drawing operation, markings look like as shown in Fig. 8.29 'B'. Distance between two marked lines is say, X. After cupping, this distance becomes Δx. This means that grains of material in X distance are compressed in a distance Δx. Had there been wrinkles, then no compression of grains would have taken place. Due to compression of grains, hardness of cup wall is maximum at the brim of cup.

It can very well be appreciated that compression of material cannot be unlimited. In case diameter of blank is too large as

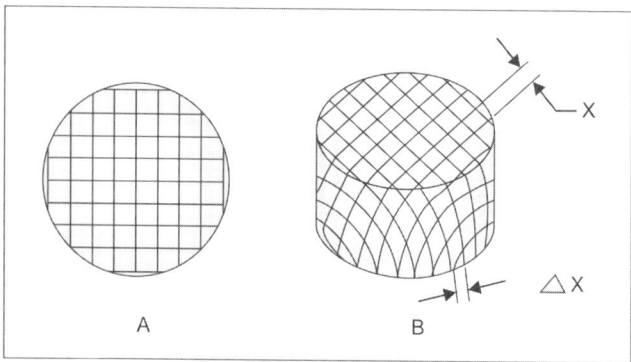

Fig. 8.29: Deep drawing stages

compared to diameter of cup, complete cupping cannot take place. Force of drawing punch would rupture the base (closed end) of cup.

In practice, maximum percentage of reduction in diameter is about 40%. Beyond this percentage, it becomes difficult to control flawless drawing of cup. Sometimes internal stress due to compression of grains, near the brim of a brass cup becomes so much that 'Season cracking' takes place. This is a very serious defect. To avoid this situation, stress relieving operation is carried out. It is done by placing cups/components in an oven at a particular temperature and for predetermined period of time. After stress relieving, some finishing operations such as 'bright dipping' may be necessary.

Deep Drawing

Deep drawing is a series of drawing operations where reduction of diameter from blank to diameter of final component is much beyond 40%, hence, deep drawing operation cannot be carried out in one draw. A typical example of deep drawing is described with the help of Fig. 8.30.

A shell of 30 mm diameter and approximately L3 mm long is to be drawn from a circular blank of 120 mm diameter, this means, a reduction of 90 mm in diameter. Percentage of reduction works out to be 75%. This is just not possible. As already mentioned earlier, reduction more than 40% offer difficulties, therefore, practically it is tried to keep percentage of reduction less than forty.

Table 8.1 shows planned percentage of reduction. After first draw from blank to cup, grains are compressed maximum near

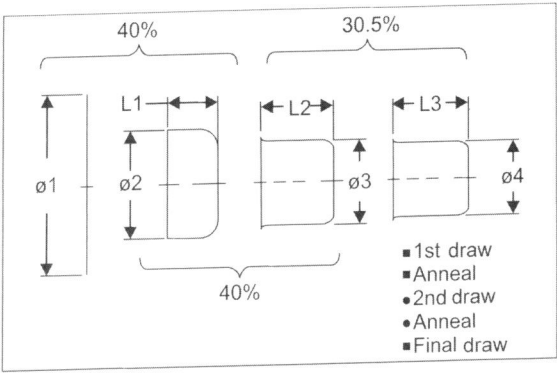

Fig. 8.30: Components at various stages of draw

Table 8.1: Deep draw, reduction percentage

S. No.	Description	Diameter Mm	Reduction in diameter	% age of Reduction
1.	Blank	120	–	–
2.	1st Cup	72	48	40
3.	2nd Cup	43.2	28.8	40
4.	3rd Cup, final	30	13.2	30.5

the brim. Consequent to this compression, brass becomes hard and stressed. For further drawing, it becomes necessary to anneal the cups to remove excessive hardness. Annealing is carried out on most suitable, practically determined temperature and period of time. Generally, temperature of annealing furnace is kept 380 to 430°C for approximately two hours. Before 2nd draw, first draw cups are washed and rinsed and then 2nd draw is carried out. Same process is carried out for third draw. Practically, temperature and time depends on shape of components, number of components in the basket and heating capacity of furnace.

Progressive drawing is the process which is carried out by a special press, progressive dies and transfer mechanism. Table 8.1 provides an idea of reductions.

Components are automatically transferred from one station to another. There is no intermediate annealing. This is due to the fact that grains are still in pliable condition near the base (Refer Fig. 8.31).

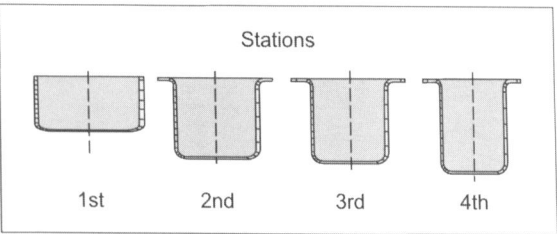

Fig. 8.31: Draw by progressive tool

Piercing

Piercing is a process by which holes are generated in a component. Multiple piercing is also carried out. Figure 8.32 shows a component which has three holes. These holes are made by a single stroke of punches.

Size of pierced holes depend on punch diameter. Pierced holes may be non-circular also. Piercing punches and dies are designed and made accordingly.

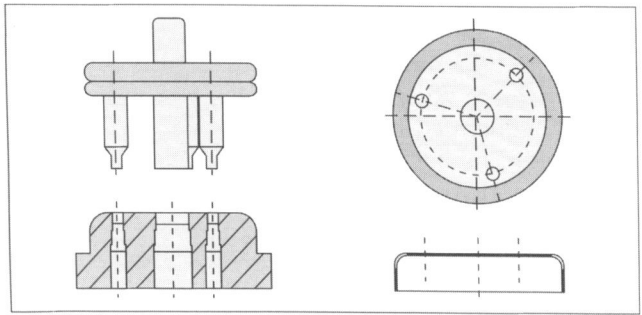

Fig. 8.32: Piercing

Beading

Beading is a process in press working of sheet metal by which certain portion of a component is raised. Figure 8.33 shows a number of components in which bead formation is there. Beads may be in a flat surface as well as circular.

Figure 8.33 'A' shows a bent strip component in which two beads are made near the edges of strip. Here the purpose of making beads is to strengthen strip against bending. As a general rule, height of bead may be two to five times the thickness of sheet. Height of bead very much depends on type, thickness, hardness of sheet and width of bead. It needs a matched punch and die to form a bead. Figure 8.33 'B' shows another flat strip on which a shaped bead is

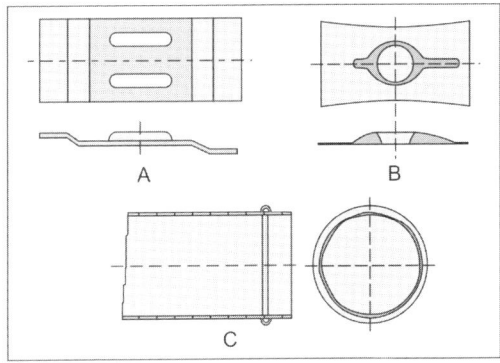

Fig. 8.33: Circular bead formation

formed. Height of bead is about four times the thickness of brass sheet which is 0.34 mm thick. Note that width of bead is about three and a half times the thickness of sheet. The purpose of formation of bead in this component is to strengthen, esthetic look and give a definite height to the component which is to be blanked out from the strip.

Figure 8.33 'C' shows a circular component, say a tube of about 36 mm bore and 0.5 mm wall thickness. It has a bead of 40 mm diameter and 3 mm wide (outside dimension), such type of beads are generally generated by mechanical means. Rotating tools are normally used. The basic working principle of rotating beading tools (die and punch) is explained with the help of Fig. 8.34.

Referring to Fig. 8.34, part (1) is a spindle of a machine, rotating in clockwise direction when viewed in direction of arrow. There is a protruding portion (5) in spindle (1). Part (2) is a bead turned

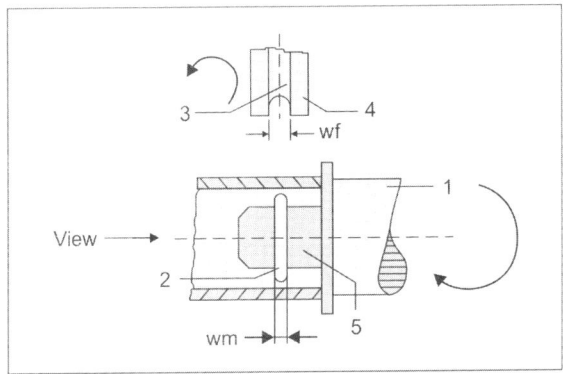

Fig. 8.34: Bead formation in lower tool

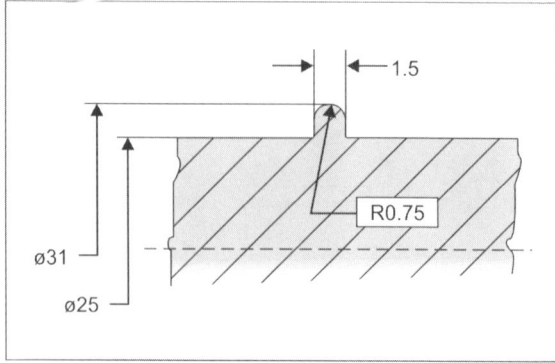

Fig. 8.35: Bead formation in upper tool

and ground in spindle. Diameter of this steel bead is about four to five millimeters less than the bore of tube so that it may be inserted over lower beading roller and taken out after beading operation. Width of steel bead in lower roller is 1.5 mm having a semi-circular shape as shown in Fig. 8.35.

The design of **beading machine** is such that upper spindle can be lowered or raised with respect to lower spindle, yet without unwanted play in machine parts or sacrificing parallelism of upper and lower spindles. Part (4) is upper beading roller with a female bead formation (3). Width of this female bead is kept so much that 2 × t (2 × thickness of tube wall) plus width of steel bead in lower roller are accommodated.

$$Wf = t + Wm + t$$
$$= 0.5 + 1.5 + 0.5$$
$$= 2.5$$

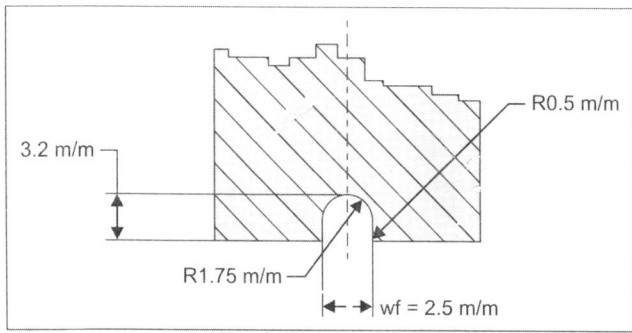

Fig. 8.36: Upper and lower beading tool on machine

A small radius R is provided on the edges of female bead of upper roller. This is to facilitate flow of material during beading operation. (See Fig. 8.35).

It is highly desirable that peripheral speed of lower roller and upper roller should be almost equal. Sometimes due to practical limitations in machine construction, upper roller cannot be lowered to touch lower roller. In such a situation, designing of upper and lower roller is done in such a way that circumferential speed of lower and upper rollers is almost equal whereas rpm of upper roller is exactly half of rpm of lower roller. Complete calculation is explained with the help of Fig. 8.37.

Let $\quad$ p min = 35 mm where p = pitch

$\qquad$ P max = 65 mm

$\qquad$ Du = mm

$\qquad$ Dl = 31

Cl = 97.4 mm, where Cl = circumference

$\qquad$ Cu = ?

$\qquad$ Nu = ?

$\qquad$ Nl = ?

Circumferential speed of upper roller ——— rpm × circumference

$\qquad\qquad\qquad\qquad\qquad\qquad\qquad$ Nu × Cu

Circumferential speed of lower roller ——— rpm × circumference

$\qquad\qquad\qquad\qquad\qquad\qquad\qquad$ Nl × Cl

Now $\qquad$ Nu × Cu = Nl × Cl

$\qquad\qquad$ 0.5 × Cu = 1 × 97.4

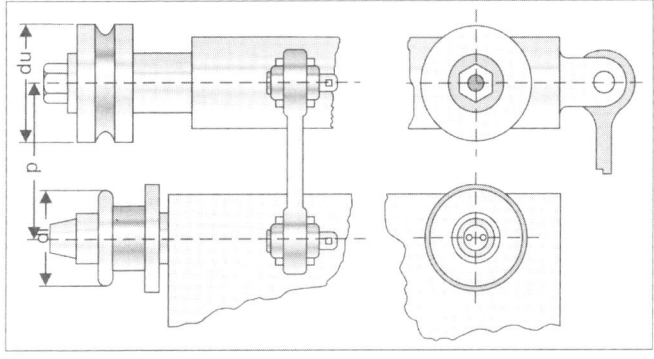

Fig. 8.37: Beading machine

Cu = 194.8 mm

Therefore,   Du = 61.98, say 62 mm

The purpose of keeping circumferential speeds of upper and lower rollers approximately equal is to avoid rubbing of upper roller on component under beading operation. In case rubbing takes place then two unwanted things may happen. Firstly, work hardening of component material at bead portion. Secondly, built up of material of component, say brass, on upper roller bead formation (steel bead). This needs frequent cleaning, causing stoppage of production.

Notching

Notching is an operation by which a small cut is produced at the brim of component. Notch can be made on flat portion of a sheet metal component or on the wall of round component. Figure 8.38 shows two typical examples of notches.

Generally, the purpose of notch is to locate component or product in a particular place. Figure 8.38 'A' shows a pre-focus miniature lamp. A notch is cut on the collar of brass cap. This notch is extended a little in the body of cap. In Fig 8.38 'B' A can is shown with a vee notch.

The basic concept of notching tool (die and punch) is shown in Fig. 8.39.

Figure 8.39 'B' shows a punch and die for generating a triangular notch. W_p is the width of punch and W_d is width of die which may be longer than W_p. Length of notch in a component depends on

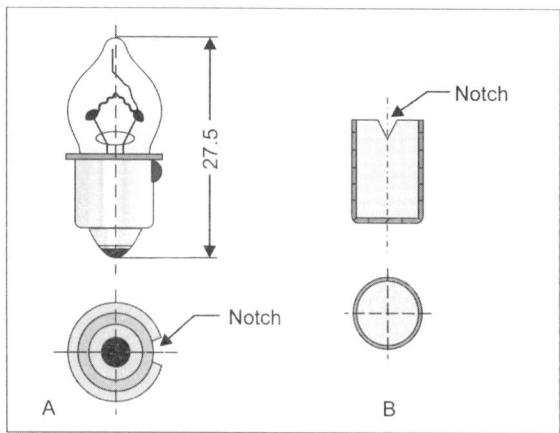

Fig. 8.38: Notching tool for circular component

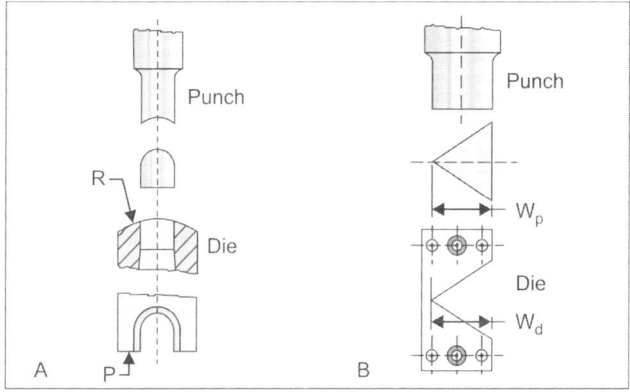

Fig. 8.39: Single and two start threads

how much deep component brim is inserted into the die, but not more than W_p, width of punch.

Figure 8.39 'A' shows a notching tool, suitable for a circular component, say an aluminum sheet tube. Radius R of die is slightly less than radius of bore of tube so that portion to be notched definitely sits properly on the curved die. Depth of notch would depend on the depth of feed of component on the length of die. P is a point from where semi-circular (may be other shape) cut opening in the die rod starts. Suppose brim of tube crosses this point only by 1.5 mm then length of notch would be 1.5 mm. If tube crosses the point P by 3 mm, then the length of notch would be 3 mm.

Thread Rolling

Many sheet metal components having rolled threads are used for general purpose and technical use. Normally, the purpose of threads in sheet metal components is to join parts, close opening or perform some function in a product assembly. Few examples of sheet metal thread rolled components are closures of a tin canister, assembly of 'ring' over the 'head' (where reflector is held) of a metallic flashlight and base which acts as closure as well as focus adjuster. By rotating the base in clockwise or anti-clockwise direction, focus of flashlight beam changes. This is an example of functional use of thread. Threads rolled in sheet metal components may be specified in the following manner:

- Direction of spiral, that is, right hand or left hand (clockwise or anti-clockwise)
- Single or multi-start   • Pitch of threads

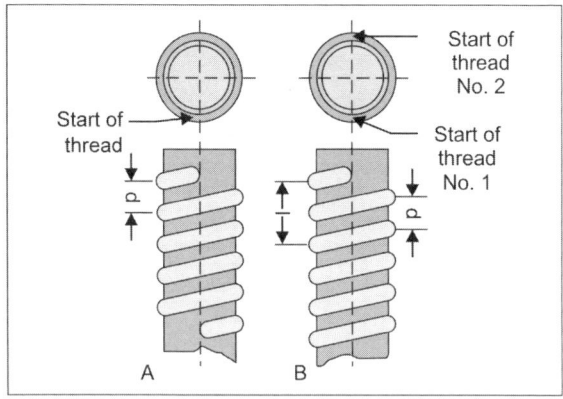

Fig. 8.40: Parameters for thread design

- Lead of threads
- **Profile of thread in bore**
- Shape of start of thread
- Thickness of sheet at threaded portion
- Matching of threads of two mating components
- Profile of thread on diameter
- Depth of thread
- Shape of end of thread

When a nut is turned in clockwise direction, it axially moves away from source of rotation, say hand of a person, such as nut or bolt is called to have a 'right hand thread'. If nut is rotating in anti-clockwise direction and moves away from source of rotation then such a bolt or nut is called to have a 'left hand thread'. Same rule applies to threads on/in a sheet metal component.

Figure 8.40 'A' shows a sheet metal component which has a single start thread having pitch 'p'. Fig. 8.40 'B' shows a sheet metal component which has a double start thread having a pitch 'p' and lead 'l' which is equal to 2p. In case of double start thread, start of two threads are at 180°. In case of a three start thread, angle between two starts is 120°. Lead in a three start thread is equal to 3p.

There are practical situations where space (length of thread portion) is not enough to generate more than one and a quarter thread. Under such circumstances, grip between two mating parts is not strong enough. Two start threads overcome this problem to a great extent. Furthermore, there are practical situations where it is desired that quick screwing and unscrewing of two mating components take place. For such situations, two or three start threads are used.

Figure 8.41 shows various parameters which are considered at the time of designing thread rolling rollers together with type and

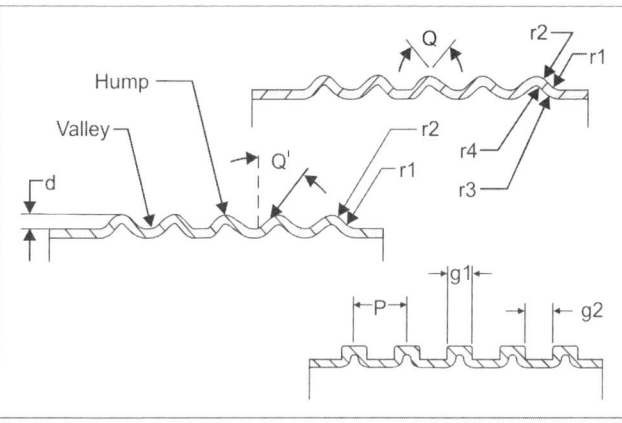

Fig. 8.41: Thread profile

thickness of sheet metal. r1, r2, r3 and r4 are radii which are desired to be maintained. Sometimes, it is not practically possible to maintain very small desired radii. Grains of sheet get 'opened' with very sharp radii, say 0.3 mm for a sheet of 0.3 mm thickness. θ° is acute angle of a 'hump' (raised portion) of thread. This varies according to design of thread. Practically, it may be 20°–40°. Acute angle of valleys (sunken portion of thread) would also be the same as that of hump. 'P' is the pitch of thread, this means distance between the center of two humps or valleys.

In case of multi-start thread, pitch 'P' is the distance between hump center of one thread and hump center of the other thread, but lead of thread is double the pitch in case of a double start thread. g1 is approximate width of hump and g2 is width of valley. In practice, it is quite difficult to measure width of hump or valley by means of a measuring instrument, say a vernier caliper. For proper measurement of depth of thread, radii, angles and width of hump and valley, it is necessary to draw a profile with the help of a profile projector. A typical thread profile is shown in Fig. 8.42.

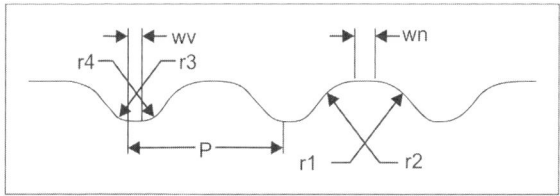

Fig. 8.42: Sunken and raised threads

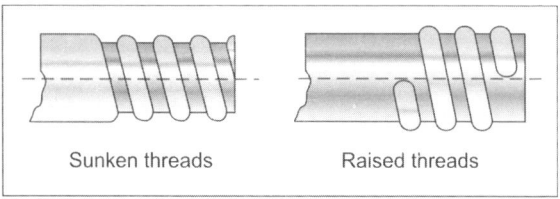

Sunken threads   Raised threads

Fig. 8.43: Thread matching in component

All measurements can be directly read on profile projector with the help of a suitable built in scale image. Alternatively, a shadow graph can be drawn by placing a tracing paper on display screen. If profile projector is computerized, an attached printer can be set to get a photo printout. Both raised and sunken threads are shown in Fig. 8.43 and thread considerations are shown in Fig. 8.44. In case working or mating side of thread is inside the bore of sheet metal barrel/tube then taking of profile is a little difficult. Either a mould is to be lifted or a narrow strip is to be cut without damaging the profile. Sometimes both internal and external sides of thread are working sides.

In Fig. 8.44, part 2 has right hand thread. On its bore, part 1 is screwed in and on the diameter, part 3 is screwed over it. In such circumstances, design of all the three components is critical especially if the pitch of thread is to be kept fine. Fine thread means less space to form thread profile. Most difficult formation of thread would be in part 1 as width of hump of thread would have to be as minimum as possible so that it screws in easily in part 2. There would be hardly any difficulty in rolling threads on part 3.

Figure 8.45 shows a theoretical **assembly** of threads of three components. Continuous threads are normally generated in sheet

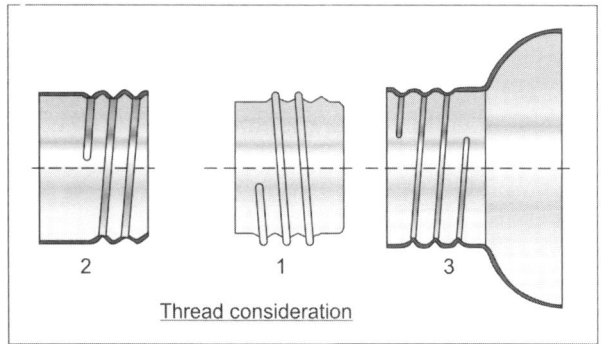

2   1   3

Thread consideration

Fig. 8.44: Theoretical assembly of three threaded components

Fig. 8.45: Assembly of threads

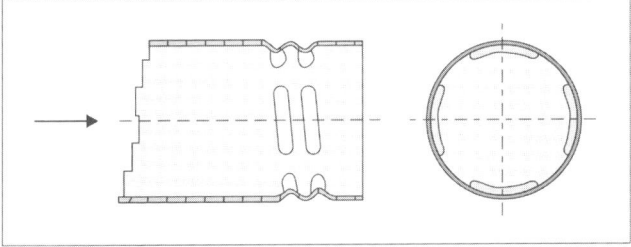

Fig. 8.46: Intermittent threads

metal components, but there are instances where thread humps are intermittent. A typical component is shown in Fig. 8.46 which has intermittent thread.

Intermittent thread design is adopted where diameter of component is so small that it becomes difficult to maintain circularity if continuous threads are generated. Furthermore, body of component remains strong and smooth if slides in bore of another component.

In this paragraph, design of **thread rolling machine** and thread rollers, quality aspect and thread rolling proceses are discussed. Figure 8.47 shows fundamental construction of a thread rolling machine where part (1) is the upper cast iron body of machine which is fitted to the main body (2) in such a way that it can swing up and down with the help of reciprocating vertical arm (6). There is one precisely assembled rotating spindle (3). This spindle has no unwanted play as thrust bearings are used to eliminate unwanted axial play. At the same time, there is such a mechanical arrangement that upper spindle can be very precisely shifted in axial direction with the help of a micrometer knob (7). Minimum axial displacement may be, say 0.02 mm by micrometer knob. Running of upper spindle diameter and roller support face (13) is highly true running. Run out is not more than 0.005 mm. Lower spindle (4) is also precisely assembled. Axial play is just enough

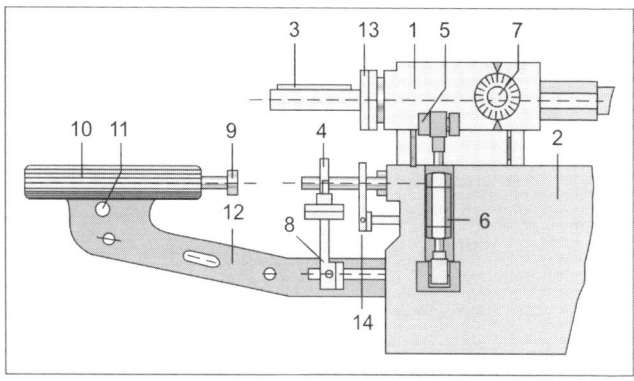

Fig. 8.47: Automatic thread rolling machine

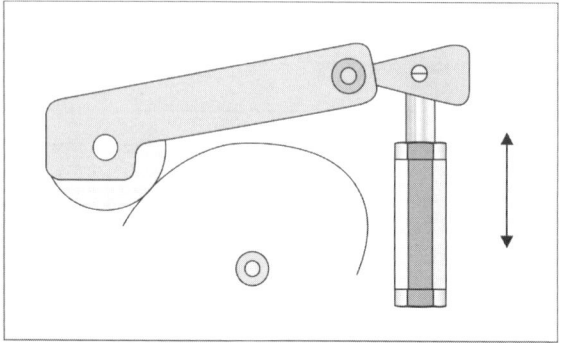

Fig. 8.48: Arbor with locating pin

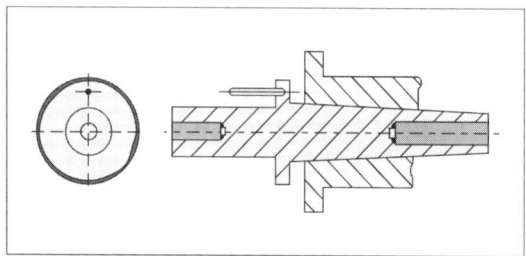

Fig. 8.49: Movement of swinging arm

for a lubricant film, say 0.02 mm. Lower spindle is hollow and has a morse taper for fitting of an arbor which carries lower thread roller disc and other auxiliary discs. Upper spindle has a key so that upper roller and other discs do not slip rotate over spindle. Arbor has a locating pin protruding from the face (see Fig. 8.49). Lower thread roller disc and other discs have holes to suit protruding pin, so there cannot be a relative motion between roller disc and arbor. Morse taper in lower spindle and arbor is so precisely machined that run out on arbor diameter and face is not more than 0.005 mm. Described accuracy is necessary for achieving good quality threads. Part (8) is an arm which swings to bring unthreaded component in front of lower spindle.

Part (9) is backrest which pushes the component over lower thread roller and remain in that position till rolling is completed. It comes back just before ejector ring (14) starts moving out over lower tool to eject threaded component. Idea of its movement is shown by Fig. 8.50.

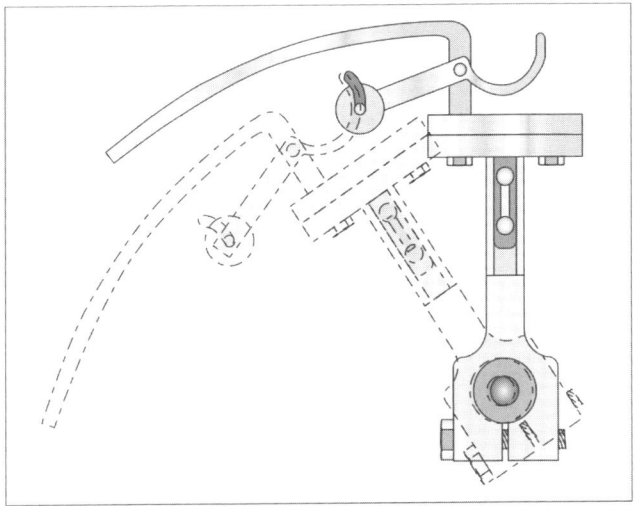

Fig. 8.50: Feeder arm

Sequence of operation of automatic thread rolling machine is as under:
- Manual or automatic feeding of component in chute of thread rolling machine.
- Switching on the machine.
- Both upper and lower spindles start rotating, lower in clockwise and upper in anti-clockwise direction.
- Backrest moves out and pushes the component over lower thread rolling tool. Stroke of backrest is adjusted according to need.
- Upper body of machine moves down till threads of upper roller sink in threads of lower roller to such an extent that desired depth of threads in component is achieved.

- Just before completion of thread rolling, feeder arm moves back.
- On completion of thread rolling, backrest moves back and ejector ring moves forward to eject threaded component from over the lower thread rolling tool.
- First few components are trial components when depth is adjusted by slightly rotating left-right hand long nut of vertical arm (6). Then profile of thread is checked to see if thread formation is symmetrical or there is need to do precision axial shifting of upper spindle.
- Machine is finally started to keep on producing threads on fed components.

Upper and lower spindles are connected to each other by a train of gears towards the back end of spindles. By changing the gears combination, ratio of rotation of spindles can be changed. 1:1 means that both spindles rotate one complete round. 2:1 means that upper spindle makes one complete round when lower spindle makes two complete rounds. Similarly, a ratio of 3:1 is possible. These ratios are mentioned as example, it may be different according to machine design.

Facility to change spindle rotation ratio is very useful for designing upper thread rolling disc diameter. As mentioned earlier, peripheral speed of thread rolling discs is kept as equal as possible. In the following paragraphs, a typical example is taken up to **demonstrate** as to how ratio of rotation of upper and lower spindles is decided.

Minimum adjustable distance between axis of upper and lower spindles —— 40 mm.

Diameter of lower thread rolling discs ——————— 28 mm.

Diameter 'D' of upper thread disc should be more than ——— 52 mm, Say, 60 mm

This means that $Nl \times Dl = Nu \times Du$

Hence, $$Nu = Nl \times Dl/Du$$
$$= 1 \times 28/60$$
$$= 0.46$$

This means that upper spindle should make 0.46 of complete one round when lower spindle makes one round. Gears for setting such type of ratio is normally not possible. Nearest possible ratio which can be set by gear train is 1:0.5. This means that on one complete round of lower spindle, upper spindle would make complete half round.

If upper thread roller is made of 60 mm diameter then peripheral speed of upper thread roller would be more as compared to peripheral speed of lower thread roller.

Peripheral speed of lower thread roller at 600 rpm ——

$$= 600 \times \pi \times 28/1000$$
$$= 52.8 \text{ meters/minute}$$

Peripheral speed of upper thread roller at 300 rpm ——

$$= 300 \times \pi \, 60/1000$$
$$= 56.6 \text{ meters/minute}$$

Hence, there would be a rubbing action of (56.6 – 52.8) 3.8 meters per minute. This is an undesirable situation. Keeping the rotation of spindle as 1:0.5, diameter of upper thread roller disc would have to be re-calculated.

$$\text{Nu} \times \text{Du} = \text{Nl} \times \text{Dl}$$
$$0.5 \times \text{Du} = 1 \times 28$$
$$\text{Du} = 56 \text{ mm}$$

Hence, diameter of upper thread roller would have to be machined as 56 mm instead of 60 mm and lower thread rolling discs 28.

This satisfies two important conditions that:

• Peripheral speeds of upper and lower thread rollers are almost the same.

• Total of two radii exceeds minimum distance between axis of upper and lower spindles.

Let required pitch of threads in the component be 2.5 mm. Hence, the pitch of threads in lower thread rolling disc would also be machined as 2.5 mm, so travel of a point will be 2.5 mm on one complete round of lower thread roller, but upper spindle has rotated only complete half round. This means that a point on thread of upper roller will travel only 1.25 mm (half of 2.5 mm). This is just not possible. Threads of both the rollers would severely strike each other and would get damaged. Solution to this problem is that two threads should be machined on upper thread roller, start of both the threads should be 180° apart and lead of each thread should be 5 mm. Hence, distance between the hump and valley of two threads would be 2.5 mm, which is pitch of threads. Figure 8.51 illustrates this explanation.

There may be cases where upper thread roller may need three start threads. The quality of threads formation very much depends on the quality of thread rollers and accuracy of rotation.

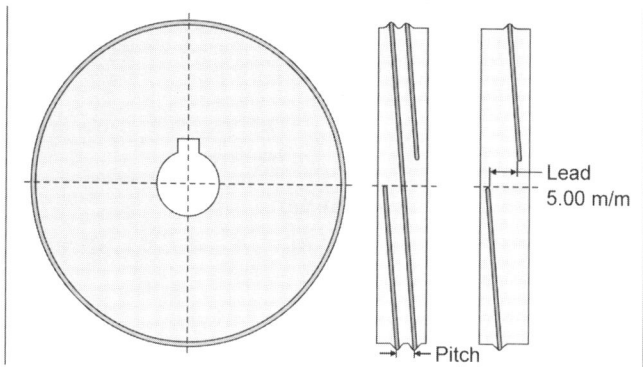

Fig. 8.51: Thread matching

Fig. 8.52 'A' shows correct matching of upper and lower thread roller's hump and valley. Symmetrical gap between hump and valley should remain unchanged if rollers are rotated. Also, vertical gap (g) should also remain unchanged. Any roller running out on diameter would change vertical gap. It will be more at one place and less at 180° opposite to first place.

Symmetry in axial gap depends on axial setting of upper spindle and true running of helix of threads. It is already mentioned that true running of diameters and faces of spindles is of extreme importance.

Diameter, bore, threads and one face is normally machined in one setting on lathe or profile grinder. Before unloading the job from

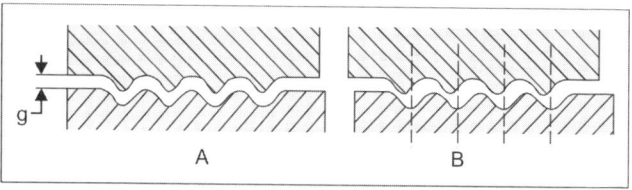

Fig. 8.52: Combination of thread rolling discs

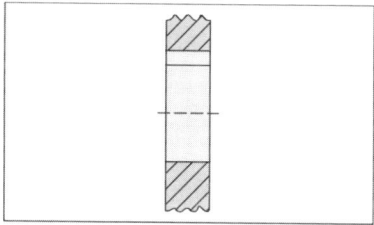

Fig. 8.53: Lower thread roller with pin hole

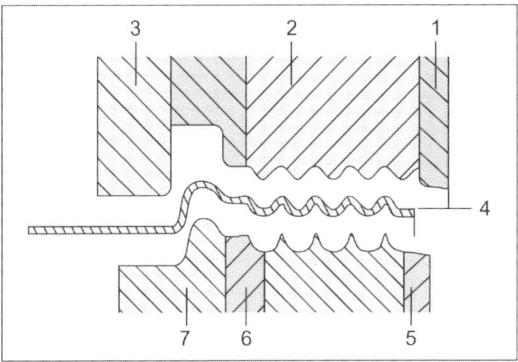

Fig. 8.54

the chuck, a chamfer is made to indicate that this side of face is true to bore and helix of the threads. While assembling thread rollers on thread rolling machine, chamfer side face should be towards spindle collar face. In this way, true running of thread helix may be ensured. Fig. 8.53 shows lower thread roller with pin hole.

In practice, some additional discs are required to support thread rolling operation to achieve correct shape of threaded barrel/tube.

In Fig. 8.54, part (1) is a plain rolling disc having a taper on diameter according to design of component (4). Normally taper angle is between 5°–15°. Smaller diameter of taper disc is kept equal to diameter of thread roller. On many occasions, it becomes necessary to slightly modify taper to achieve best results. This is very important that thickness of disc is equal on all points, say within 0.01 mm. Similarly, thread rollers (2) and (6), beading rollers (3) and (7) should have parallel faces very accurately. These accuracies ensure generation of good quality threads on sheet metal components. Inaccurate rotation of thread rollers and other discs cause unequal depths of threads or squeezing of sheet material, causing pin holes or cracks in threads.

Forming

Sheet metal forming involves a wide range of processes that manufacture parts for a vast amount of purposes. Sheet metal refers to metal that has a high surface area to volume ratio. Sheet metal work stock, used for sheet metal processes, is usually formed by rolling and comes in coils, in which force is applied to a piece of sheet metal to modify its geometry rather than to remove any material. The applied force stresses the metal beyond its yield strength, causing the material to plastically deform, so the sheet can be bent or stretched into a variety of complex shapes.

9

Cost and Productivity Considerations

This chapter consists of the following aspects:

- Material saving
- Multi-impression tool design for enhancing productivity
- Calculation for percentage of scrap
- Stamping of reflector for attaining good parabolic surface
- Fast and economical tool assembly example

MATERIAL SAVING

Manufacturing **cost** of components basically depends on material, process and **tooling**. As far as material is concerned, there should be least possible wastage of raw material such as sheet metal. Wastage may be of two types, one due to erroneous process, mismanagement, inconsistent quality of sheet metal strips, coils or rounds and the other is tool design and its actual construction. No matter how best a tool is made, it will produce higher percentage of scrap together with component if **economical layout** of blank in a blanking tool is not taken into account.

In case of a cut and cup tool, blank diameter may be kept more than just needed, because designer may 'play safe', taking into consideration that **shell height** may be uneven and the minimum wall height may not get trimmed. Consequently, produced cups may be a reject. Here comes the accuracy in tool making. If the parts are precisely made and assembled then there is no reason why a cup with uniform wall height may not be produced. Here most economical blank diameter, more number of cups would be produced per unit weight of sheet metal.

Further, economical production of component by using a **multi-impression tool**, say a two or three impression cut and cup tools are produced.

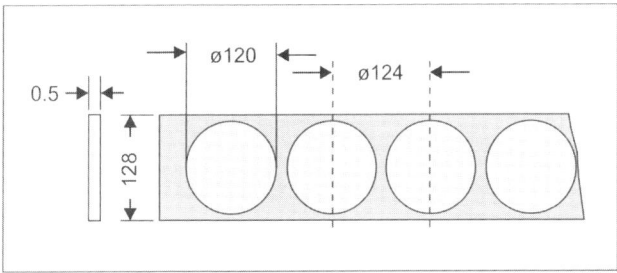

Fig. 9.1: Strip scrap after blanking

Comparative **scrap percentage** for single, two and three impression cut and cup tools is given below to provide an idea of reduction of scrap percentage (Refer Fig. 9.1).

Single Impression
- Length of strip required to produce 10 cups ───────────
  $120 \times 10 + (4 \times 11)$
  $= 1244$ mm
- Area of strip ─────────────────── length × width
  $1244 \times 128/100$
  $= 1592.32$ cm^2
- Area of one blank ───────── $\pi/4 \times D^2$
  $= 3.14 \times 12 \times 12/4$
  $= 113.04$ cm^2
- Area of 10 blanks ───────── 1130.4 cm^2
- Hence, scrap generation ───────── $1592.32 - 1130.4$
  $= 461.92$ cm^2
- Percentage of scrap ───────── $461.92 \times 10/1592.32$
  $= 29$ %

Two Impression (Refer Fig. 9.2)
- Length of strip required to produce 10 cups ───────────
  $120 \times 5 + (4 \times 6)$
  $= 600 + 24$
  $= 624$ mm
- Area of strip ───────────── 235.31×624
  $= 1468.33$ cm^2

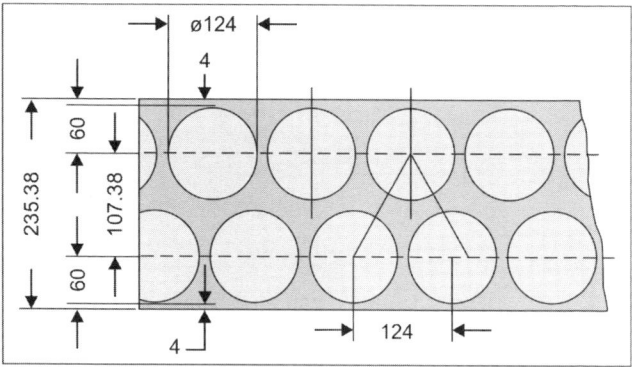

Fig. 9.2: Two impression strip layout

- Area of 10 blanks ——————— 1130.4 cm²
- Hence, scrap generation ——————— 1468.33 – 1130.4

  = 337.93 cm²

- Therefore, percentage of scrap ——— 337.93 × 100/1468.33

  = 23 %

Three Impression (Refer Fig. 9.3)

- Area of strip ——————— (1244 × 342.36)/100

  = 4258.95 cm²

- Area of 30 blanks ——————— 1130.4 × 30

  (29 + half + half) = 3341

- Area of scrap ——————— 4258.95 – 3341

  = 917.95 cm²

- Scrap percentage ——————— 917.95 × 100/4258.95

  = 21.5 %

The above comparison of percentage of scrap indicates that in having two impression cut and cup tools, there would be reduction of six percentage of scraps as compared to single impression. But the reduction of percentage of scrap with three impression tool is insignificant as compared to two impression tool. Hence, it may not be advisable to go for a three impression tool (see Fig. 9.3). More so, **cost of three impression tooling** would be much high as compared to single impression tool and much higher tonnage

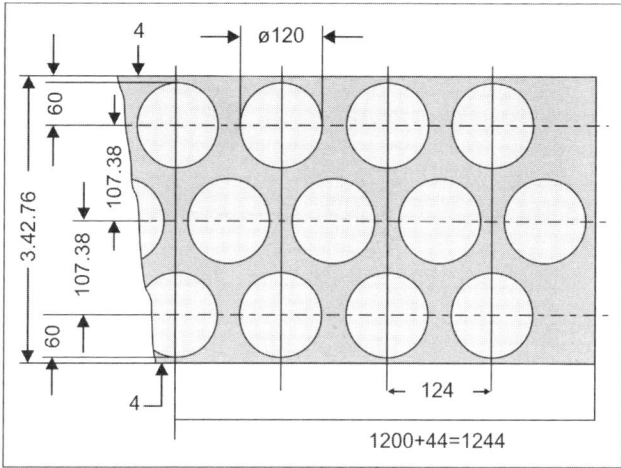

Fig. 9.3: Three impression layout

power press would be needed as the size of tool would be such that a 200 tons (say) press may provide desired bolster plate size.

Comparing percentage of scrap from single impression tool with two impression tools, there is a reduction of 6% and it is worth while to go for a two impression cut and cup tool after other factors affecting overall **cost**. Let us now examine that six percent of reduction in scrap would yield how much saving in **material cost**. Let the material of cup is 0.5 mm thick and brass sheet of 8.4 density (8.4 grams per cubic centimeter) and prevailing cost of brass sheet (assumed) be rupees two hundred twenty five (₹ 225/kg) per kg.

Reduction in scrap generation in terms of area (cm²) ————

461.92 – 337.93 = 123.94 cm²

Volume of this area = 123.94 × 0.05 = 6.197 cu cm

Weight of this scrap saving ———————— = Volume × Density

= 6.197 × 8.4

= 52.05 gm

Cost of 52.05 gm of brass sheet for 10 components ————

= 0.225 × 52.05

= ₹ 11.71

Saving for 1000 components = ₹ 1171

Increase in cost of tooling from one impression to two impressions is rupees two lakh, then number of components to be produced to recover these rupees, two lakh would be

200000/1171 × 1000 = 170794 number

Supposing 8000 components are produced per day then 21 days would be needed to recover two lakh rupees investment.

For a single impression tool, a hundred tons power press may be suitable, but for a two impression, a 200 tons power press would be needed. Hence, the increase in processing cost due to change of press may be examined to decide if going for a two impression tool would be beneficial, and so a decision could be taken accordingly.

Cost of tooling (design and making) depends on many factors such as

- Open tool (not in a die set), simple
- Tool in a **die set**

Open tools for operations such as stamping and bending may be used.

Figure 9.4 shows a basic **stamping tool** for a reflector cup. It is stamped to precisely maintain parabolic surface. In this case, first of all, a die is held in power press RAM. Punch part assembly is placed over the bolster plate. Punch is covered with an already stamped component which is preserved for tool settling purpose. RAM is brought down so much that there is a little gap between die parabolic surface and setting component's upper surface. Now the RAM is further brought down by unscrewing pitman screw so much that it exerts some forces on the punch. Now the lower portion of tool is secured (tightened) over bolster plate. It is worth appreciating that a play of say, 0.03 mm in RAM will not affect tools, performance because punch and die would align themselves. Therefore, there is no necessity of using a die set, hence, reduction in tool cost.

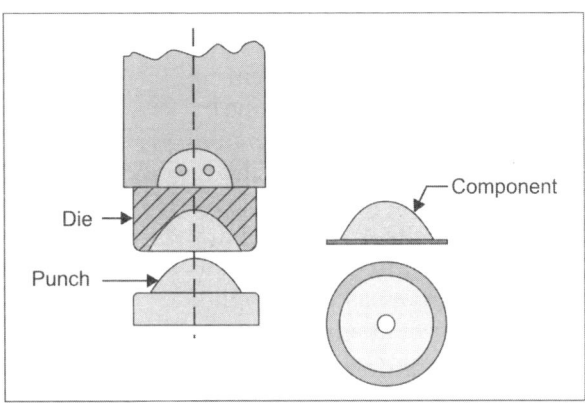

Fig. 9.4: Basic stamping tool

Situation with a cut and cup, blanking, trimming and piercing tool is quite different. Here **cutting clearance** may be as small as 0.02 mm. If an open tool is used then there is a likelihood of saving of punch die or both. It is therefore necessary to design and make a tool in a die set.

Inclusion of a die set would certainly increase tool cost.

For heavy and critical tools, die set with **linear cage ball bearing** is used instead of plane bush bearing. Consequently, tool cost further increases. It is not only the increase in cost of pillar type die set (with linear cage ball bearings or plain bush bearings). Machining cost of die set appreciably increases. All these factors are taken into consideration at tool designing stage. Hence, tool designer has a great responsibility to offer a design which is very precise, reliable and cost effective.

The cost of tooling may not be considered in isolation. It has to be thought that what arrangements would be necessary for **feeding of strips** or rolls into the tool and safe ejection of components and scrap, since this arrangement may be considered together with tool cost.

Frequency of **maintenance of tool** may cause expenses which add to tool cost for a particular volume (number of components) of components production. Most of the time, frequency of maintenance of tool depends on tool parts making and assembly rather than tool designing. Precise selection of steels, machining of parts and heat treatment adds to good performance of tool in terms of larger number of components produced between two maintenance, but at the same time, it increases the making cost of a particular tool.

Another important factor of tooling cost consideration is **over head** of a tool room of an organization. Overhead of a tool room increases if various types of sophisticated machine tools are installed. Advanced cost management techniques advocate that maximum possible tooling may be done by off loading. Obviously, essential machine tools such as lathe, milling and surface grinder may be in-house. Other operations such as engraving, spark erosion machining or laser itching may be off loaded. The advantage of off loading is that over heads of a tool room remain low. Secondly, it becomes the responsibility of the party (to whom off loading is done) to machine the part as per supplied drawing and to affect delivery of machine part at agreed dates.

If the same operation is done at in-house facility then the rejection if any in machining, would be a loss as the defective component may be thrown in junk. Such a loss being small goes unnoticed in

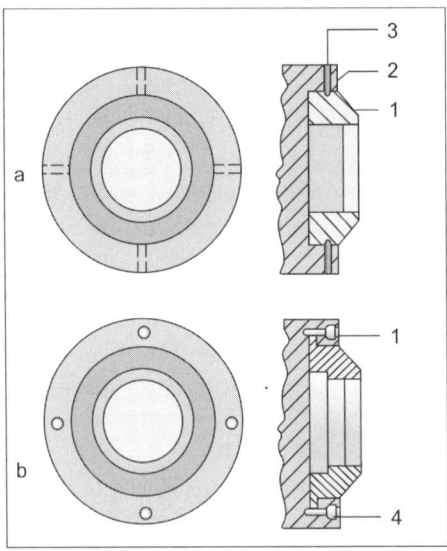

Fig. 9.5: Cutting ring holding option

a big organization, but the party on which part machining is off loaded, cannot afford to machine an erroneous part as he may not get paid for defective part as per agreement between the work providing party and the party takes up the work to complete it.

The cost of tool also depends upon design of joining techniques of mating part. Figure 9.5 illustrates two methods of fitting of a cutting ring in a C.I. die block.

Figure 9.5 'a' shows that cutting ring (1) is held in C.I. die block by means of four grub screws. Cutting ring is pressed inside die block with transition fit and then grub screws are tightened. This arrangement is tedious when a cutting ring is to be replaced by a new one. Difficulty is that how to take marking of grub screw as the diameter of cutting ring may have a grinding allowance for grinding after hardening. Hence, round grooves for grub screws are made by marking before heat treatment. There is a likelihood that round grooves do not match properly with grub screws. Due to all these reasons, die maker has to devote much time to do assembly, still having a chance of errors.

Methods shown in Fig. 9.5 'b' is straight forward. Cutting ring is placed inside the die block with precision fit and it is tightened by tightening ring (4). Now in these two alternatives, cost of 'b' alternative may be a bit high as compared to 'a' but it may be compensated by the time saved in assembly process by die maker.

10

Types of Sheet Metal
Press Tools

This chapter consists of the following aspects:

10.1 Classification of tools

10.2 Simple, progressive, compound and combination tools

   • Multi-impression tools

10.3 Multi-impression progressive tool

10.1 CLASSIFICATION OF TOOLS

Most modern sheet metal press tools are used today as a result of long quest of engineers, technicians and workmen for developing the tools design and making so that good quality tools could be designed and made. Sheet metal press tools designing and making took about a century to reach a stage as it exists now. It may be said that a time span of say, hundred years points towards a slow progress. It may be reason out that it is not a slow progress as the development in sheet metal press tools design took place mainly due to two reasons. Number one is the desire and urge of tools development, secondly developments in the field of metallurgy, machine tools and infrastructure.

10.2 SIMPLE, PROGRESSIVE, COMPOUND AND COMBINATION TOOLS

High performance steels as available now were under development stages, say about sixty years back. Die sinking machines got developed to present status during last forty to fifty years. CNC, CAD and CAM are the latest developments. All these developments took place gradually over period of years. Consequently, took so long to reach present stage of fineness in tools design and making. Since so many varieties of sheet metal tools got developed, it became necessary to classify them.

The broad classification of sheet metal press tools may be as under.

- Simple tools
- Combination tools
- Progressive tools
- Multi-impression tools
- Compound tools

Simple Tools

Simple tools may be defined as those which perform one of the following operations. This means one tool for one operation.

- Blanking
- Piercing
- Trimming
- Cut off
- Lancing
- Bending
- Forming
- Drawing
- Deep drawing

Any of the above simple tools may be an open tool or a die set tool. This depends on the size, shape, material, its thickness, accuracy in dimension required and productivity targets. Examples of such tools are illustrated in Fig. 10.1.

Figure 10.1 'A' shows a component blanked out of a 0.34 thin brass strip. The outer contour of this component has no functional importance. Moreover, component is 45 mm long and 18 mm wide, hence, the size of tool may be such that it may either be operated on a screw hand press or a 5 tons power press. Batches of this component are tumble polished on the brim (cut contour). Due to all these reasons, it may be an open tool, a tool without a die set. Figure 10.1 'B' is a strip bending tool. There is no precision matching

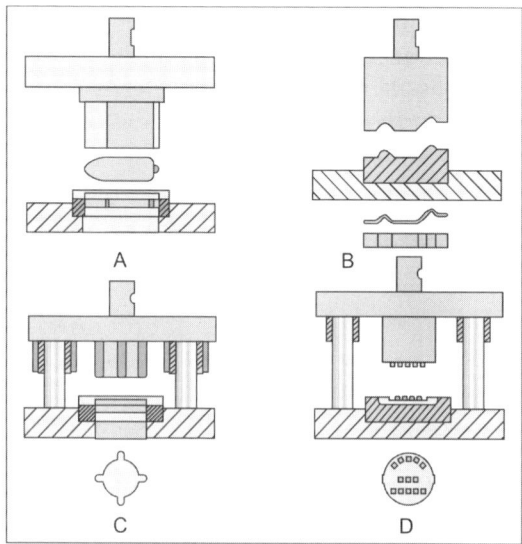

Fig. 10.1: Showing basic tools with components

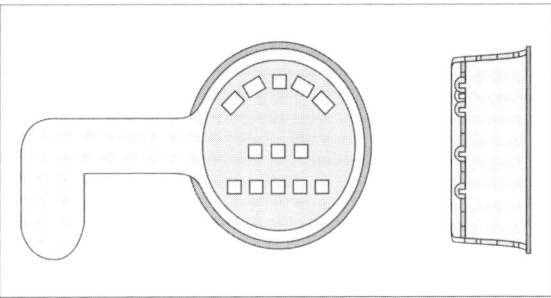

Fig. 10.2: Enlarged view of embossing

of cutting edges, therefore, matching of upper and lower parts of tool is not critical. Any looseness of say, 0.1 mm in press RAM may not affect bending angles in component or any damage to upper and lower bending inserts is not likely to take place because bending angles (slant surfaces) will shift the RAM so that both the bending inserts sit together properly. This may happen as this 'shift' is very minute, say 0.05 – 0.1 mm. Due to this reason, tool may be an open one.

Figure 10.1 'C' shows a stainless steel sheet component of 0.46 mm thickness. It has a complicated inside profile with critical dimensions. It has a functional importance in a vacuum system. Therefore, it should be free from burr. Moreover, this component is to be produced in large quantities. A round disc of stainless steel is to be fed in the guide ring of lower portion of tool. Stroke of press would blank out a slug and component would remain in the guide ring for manual or automatic removal. Manual removal of component may be by means of a probe. Automatic removal may be a robotic vacuum picker. This type of tool must be built in a precision die set with linear cage ball bearing.

Figure 10.1'D' is another example of fine **embossing** on a sheet metal component which is shown enlarged in Fig. 10.2.

Here the matching of upper and lower inserts should be very precise and there should be no chance of relative shifting of upper and lower inserts, hence, tool should be made in a die set.

Progressive Tools

Progressive tool may be defined as a tool in which more than one operation takes place progressively. This means that strip is fed to a certain length and then a working stroke is given to tool. In this first stroke, certain features are generated. Take example of a component as shown in Fig. 10.3.

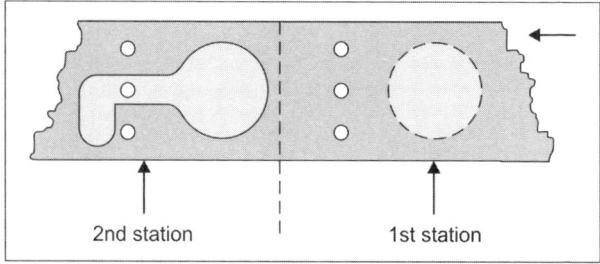

Fig. 10.3: Example of progressive operation

A progressive tool for the component (Fig 10.3) may have two stations, station first and station second. Strip would be fed up to station first. This station has a pair of embossing inserts (upper and lower) and piercing arrangement for two guide holes near the edge of strip and a third hole for component, which is feature of component design. First stroke would produce embossing feature, pierced hole and two guide holes. After first stroke, strip would be moved forward equal to the pitch of two stations. On second station, there are two guiding pins with rounded off protruding ends. Guide holes produced at first station would be roughly on guide pins. On second stroke, a pressure pad first presses the strip over guide pins, consequently, the first embossed and hole feature would be in true symmetry of blanking contour. Further, down stroke of upper portion of tool would blank down the component from die and would be collected under the table. After two strokes of press, the processed portion of strip would look like as shown in Fig. 10.4.

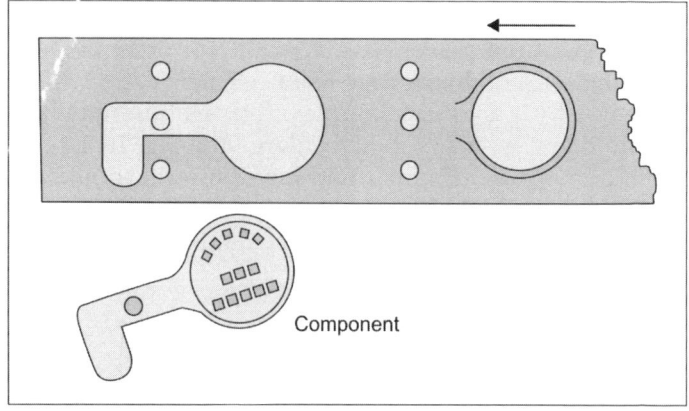

Fig. 10.4: Strip and component after two strokes

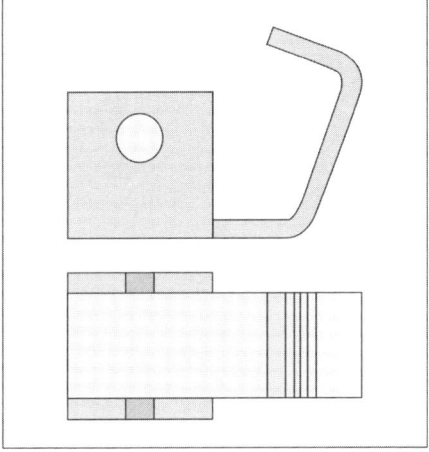

Fig. 10.5: Component requiring progressive operation

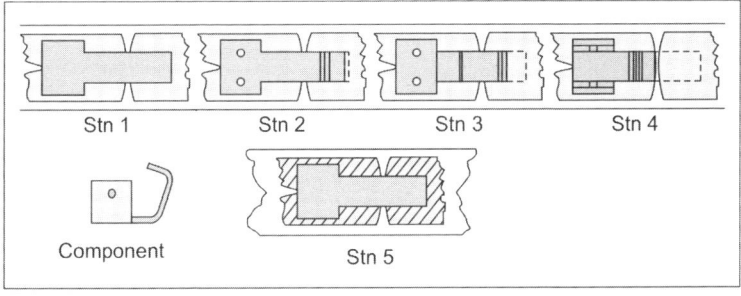

Fig. 10.6: Progressive operations briefly displayed

Another example of progressive tool for component shown in Fig. 10.5 is briefly explained with the help of Figs 10.5 and 10.6.

Tool shown in Fig. 10.6 is a four station progressive tool. **Sequence of operation** would be as follows:

1. Strip is fed up to the stopper at station 1.

2. First stroke is given. Three blanking punches blank down material beyond component geometry, leaving three hanger ribs which hold the formed blank attached to strip.

3. Strip is moved forward. Formation from station 1 reaches station 2.

4. Second stroke is given. Piercing and first bend takes place.

5. Strip is moved forward, formation from station 2 reaches station 3.
6. Third stroke is given, second (final) bending takes place.
7. Strip is moved to next station with complete component still attached to strip.
8. Forth stroke is given. Here three holding ribs are sheared near component edge, thus finished components get detached from strip and ejected.

On each stroke, component features are made at each station. Since the strip with components is moved forward for each stroke, therefore, one completed component is produced on each stroke.

Compound Tools

Compound tools are those which perform more than one operation. It may have two stations, but in this case, a blank or component is fed manually at first station. Manual feeding means that operator does not bring his/her fingers in between upper and lower parts of tool. Instead, a kind of gripper is used to hold the part and put it at station one. Let it be a piercing cum nibbling station. Stroke is given and piercing and nibbling operations get completed. Now the component is lifted by operator and placed on other station in close vicinity of station one. Second stroke is given and next operation takes place, say bending.

Above described basics of a compound tool is illustrated with the help of Figs 10.7 and 10.8.

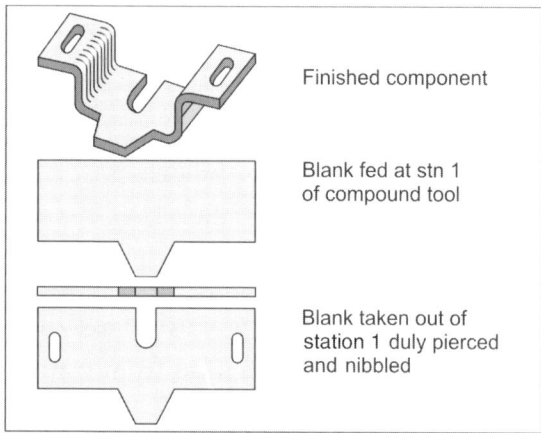

Finished component

Blank fed at stn 1 of compound tool

Blank taken out of station 1 duly pierced and nibbled

Fig. 10.7: Component and basic operation described

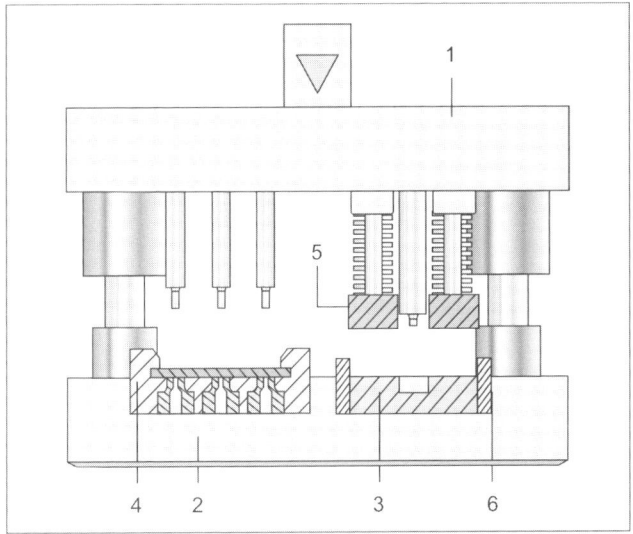

Fig. 10.8: Basic construction of compound tool

Figure 10.7 shows finished component which would come out of second station. Blank 'a' is shown which is to be fed to station 1 of compound tool. At this station, piercing of two elongated 'holes' takes place and nibbling of open slot (which is rounded inside) also takes place simultaneously with piercing. Component is now taken out and placed at station 2 where bending takes place and finished component comes out.

The basic construction of compound tool for already mentioned component is shown in Fig. 10.8.

Combination Tools

Combination tools consist of tooling to perform two or more operations at a time in one stage only (see Table 10.1). Two

Table 10.1: Parts of combination tool

Parts	Description	Remark
1.	Upper part of dieset	Fitted with guide bushes
2.	Lower part of dieset	Fitted with pillars having linear ball bearing
3.	Bending die insert	
4.	Blank stripper plates	For station 1
5.	Stripper cum under holding of component from 1st station	Spring loaded
6.	Input component guide plate	

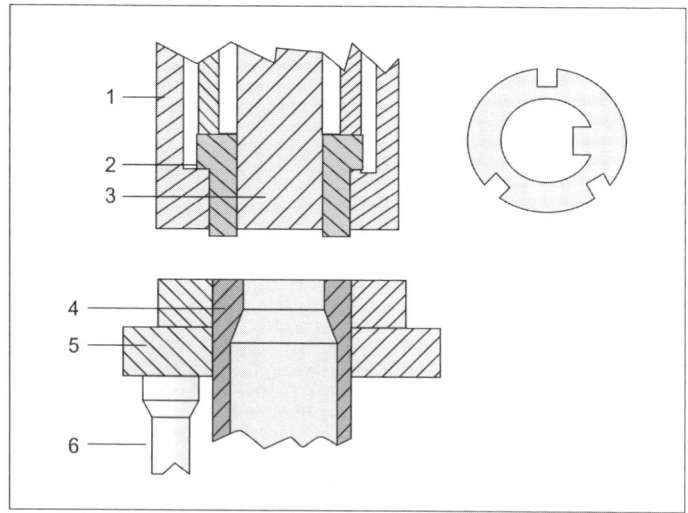

Fig. 10.9: Example of combination tool

examples of combination tools are given in this chapter. Only basic construction is illustrated to give an idea of combination tools.

Figure 10.9 shows a special purpose washer of brass sheet of 0.38 thickness. Combination tool to produce this washer is also shown. A brief description of tool is as under.

Part (1) is blanking die having internal shape to blank out outer space and size of washer.

Part (2) is a circular ejector member of upper portion of tool. After operation, upper portion of tool moves up, it carries the component inside it. It is then the ejector which pushes the component out of blanking die and piercing punch.

Part (4) is the lower portion of tool.

Part (3) is the blanking punch cum piercing die.

Part (5) is the ejector ring to push the strip out of blanking punch. So each stroke of press gives completely finished component since cutting clearance between punch and dies are very important, therefore, this tool should be built in a robust die set with linear ball cage bearing between die pillars and their respective bushes.

Another example of combination tool is provided herewith where three operations are combined that is cut, cup and pierce. Figure 10.10 shows component and combination tool to produce it.

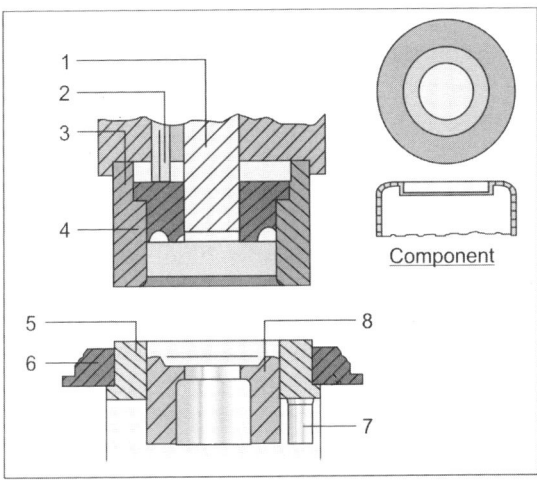

Fig. 10.10: Component and its tool

In upper portion of the tool, part (1) is a piercing punch to create a hole in the component. Part (2) are three ejector pins which push part (4) down. At the same time, it acts as a forming die when upper tool portion reaches its critically set portion. Part (5) is a pressure pad in the lower portion of tool. Before draw starts taking place, strip is held between the faces of part (3) and (5) with desired under holding pressure. The sequence of operation for complete making and delivery of component is given in Fig. 10.10.

10.3 MULTI-IMPRESSION PROGRESSIVE TOOL

Multi-impression tools are those tools which perform same operation at more than one portion in a single stroke. Multi-

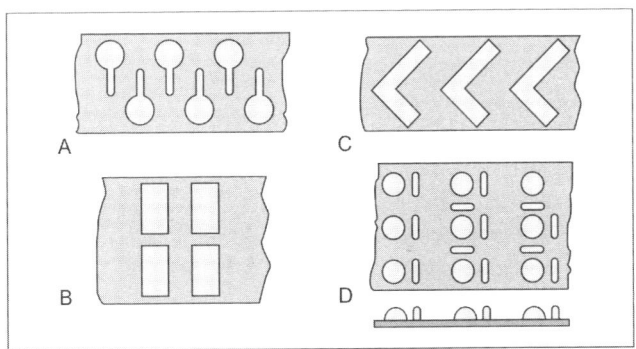

Fig. 10.11: Examples of multi-impression tools

impression tools may be simple or progressive tools. Few examples of multi-impression tools are shown in Fig. 10.11.

Figure 10.11 'A' shows a blanking tool having two impression blanking. Punches and respective dies are so placed that resultant ties between two blanked hollow and edge of strip is not more than four times the thickness of sheet metal. Two punches cannot be placed near to each other so much that tie of not more than four times thickness of sheet remains after blanking. Fitting of punches in punch holder plate is a constraint, so the punches are placed with sufficient thickness of steel between two die openings. Figure 10.12 shows strip layout and placement of two punches in punch holder plate. Crs (distance between the center lines of punches) is also shown. Now the stroke would be given and a blank would be cut by punch P_1 and pushed down through die. Punch now moves up. Sheet is now indexed forward by a value of S, which is equal to Ø. This brings already punched space (from where blank is pushed down at position under P_2) as shown in drawing. Next stroke would punch down two blanks, P_1 the second time and P_2 the first time. On next indexing and stroke, two more blanks would be pushed down through dies. In this way, two blanks would be produced on each stroke.

Please note that steel thickness 't' is of sufficient strength. In fact, it may be calculated at design stage where factor of safety is also taken under consideration. The process sequence of a two impression tool would be as follows:

- Move strip from left to right.
- Place leading edge of strip against stopper pin.

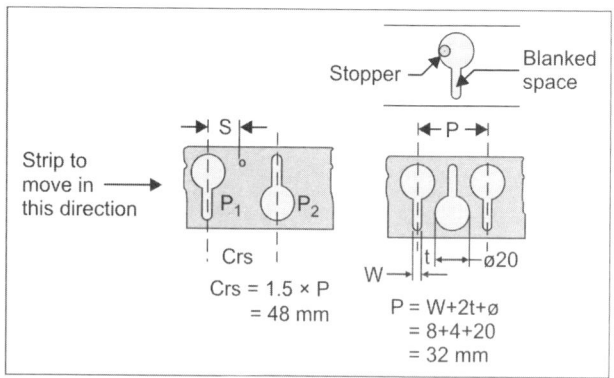

Fig. 10.12: Punches placement (crs)

- Give blanking stroke and move the punch up.
- Push the strip in such a way that inside of blanked edge of strip stops against stopper pin.
- Give blanking stroke. First blanked portion of strip would be partially blanked. It would be a one time waste.

·Subsequent indexing would produce two blanks in each stroke.

In this way, two components would be blanked down at each stroke of tool.

Another example of a sheet metal eyelet is given here. The shape and approximate sizes of eyelet are shown in Fig. 10.13.

Normally eyelets of various descriptions are produced in bulk quantities. To meet market demand, eyelets are produced by using multi-impression tools. Incidentally these tools are progressive tools. Such tools are combination of multi-impression and progressive tools' formation. A typical arrangement of tool and its effect on sheet strip/roll is shown by Fig. 10.14.

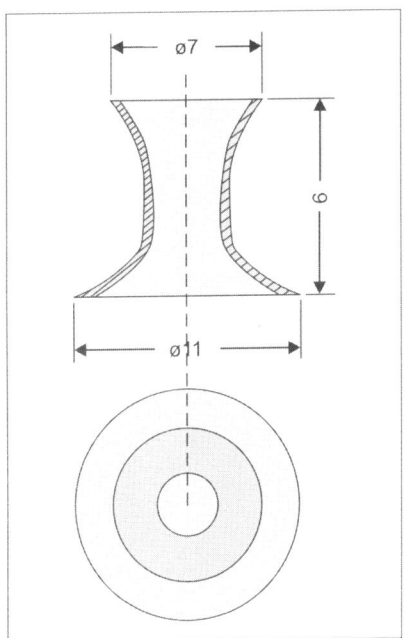

Fig. 10.13: Sheet metal eyelet

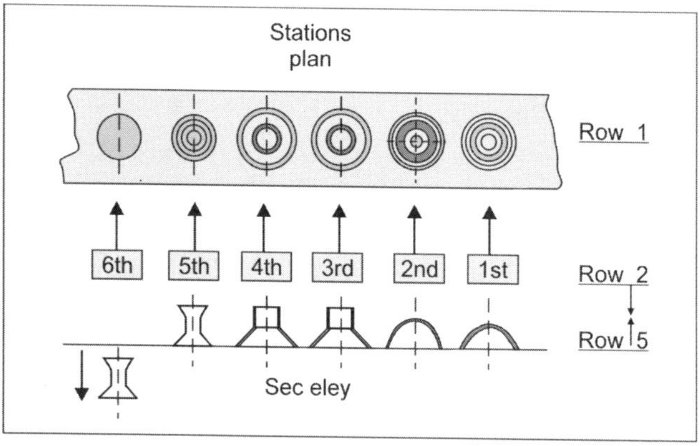

Fig. 10.14: Progressive operation

Figure 10.14 shows progressive formation of eyelet. There may be five or seven such rows, depending on design of tool. Progression of one row is briefly explained herewith which is the same for other rows as well.

At 1st station, a dome is formed, on 2nd station, this dome is drawn to increase in height and with reduced diameter. As explained earlier, percentage of reduction for each station is determined by tool design. On third station, final length and diameter of draw is done. On 4th station, reverse piercing is done to open the eyelet. On 5th station, flaring of brim of eyelet is done and finally eyelet is trimmed out of strip. Assuming that there are five rows for eyelets formation then the production per stroke would be 5. If tool is being run at 120 strokes per minute then production per minute would be 600 eyelets per minute. Hence in one shift of eight hours, production would be 2,88,000 (two lakh eighty eight thousand).

11

Tool and Die Designing

This chapter consists of the following aspects:

11.1 FEATURES OF TOOL DESIGNING

Designing a sheet metal tool is an engineering science as well as personal trait of a designer. This means ability to do deep imagination and draw mental pictures of a design. Many times, a designer has to complete his design in imagination bit by bit. After doing some thinking, he draws ideas on paper to review it at some other convenient time. In due course of time interval, designer comes up with more ideas to refine his initial thinking.

A freshly qualified designer must know various **design principles**, calculations and use of design tools such as **AutoCAD or other software**. What he needs more now, is to have a deep look on various types of sheet metal press tools which are not in action and in action as well. Actually designer should have opportunities to visit places where tools are designed, made and used. The more he will see, the more he would have designing options as model. In this way, a designer may go ahead with the design of a particular tool for a given component.

A tool designer has a great responsibility for designing a tool which has the following features:

- Produces components with specified accuracies and reliability
- Maximum possible components between two maintenance operations
- An intelligent compromise between tool design and its cost
- Safe operation
- Ease of loading on press
- Drawing of individual parts and assembly to be AutoCAD generated so that computer aided machining (CAM) is possible.

To achieve above mentioned features in a sheet metal press tool, a designer has to consider the following aspects when designing a tools

- Study of component design
- Sheet metal process "thumb rules"
- Application of "strength of material" rules
- Availability of suitable steels
- Press selection
- Productivity
- Availability of machining facilities
- Reliability
- Tool life
- Ultimate tool cost
- Tool completion time
- Available CAD, CAM and CNC **machining facility**

11.2 STUDY OF COMPONENT DESIGN

There may be hundreds and thousands of sheet metal components used in various fields of engineering, utility and domestic products. Many components look alike but there may be minute difference in their features. Not only features but the material may be different and of varying thickness. All these factors may be the cause for changes in tool designs. Designer may come across with a situation when he is given a sample component where certain modification in design is desired. Consequently, a new tool would have to be designed. Alternatively, designer may be provided with a component drawing for designing a tool. In such a case, tool designer has to examine the component drawing critically to see if tolerance in various sizes are provided and if it is possible to design a tool which can provide specified tolerance in component dimensions. Further, many a time, radii at bending points or cup or shape drawing point are specified so much sharp that it might not be possible to bend a portion of component or draw. If that is the case, then tool designer or user has to point out the possibility of increase in the percentage of rejection of components due to cracking or breaking at bending point or becoming weak. Most likely a component designer or user would agree to tool designer's recommendation for modifications.

Another thing is that specification of sheet metal should be studied for tolerance in material composition and width of strips or rolls. If the tolerance is too wide then tool design and dimension would be affected. At this stage, tool designer may foresee the consequences. He may hardly have a chance to accommodate wide tolerance. For example, in case of cup drawing, cup may have slight wrinkles when thinner portion of sheet is drawn. There may be ironing on walls of cup as thickness is more. He may have to discuss the matter with the authority, who has assigned the job of tool designing. Authority may be asked for having sheet metal with close tolerance or to accept certain percentage of defective components.

Sometimes, views given in the drawing are not correct or some sectional view is missing. Tool designer must pick up such errors or omissions in component drawing and must ask for a new drawing with errors and omissions removed.

11.3 SHEET METAL PROCESS, 'THUMB RULES'

For a sheet metal press tool designer, it is almost necessary to have knowledge of sheet metal behavior under various operations such as cutting off, bending, piercing, drawing, etc.

Persons with long practical experience of shop floor and research organizations especially for sheet metal production have come up with certain 'thumb rules' for various operations carried out on sheet metals to convert it to desired component.

'Thumb rules' are briefly described in the following paragraphs:
- Use outside dimension of components in drawing unless inside dimension is specifically given as per example in Fig. 11.1.
- Dimensions as general 'thumb rules' should be given on the same side of the sheet. An example is given in Fig. 11.2.

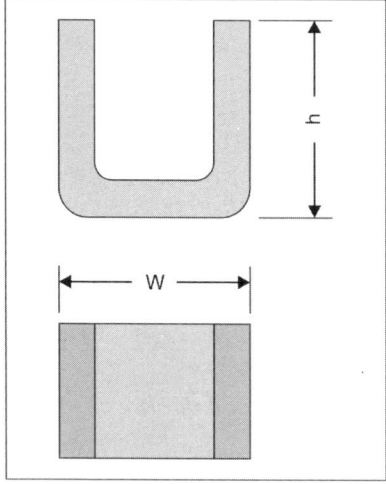

Fig. 11.1: Showing outside dimension

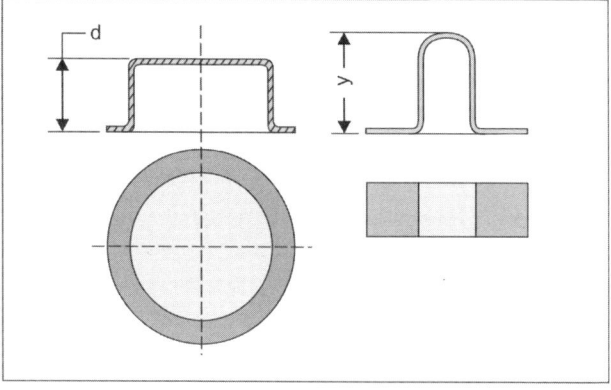

Fig. 11.2: Dimensions on same side

- Inside bend radius equal to the thickness of sheet metal should be preferred.

  Figure 11.3 shows three conditions of bend.

  Bend condition shown in (a) is highly undesirable as tearing of material takes place at bending point.

  Bend condition shown in (b) is OK from bending point of view but size Y has increased to $(Y + R + T)$. It may not be acceptable from component function or assembly point of view.

  Bend condition shown in (c) appears to be much more acceptable because there is no tearing of material or increase in size Y to

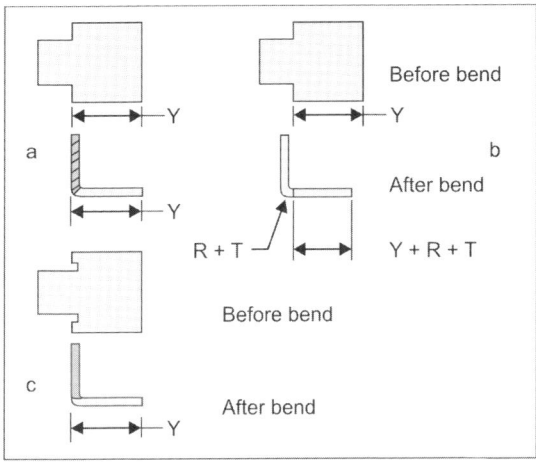

Fig. 11.3: Conditions of bend

(Y + R + T). The only point not in favour of this bend option may be additional feature in the shape of blanking punch and die.

- **Bending near a hole or slot features**

Figure 11.4 (a) shows that hole shape is distorted as bend of radius 'r' is very near to hole. (b) Shows that brim of hole is (2T + R) away from bent surface. Hence, there is no distortion.

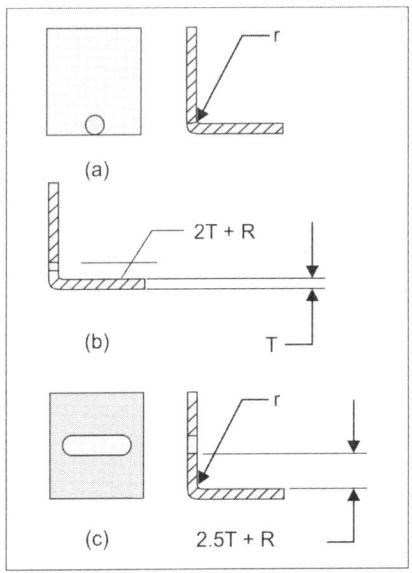

Fig. 11.4: Bending near a hole

- **Piercing and edge relationship**

In Fig. 11.5 'a', thickness X between hole brim and edge of material is less than sheet thickness. As a result of this, on piercing, a bulge would be formed on the edge of sheet. In 'b', value of X is more than sheet thickness. Consequently, there would be no bulge.

- **Punch and die clearance** in trimming and blanking operations, normally kept as 10% of sheet metal thickness.

- In a particular sheet metal component, a hole less than sheet thickness is not recommended. At the most piercing punch, diameter may be equal or more than sheet thickness. This 'thumb rule' may hold good for quarter to spring hard brass sheet. If the same hole is to be pierced in a stainless or titanium sheet then piercing punch and die diameters have to be increased to achieve strength in punch so that it does not bend or break.

- In case of multiple impression embossing, after first stroke, strip has an embossing. It is then moved forward in embossing tool. If proper pitch of movement of strip is not provided then there is a chance of first impression getting damaged on second stroke of the tool.

- **Counter sinks** in sheet metal may be created in different ways. For sheets up to 18 gauge (1.214 mm), counter sink may be created by forming tool. In case of heavier gauge, punching and machining may be resorted to.

- In blanking, size to be checked from punch size rather than the die side, because it is the punch which determines the size of hole. Hole size towards die gets a little enlarged as about one third of sheet thickness is torn down. This situation is illustrated with the help of Fig. 11.6.

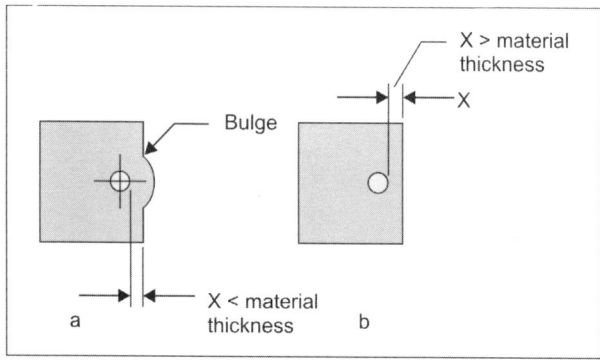

Fig. 11.5: Thickness between hole brim and edge

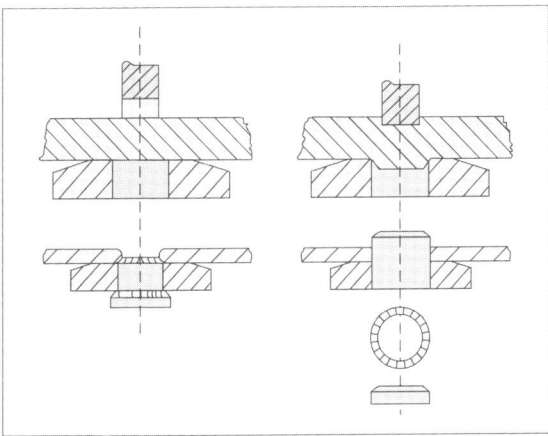

Fig. 11.6: Punching action

- In trimming and blanking, punch corner (not cutting edge) should not be razor sharp unless specified. A radius of half the material thickness may be given. When punch corners have radii then obviously die would also have matching corner radii.

- **Notches and tabs**

  Figure 11.7 shows a piece of sheet metal having notches and tabs. As a general 'thumb rule', width 'w' of notch or tab should not be less than 1.5 times thickness 't'.

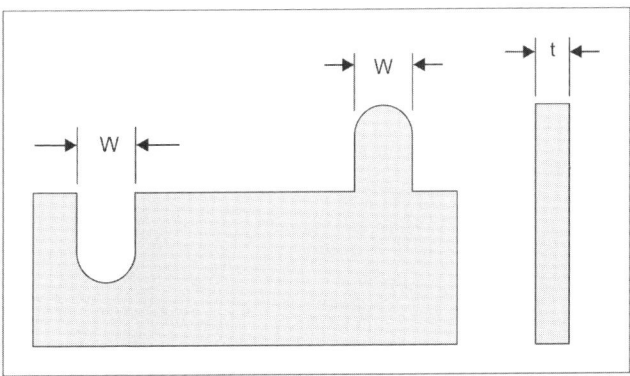

Fig. 11.7: Notches and tabs

- **Cut outs**

  Three types of cut outs are shown in Fig. 11.8.

  Rounded cut outs are not suitable due to presence of thin 'needle sharp' edges.

- **Drawing**

  Drawing of circular shapes is easy to achieve as compared to square or rectangular shapes. Quite generous radii of punch and die are required to get a component free from wrinkles.

- **Radii**

  Punch and die radii may be about four times the material thickness.

- **Drawing clearance**

  Drawing clearance between punch and die diameters should be two times the higher value of sheet thickness (See Fig. 11.9).

  So, $\varnothing 1 - \varnothing 2 = 2 \times (t + 0.015 \text{ mm})$. Axis of punch and die must be concentric so that clearance all around is same.

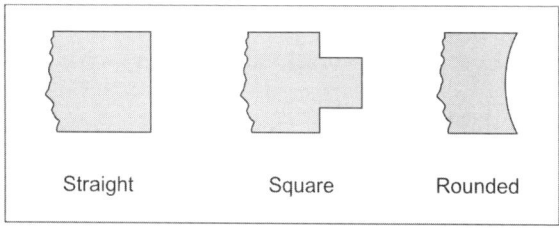

Fig. 11.8: Cut outs

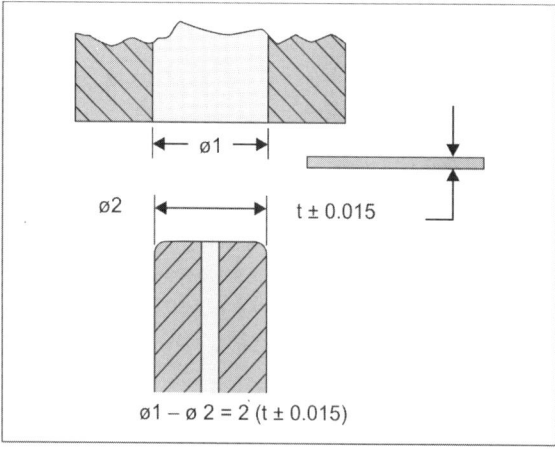

Fig. 11.9: Drawing clearance

- **Grain direction**

    Orientation of grain takes place at the time of rolling of sheets in calendaring plant. Grain direction is almost parallel to sheet, strip or roll length. In components where bending of same portion has to take place then that should be along grain direction. This 'thumb rule' is illustrated with the help of Fig. 11.10.

    Bend tail is along grain direction. Chances of breakage/crack at bending line is very little as compared to if grain direction is at right angle.

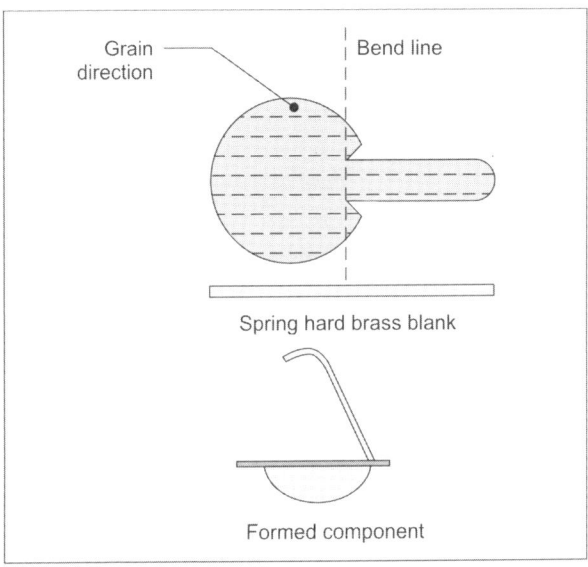

Fig. 11.10: Grain direction

11.4 APPLICATION OF 'STRENGTH OF MATERIAL' RULE

The objective of application of 'strength of material' rule is to reasonably ensure that object, in our case, a sheet metal press tool, does not fail or perform erroneously. Behaviors of tool is very much affected by many physical effects on various parts of tool, such as

- Mechanical stress
- Fatigue
- Heat transfer
- Wear
- Mechanical vibration
- Motion
- Shock loads

An effective designer of sheet metal press tools would have knowledge and experience to carry out finite elements analysis by using suitable software such as full version of autodesk software for student or professional software. Moreover, the designer may have a 'feel' if sizes of a particular part of a product are safe, unsafe or 'over safe'. To illustrate this point, Fig. 11.11 is provided and a brief discussion is given.

Figure 11.11 'A' depicts wheels of a motorcycle. Wheels may have axial of, say 6 mm, 12 mm or 18 mm diameter. Wheel with 6 mm diameter certainly looks weak. A rider would not opt for this axial. Wheel in the middle has a 12 mm diameter axial which appears to be safe. Third view shows a wheel with 18 mm diameter axial. It seems certainly oversized, thus waste of material and other oversized components.

Another illustration is 'B' which is a **die set**. On left hand side is a die set having **pillars** of say, 10 mm diameter and on right hand side 18 mm diameter pillar. Which die set is safe to use? This can be answered by a designer who has seen many die set tools functioning and by seeing has attained a 'feeling' as to which diameter of pillar is safe.

The third illustration is of two ladders having 10 mm and 25 mm diameter cross bar. Which one is safe to use? Most likely the answer is ladder having 25 mm diameter cross bar.

The above discussion is based on the feeling of individuals. Similarly, design of a tool cannot be completely based on 'feel'. 'Strength of material' principle has to be applied to design a safe,

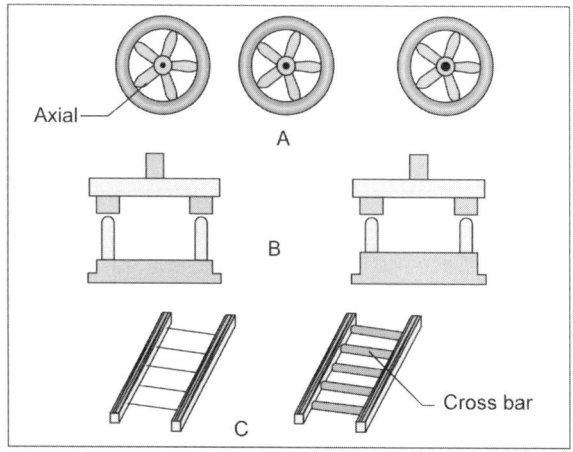

Fig. 11.11: Under safe, safe, over safe dimensions

accurate and durable tool. It is therefore necessary to go into some details of 'strength of material' rule.

What is 'strength of material'? When a material is subjected to a load, tensile or compressive, its grain/molecule tend to deform or displace, consequently, resists applied force equally. It happens till deformation remains in plastic limits. Once plastic limit is crossed, resisting force of grains/molecules become less than deforming external force. Result of this is that material gets stretched or compressed to such an extent that material fails due to breaking. Resisting force by material before crossing plastic limit is 'strength of material'. The study and determination of tensile, comprehensive, tensional, bending and shear strength come under the preview of 'strength of material'. Tensile strength of a material may be different at two stages of loading. One, when plastic limit has just arrived, second, when material fails or breaks. Tensile or comprehensive strength is measured in force per unit area.

Units in metric system may be N/cm^2 Or $kgsf/cm^2$.

In British system, it is Lbsf/sq inch.

As already stated earlier, tensile, comprehensive and bending strengths are measured by universal testing machine and shearing strength is tested by equipment which is based on application of kinetic energy of swinging pendulum.

Let us consider few 'real life' situations in press tools where dimension may be calculated to achieve non-failure condition. Figure 11.12 shows a draw punch fitted to a **cast iron die block**.

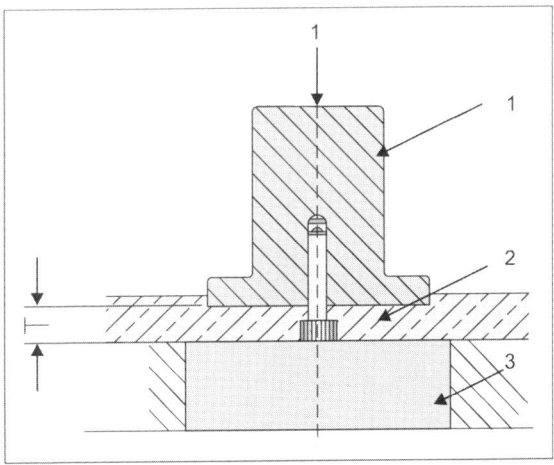

Fig. 11.12: Draw punch fitting

Draw punch (1) has a base diameter of 40 mm. T is supported on a C.I. block (2) of thickness T. Draw punch is made of tool steel which has a tensile strength of say, 1200 MPa and shear strength of cast iron is 124 KPSI (for iron ASTM number 35) or 71.7 kg per square centimeter.

During draw of 0.38 mm thick brass sheet into a cup would roughly exert a force of 3 tons (3000 kg).

Safe thickness T of cast iron which is supporting the punch may be calculated in the following manner:

Formula:

$$F = F_s \times C \times T$$

where,　T = C.I. thickness in cm

Or

$$T = \frac{F}{F_s \times C}$$

F = Force in kgf

F_s = Shear strength in kgf per cm square

$$T = \frac{3000}{7.1 \times \Pi \times 4}$$

$$= \frac{3000 \times 7}{71.7 \times 22 \times 4}$$

C = Circumference of punch base in cm

$$= 3.32 \text{ cm}$$

Considering a factor of safety as 1.5 then the thickness T may be maintained as

$$T = 3.32 + 1.66$$
$$= 4.98 \text{ cm}$$

For iron, ASTM number 20 (for example), value of T works out to be 4.97 cm plus safety margin (4.97 + 2.48) 7.45 cm.

Again, for iron ASTM number 60 (for example), value of T works out to be 3.3 cm (2.2 + 1.1), so it may be seen that punch supporting thickness T of cast iron varies according to shear strength of iron (cast iron). Results of calculations are summarized in Table 11.1.

	Table 11.1: Calculated value of thickness		
Iron ASTM Number	*Shear compressive strength*		*Calculated value of T in cm with factor of safety*
	KPSI	*kg per cm square (calculated)*	
20	83	48	7.45
35	124	71.7	4. 98
60	187.5	108.4	3.3

As already discussed, cast iron is weak in tension. It is therefore almost necessary to workout the sectional area of a portion of a casting which undergoes a direct tension or compression/tension due to bending.

Figure 11.13 shows assembly of two cast iron: Part (1) and (2). Part (1) is a protruding tab of thickness 't' and width 'w'. Parts (1) and (2) are tightened by means of a Ellen screw. XX' is an imaginary plane passing through the middle of thickness of tab. Pull of Ellen screw is creating compression above XX' plane and tension below XX' plane. A line (3) is a probable crack location.

It is required to be calculated that what should be the safe value of tab thickness 't' and width 'w' so that tab does not break at line (3) when M10 Ellen screw is fully tightened. First of all, it should be known that how much pulling force is exerted by screw when fully tightened with specified torque. Let the pulling force created by Ellen screw be 'F' and line of application of this force from part body be C, and there is a bending tendency of tab. It is the line (3) below imaginary plane XX', where the grains of cast iron are in tension due to bending tendency.

Hence, So to calculate safe cross-section of tab, values of the following should be known:
- Pulling force F in kg
- Value of 'C' in cm
- Tensile strength of C.I. in kg per centimeter square
- Value of 'W' in centimeters

Formula for calculating thickness 't' of tab may be as follows:

Figure 11.14 shows relationship of force F, distance C, t (the thickness of tab) and f_t is shown. Mathematically, it would be

$$F_t = \frac{F \times C}{t}$$

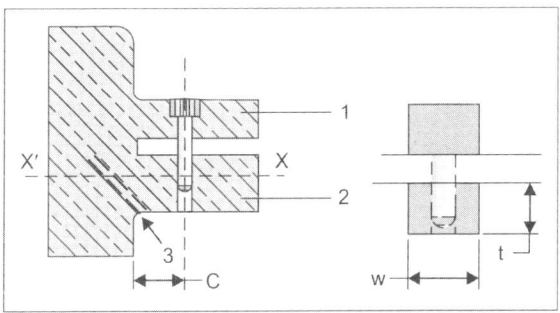

Fig. 11.13: Assembly of cast from parts

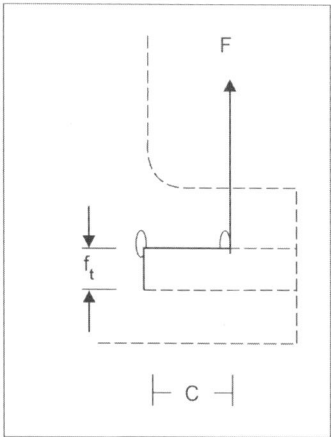

Fig. 11.14: Relationship of force, distance and thickness

Area of tab under tension

$$\frac{t}{2} \times W$$

Assuming width 'w' of tab as three times the thickness t, then
$$W = 3t$$
Hence, area of tab under tension

$$= \frac{t}{2} \times 3t$$

Again,

$$F = f_t \times area$$
$$= f_t \times 3/2 \; t^2$$

Therefore,

$$t^2 = \frac{F}{f_t} \times \frac{2}{3}$$

Let us take a numerical example where pulling force F by screw is taken as 180 kg (assumed) and tensile strength f_t is taken as 300 Newton per square millimetre which equals approx 306 kgf/cm^2.

$$t^2 = \frac{180}{306} \times \frac{2}{3}$$

$$\approx 9 \text{ cm}$$

$$t = 3 \text{ cm}$$

Taking a factor of safety as 1.5

$$t = 4.5 \text{ cm}$$

Fatigue

There may be an example of axial of a car or a truck braking while vehicle is in motion. Such cases may also be found in railway axial failure. Why does this happen? This happens due to a phenomenon which is known as fatigue. It is caused by cyclic exertion of a force on a machine or tool part. In case of an axial, it keeps on rotating while the bending force (load) remains acting downwards, that is, in one direction, so the grains of material alternate rapidly into compression and tension.

In case of sheet metal press tools, there are instances where tool members are subjected to compressive or tensile force and no force in cyclic manner with changing frequency as high as 200 cycles per minute.

Fatigue in a tool component or in any component of any other system promotes minute cracks in material which gradually keeps on increasing till component breaks.

Under fatigue conditions, the ultimate tensile strength of metals, such as ferrous and non-ferrous is adversely affected, causing it to lower down. Therefore, it may be a general practice to design sizes of component with about 15% lower value of specified ultimate tensile strength of steel to be used for making a part. Due to this practice, failure of sheet metal press tool component is a rare happening.

Diameter of a Die Set Pillars

Designer of a sheet metal press tool need not worry much about the diameters of a die set pillars. Makers of pillar **die set** make **various types** of die set suitable for certain tonnage of power presses. Accordingly, type of steel, heat treatment, surface hardness, accuracy in diameters and surface finish are considered. Diameters of pillar are so designed and made to withstand some bending and shearing forces which may act on it due to inaccurate movement of RAM.

The other reason for generation of bending and shearing forces on die set pillars may be incomplete feeding of sheet metal in the tool, as shown in Fig. 11.15.

While designing tool, stem is placed at such a position on top of die set that punching force is equally distributed in two rows, one having two punches and the other having three punches, but at the start, only three punches at right hand side would be blank. Hence, there would be unbalanced force pressing upper member of die set. Consequently, there would be tilting tendency in upper

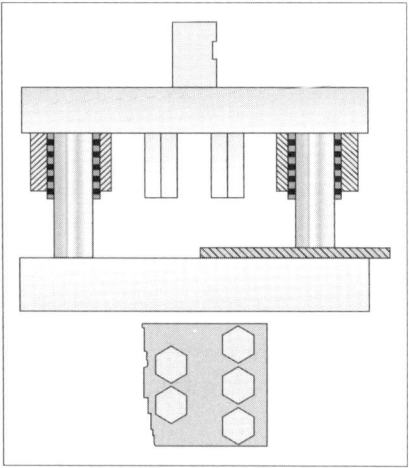

Fig. 11.15: Unblanced force on die set

member resulting in bending and/or shearing forces working on **pillars** of die set, but the diameters are so calculated and made that they sustain forces and do not allow upper member of die set to tilt even if it is of small value.

Refering to Fig. 11.16, in case Y is equal to 2X then value of force F at point P_2 would be F/2. Bending tendency is illustrated in Fig. 11.17.

If values of Y and X are different then resultant force at P_2 may be calculated as per following formula:

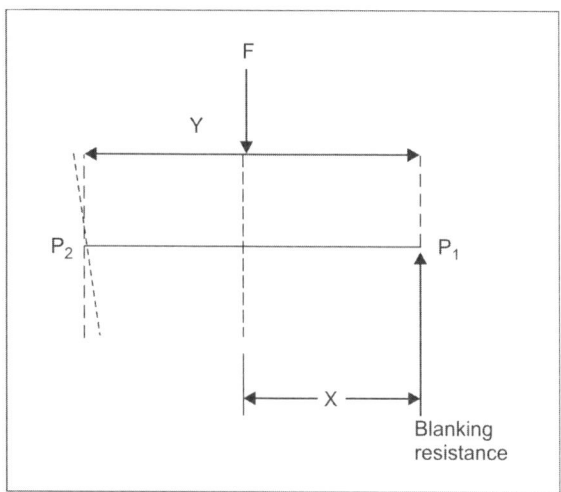

Fig. 11.16: Effect of force and distance on die set pillars

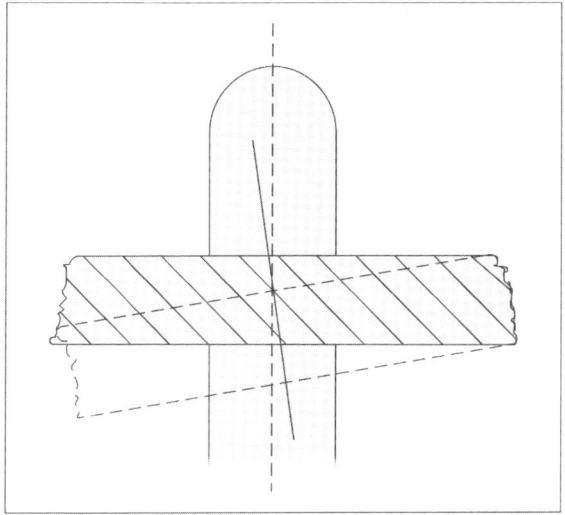

Fig. 11.17: Bending tendency is illustrated

$$F \times X = f \times y$$

Therefore, f

$$f = F \times \frac{x}{y}$$

Where   f = Resultant force at point P_2 in tons

F = Force applied by RAM in tons

X = Distance between line of force and point P_1

Y = Distance between point P_1 and P_2

Thickness of Punch Holding Plates

Generally, there are two common ways to hold and fix punches under the upper member (plate) of a die set. Most common is to fix punches in a steel plate having suitable openings (holes or profile) so that punches are assembled with interference fit to achieve rigidity in fitting. This practice is adopted for small to medium size press tools. In large press tools, punches may be screws on the inner face of upper die set members. Such punches are firmly located by providing location dowel pins. This practice is mostly adopted for drawing and stamping station for blanking or trimming operations the use of punch holding plate system is preferred.

The basic arrangement of both the systems is shown in

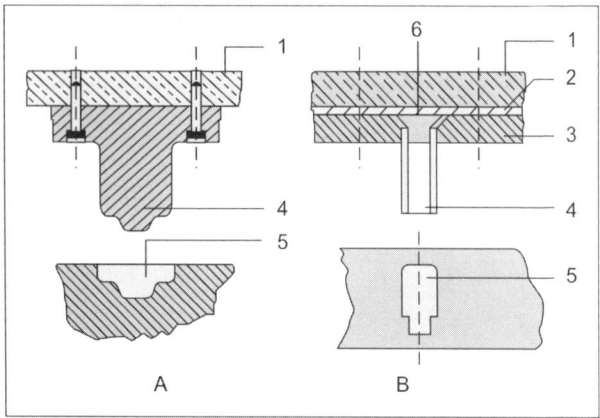

Fig. 11.18: Punch holding possibilities

Fig. 11.18

In Fig. 11.18 'B', Part (1) is upper plate of die set which is generally of cast iron. Part (2) is a hardened and tempered back plate which is fitted in between (1) and punch holder plate (3). The object of using plate (2) is to provide a strong and hard backing for punch face (6) so that it does not shift slightly up in lower face of upper die set member (1). It may happen due to repeated force and no force cycle during production runs. Plate (2) may be of carbon steel having a hardness of 52–55 HRC after tempering. Thickness of plate may be calculated if compressive strength of hardened plate is known. For example, AISI – SAE steels 1050 may have shear compressive strength of approximately 140 KPSI and Brinell hardness as 370 BHN.

Referring to Fig. 11.18 'A', Part (4) is a big size draw cum forming punch such as used for drawing and forming a stainless steel sheet washbasin for kitchen.

Figure 11.19 and Photo 11.1 of washbasin would give an idea of size of draw cum forming punch. In such cases, punches may be built by suitability joining a number of portions. Ultimately total area of back of punch would be fitted on a big plate (which may be a spacer block). Common practice is to enlarge back face area of punch so that force per unit area does not exceed compressive strength of top plate.

Thickness of punch holding plate (3) in cases such as shown in Fig. 11.18 'B' is kept 1.5 to 2 times the size 'W' of punch where it is held in plate. In practice, it is found that punch holding plate of 212–235 Brinell hardness is suitable because it very well resists

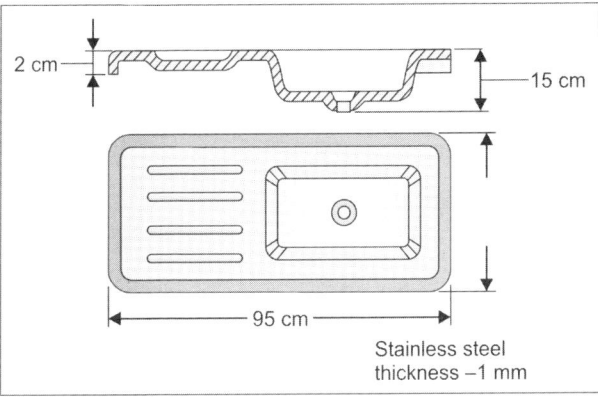

2 cm

15 cm

95 cm

Stainless steel
thickness –1 mm

Fig. 11.19: Kitchen wash basin

Photo 11.1: Kitchen sink

(Source: Google-write kitchen sink images), as on October 30, 2013)

deformation force which might get generated due to unbalanced trimming or blanking load on punch.

11.5 AVAILABILITY OF SUITABLE STEELS

In selecting type and grade of steel for various tool parts, a designer of sheet metal press tool has a number of options, but what is actually available or may be made available is a matter of information collection. In an organization, purchase department may have the responsibility of having a database for various steels from different producers, suppliers, rates, availability of technical data sheet and time required to procure steel.

Equipped with the above information, a designer may start

designing tool accordingly.

Nowadays, collecting techno-commercial information is not a problem. Most of the information may be gathered on internet by visiting sites of steel makers, suppliers and marketing agencies such as 'India mart'. Designer may gather information himself.

11.6 PRESS SELECTION

Press selection and tool design are very much dependent on each other. Designer generally takes into account the power presses which are already installed in a manufacturing shop. It rarely happens that a new power press of a definite tonnage has to be purchased for a particular design of press tool. In such a case, designer has to consult an appropriate official of the organization, say a general manager to obtain a decision if a new power press is really necessary for benefit to organization. A press tool designer may have to explain to general manager various decision making aspects such as productivity, tool cost, time period for recovering investment to be done in purchasing a power press and future possibilities of utilization of power press of proposed tonnage.

Sometimes, it may be a compulsion to purchase and install a high tonnage power press because company has planned to introduce a new product which definitely needs a higher tonnage than what is already existing.

For example, a manufacturing company is producing light weight and size kitchenware by using existing 40 and 60 tons power press. In case, company decides to produce a pressure cooker then power press of 150 tons and above might be necessary on account of required tonnage for draw, trimming operations and sizes of tool.

In case, company further decides to diversify and produce items such as washbasins, cover plates of earth moving machine, truck engine bonnet or front panels then power presses of even much more tonnage, say 400–600 tons, might be necessary.

So after in-depth discussion, a press tool designer may go ahead with tool design, knowing that a power press of a particular tonnage would be available for running the tool on it.

Productivity

Number of components produced per unit time may be considered as productivity of overall set up including tool design.

For small and medium size component, press tool may be designed to produce more than one component per stroke of the press. Tools may be two, three, four or more impression tools. As already discussed earlier, a balance has to be struck between productivity and cost of tool. Again, tool designer may have to discuss the matter with general manager as the element of financial consideration is involved.

11.7 AVAILABILITY OF MACHINING FACILITIES

A tool designer normally has basic knowledge of machining operations which are required to be performed to make a particular part. Let the part be a blanking punch as shown in Fig. 11.20.

In Fig. 11.20, small working length is shown, but how the punch would be held in punch holder plate is shown in Fig. 11.21. There may be more than one option as shown in Fig. 11.21.

In option 'A', punch profile is throughout the length and it is easy to machining, but making of punch holder plate to achieve a sturdy fitting is difficult.

In option 'B', top of punch (fitting end) has a circular shape which makes the fitting into punch holder plate easy and sturdy. If designer opts for option 'B' then he must know that tool room in which punch would have to be machined has a punch shaper. It is

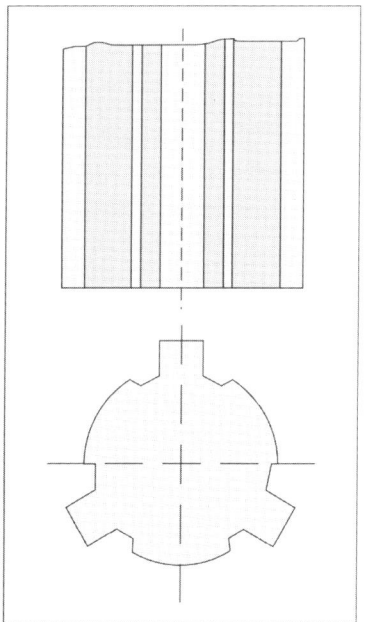

Fig. 11.20: Blanking punch

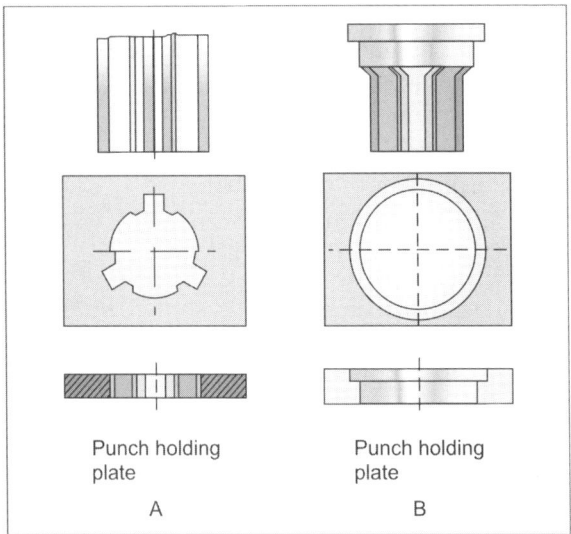

Fig. 11.21: Punch holding possibilities

a special purpose machine suitable for shaping punches similar to the one shown in 'B'.

The above example highlights the necessity of having knowledge of availability of suitable machine tools.

11.8 RELIABILITY

Most of the designers of sheet metal press tools are conscious of the need of inbuilt reliability in the design of a press tool. There are many factors which contribute to the reliability. These are as follows:

- Design of parts for sturdiness
- Selection of most suitable steels
- Proper heat treatment
- Proper machining operation after heat treatment, such as grinding, lapping and polishing
- Specifying most suitable fits and tolerance
- Use of proper fasteners and dowel pins
- Use of very accurate and reliable linear ball bearings

Another important condition of reliability is that power press to be used is accurate to specified accuracy limits. This condition may be mentioned in design sheet when it is handed over to tool room, may be with a copy to manufacturing shop.

Tool Life

From design point of view, tool life has to be good if all 'reliability' points are taken care of, but tool life does not only depend on tool design, its use and maintenance are also factors which affect tool life. There are two aspects of tool life, which are as under:

- Production or number of strokes between two maintenance
- Total period of time in months or years till the tool becomes unserviceable and to be discarded.

11.9 ULTIMATE TOOL COST

Costing of a tool is normally done by a cost accountant because he is supposed to have all necessary data required for costing. Such data are

- Fixed cost
- Variable cost

Fixed costs are those which are made up of salaries, overheads, tool room maintenance, etc. In this, tool designer has hardly any role to play.

As far as variable costs are concerned, this has **elements of steel** cost, fasteners, machining, heat treatment, again some machining and assembly cost. Here is the role of a designer who decides on the most suitable and economical design of parts and assemblies. The designer may provide the following information to cost accountant:

- Starting weights of various steels specified in tool design
- Machine wise approximate machining time for various parts
- Die maker's hours
- Bought out items with specification and quantities, such as die sets, fasteners, dowel pins, spring, etc.
- Estimated tool completion time period
- Purchase of standard coil feed in indexing system, complete specification
- Strip coil holding and haul up system, complete specification (if already not available in manufacturing shop)
- Specification of any robotics and availability sources. In this case, cost accountant may not have necessary information, hence, tool designer may provide cost accountant with relevant information.

Tool Completion Time

Determination of tool completion time is a necessary exercise.

Management requires this information for planning further activities which may follow after completion of tool. Moreover, cost accountant also needs this information because tool completion time is an important factor for costing.

Estimation of a tool completion time is a combination of the following:

- Theoretical calculations for machining time for each and every part of the tool. Calculation from first principle may not be needed as standard ready reckoner charts may be available, if not then may be developed once, to be used time and again.

  Although theoretical calculations or charts carry an allowance for fatigues, necessary machine stoppages, etc.

- In practice, most of the time, it is found that estimated time fall short of actual time taken for complete machining of parts. This usually happens due to unforeseen reasons such as operator who is machining a particular job fell sick and goes on leave. If another operator is deputed to complete the job then obviously would take time to check the job for operations already done so far and then to reset the machine for completion of job.

  Although some percentage of time is taken into account for such eventualities, still there is a time out.

  In practice, it is found that increase in time for completion of job ranges from 10–30% of estimated time.

  More realistic time estimation may be arrived at through discussion among tool designer or time estimator and tool room manager.

11.10 REFERENCE TO SOME TOOL DESIGN SOFTWARES

Availability of CAD, CAM and CNC Machining Facility

CAD: Computer aided design

CAM: Computer aided machining

CNC: Computerized numerical controlled

Above are the sets of modern machining systems for machines and tool parts. These systems have great advantages in terms of machining accuracies, much less time is required to complete a job as compared to manual operations.

Most of the modern tool rooms are equipped with machine tools having mentioned systems. Use of these systems need the following inputs for best utilization:

- Trained tool designer who is capable of generating drawings of individual parts on AutoCAD in such a way that computer aided machining is possible from generated drawing.

- A variety of **softwares** may be needed to operate the system.
- Specific sheet metal press tool design software is almost necessary.

Few of the softwares located on internet are as follows:

- Product from Autodesk, a renowned software making company. Website address is as below (as on June 20, 2013)

  http://usa. auodesk.com
- Product from Solid works (DASSAULT SYSTEMS)

  www.solidworks.com

Softwares Included

- Autodesk inventor
- Professional
- Autodesk simulation
- Multi-physics
- Autodesk simulation mould flow
- Advisor ultimate
- Autodesk robot structured

Analysis of Professional Software

It has to be investigated that which one of the above would be suitable for sheet metal press tool designing.

Software, such as AutoCAD 2000 is suitable for general mechanical design work. It has a library of standard machine elements such as screws, nuts, bolts, rivets, bearings, shafts, gears, sprockets, etc. While designing a mechanical system, standard elements from 'library' may be picked up and suitably used in designing.

Similarly, software suitable for sheet metal press tool design may have the following elements in its 'Library'.

- Tools elements such as fasteners, dowel pins
- Various types of die sets
- C.I. blocks for open tools
- Standardized stems, pillars, linear cage bearings
- Variety of pressure pins
- Ejector pins of various shapes and sizes
- Stripper plates of various shapes and sizes
- Die block blanks of various sizes and types of steels
- Stopper pins

- Formulae to determine draw clearance blank diameters, bending radius draw radii
- Formulae to determine tonnage of power press for a given compound design, sheet thickness, its material
- Various possible layouts for multi-impression drawing, blanking, etc.
- Various designs, shapes and sizes of sheet or coil feeding indexing systems.
- In 'follow on' tools to provide information such as number of draws
- To get advice if tagged blanking be done prior to subsequent draw, in a 'follow on' tool
- And many other design support actions, selective or automatic.

A computer equipped with a good software would enable the designer to produce a good tool design in much less time as compared to designing a tool without a software.

12

Sheet Metal Press Tools Making Procedure and Practice

This chapter consists of the following aspects:
- Details of tool design and making for a number of components
- Tool height and press 'Daylight' relationship
- Making of forming punch and die
- Alignment of forming punch and die
- Strip stopper
- Shearing of formed portion of strip
- Setting of stopper pin in a multi-impression progressive tool
- Punch trimming
- Complicated piercing tool
- Bending cum forming operation
- Progressive tool of multi-operations, including curling
- Progressive tools
- Zinc calot blanking tool
- Press tonnage requirement for a big size component 356
- Blanking cum piercing of component 6
- Dome draw tool
- Bench for long drawn tube
- Panel rolling system

DETAILS OF TOOL DESIGN AND MAKING FOR A NUMBER OF COMPONENTS

In this chapter, actual making of tools for various components is explained with the help of figures. A number of components are selected to explain making of tool. Selection of sheet metal components is done in such a way that almost all **types of operations** such as bending, trimming, piercing, drawing, deep drawing, blanking, circular trimming, bulging, lancing, thread rolling, embossing, forming, stamping, etc. are covered. First of all, drawing of a component is given followed by brief description of its use and finer quality requirements for the component. All examples taken up are based on actual practical experience of the author of this book. Tool construction as explained in this chapter should not be used in practice for similar components. Slight variations in component design or dimension may need modifications in tool design.

Examples given here may be taken as tool design ideas generation. The study of component and its tools design may help readers to design their own sheet metal press tools.

COMPONENT 1

This component is used in assembly of a switch. Legs are inserted in other two components, placed one over the other. Legs are then curled to complete the assembly (Refer Fig. 12.1).

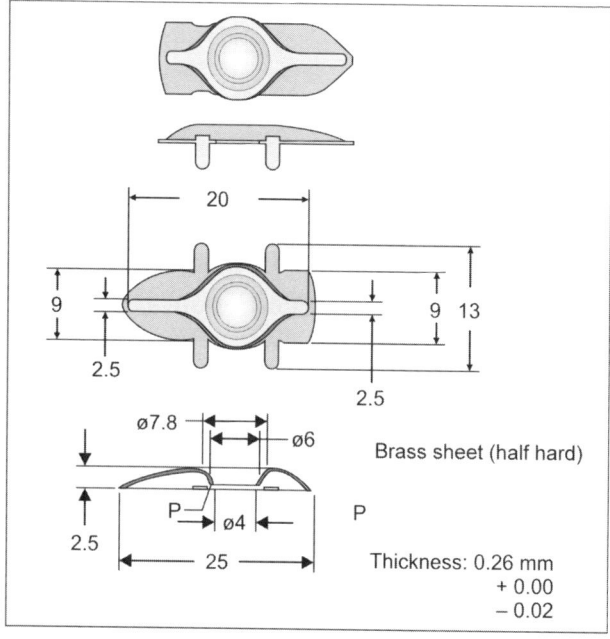

Fig. 12.1: Component of a switch

It is important that there should not be any burr at any point P on trimmed edges. This component is to be produced in the following operations:

- Forming/stamping
- Piercing
- Trimming
- Legs **bending**

At earlier development of tooling, all the above operations are performed by separate tools. Since the quantity requirements are low and component size small, it is found appropriate that operations are to be performed by hand screw press. Therefore, the tool has to be designed to suit hand press.

Figure 12.2 shows the tool construction in a die set with bush and pillars.

Part (1) is upper C.I. plate of die set and (2) lower member (plate). Part (3) is upper male forming steel block and (4) is lower female forming steel block. These blocks are dimensioned for length and width. From strength point of view, sizes of length and width are quite safe as 'bursting' force due to forming/stamping force which is about ten percent of forming force. Length of blocks (upper and lower) is kept as 52 mm, while width is 40 mm. These sizes are kept to provide space for socket cap screws M6 and dowel pins of 5 and 6 mm diameters. Thickness T_1 and T_2 of blocks are to be decided according to minimum and maximum space 's' available

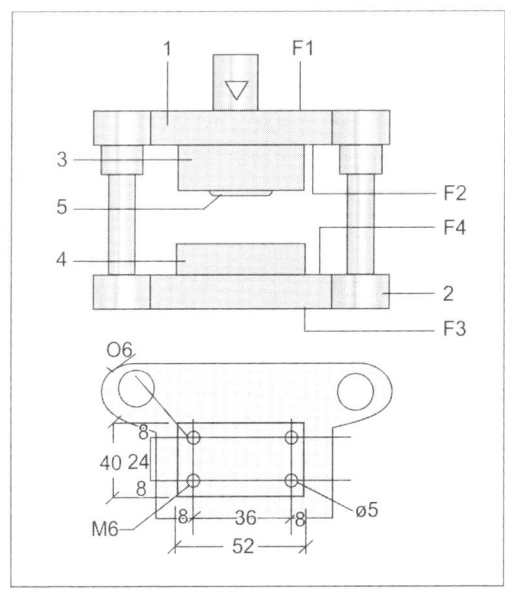

Fig. 12.2: Tool construction

between RAM face and tool fixing surface of hand screw press (see Fig. 12.3). Hand screw presses are generally specified by numbers, say No. 2, No. 3 and No. 4 as shown in Table 12.1. Figure 12.4 shows minimum and maximum space available.

Table 12.1: Data for tool space

Press No.	's' Minimum mm	's' Maximum mm	Approx angle of wheel rotation	Remark
2	80	110	90°	Angle based on ease for operation
3	100	135	80°	
4	130	170	75°	

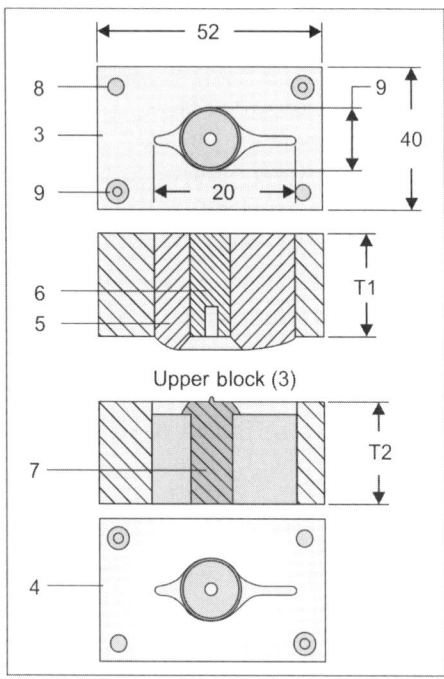

Fig. 12.3: Dimension for calculating tool height

Total thickness of tool should be more than minimum value of 'S' (Fig 12.4) say, 10 mm more.

Total thickness/Height of tool (Refer Fig. 12.2)

$$= \text{Thickness of (1)} + (3) + (4) \text{ and plus (2)}$$
$$= 20 + 24 + 28 + 20$$
$$= 90 \text{ mm}$$

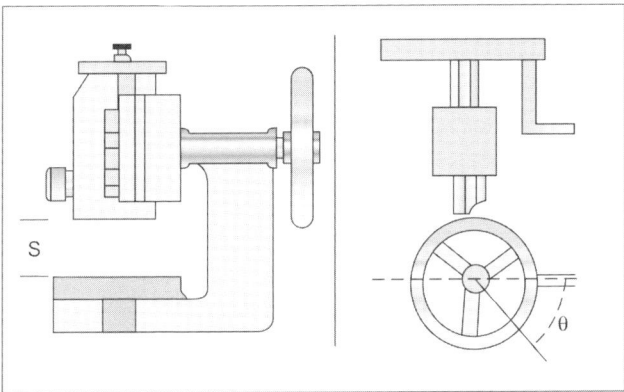

Fig. 12.4: Hand press, showing space for tool

Hence, the thickness of die and punch blocks are okay.

First of all, making of blocks of tool steel in old conventional way is described. Prepare both blocks by grinding the surfaces smooth and right angles to each other.

The following stepwise **procedure** for completing the punch block (3) and die block (4) may be adopted (Refer Figs 12.2 and 12.4).

- Apply copper sulphate solution on the top surface die set plate (1).
- Do the marking as shown in Fig. 12.2. Marking should be centralized on the plate.
- Drill dowel pin holes of 5 and 6 mm diameter as shown, through full thickness of C.I. plate 1.
- Drill tapping drill size for M6 socket cap screws. Tapping drill size for M6 is 5.6 mm.
- Make counter sunk for socket cap screw heads which is 10 mm in diameter and 10 mm deep.
- Clamp block (3) on the opposite face F2 of plate (1). Clamping should be done in such a way that position of block lies under the marking already done on face F1.
- Drill tapping hole (5.6 mm) through one tapping hole already done in plate 1. Depth of hole in block 3 should be approximately 10 mm deep.
- Remove the block and do M6 tapping in 'blind' hole. Remove burr, if any from die set plate 1.
- Increase diameter of 5.6 mm hole to 6 mm which lies over the tapped hole of block (1).

- Now tighten block (3) over face F2 of plate (1) by means of M6 socket cap screw.
- Drill other tapping hole in block (3), 10 mm deep.
- Dismantle block (3), remove burr if any and clean up.
- Increase diameter of 5.6 mm hole to 6 mm in C.I. plate (1).
- Fix block (3) to plate (1) by means of two M6 socket cap screws.
- Now this much assembly is ready for drilling and reaming 5 mm and 6 mm dowel pin holes. Do it.
- Fixing of block (3) on plate (1) is now completed. Mark location 'zero' before dismantling.
- Marking would be repeated on face F3 of lower die set plate (2). It should be done so that marking lies under previous marking on face F1 of plate (1).

No further drilling work would be done. First of all, making die cavity in block (4) and punch fitting in block (3) would have to be done.

In Fig. 12.2, punch has to be fitted in block (3). First of all, die profile is to be marked as per Fig. 12.3 by die maker. He would drill a hole of 8.5 mm diameter and would file profile by needle files. A highly skilled die maker can file out the profile nicely and polished out, but this hand work would take a long time, say 16–20 hours.

MAKING OF FORMING PUNCH AND DIE

Forming Punch

Procedure for making would be as follows:

- Prepare a rectangular block of tool steel as per dimension shown in Fig. 12.5.
- Do marking of profile on both the faces in same direction meaning that small form remain in same direction on both faces.
- Remove excess material around marking by shaping or milling, leaving filing margins.
- Remove remaining material by filing and to maintain specified dimension. Also keep on trying if punch gets fitted in the profiled hole in block (3).
- Fitting should be a transition fit. This means that punch gets in completely by slight tapping by means of a plastic hammer. Figure 12.6 shows forming punch which is to be fitted in block (3).

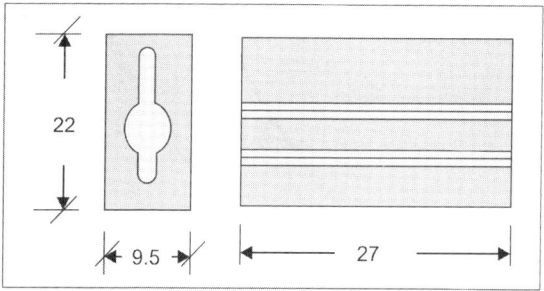

Fig. 12.5: Forming punch

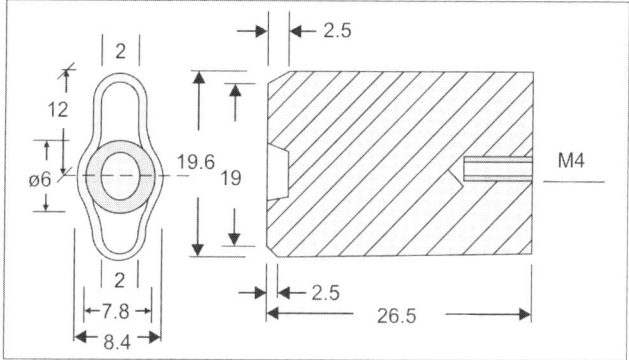

Fig. 12.6: Forming punch blank

- Working end (forming) of punch should protrude over the working surface of block (3) by 2.5 mm. Punch may have M4 threaded hole to hold it firmly against face F2 of die set plate (1).
- A tapered recess of 2.5 mm deep is to be turned for specified diameters and angle.
- Edges of raised profile should be left sharp. During trials, if a component cracks due to sharpness then sharpness of edges may be removed gradually by emery stone or emery paper, till no cracking takes place.

Forming Die

Profiled hole in block (4) would be made the same way as in **block** (3). The only difference is slight changes in sizes as compared to sizes in block (3). Sharpness of edges of profiled hole towards working face is removed by rubbing emery stick or fine needle file (Refer Fig. 12.7a).

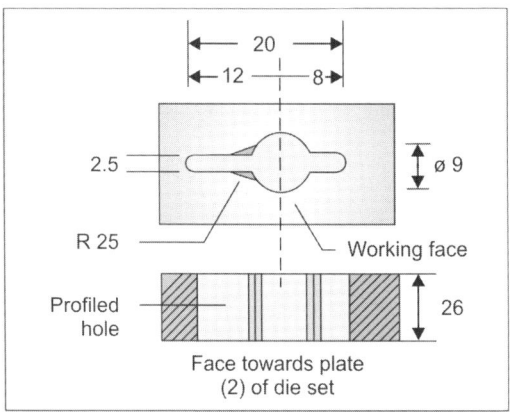

Fig. 12.7 A: Forming die

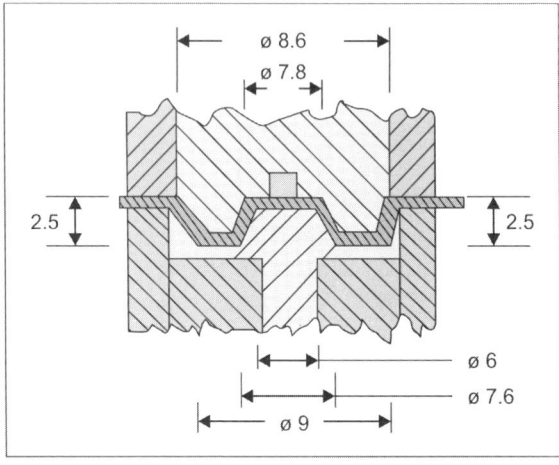

Fig. 12.7 B: Punch and die alignment

Please note that no hole is drilled so far in block (4). Certain procedure has to be adopted before any hole is drilled. Procedure is as follows:

- Cut a blank of 52 × 40 mm from brass sheet of 0.24 mm thickness, 0.24 mm is purposely chosen, instead of 0.26, + 0.02 – 0.02.

- Keep the blank over die block (4) inside the jig so that edges are matching with those of block (4).

- Put block (3) with punch fitted in such a way that edges of both the blocks match. It may be done with the help of a rectangular shaped jig, made out of MS strips. Jig is shown in Fig. 12.8.

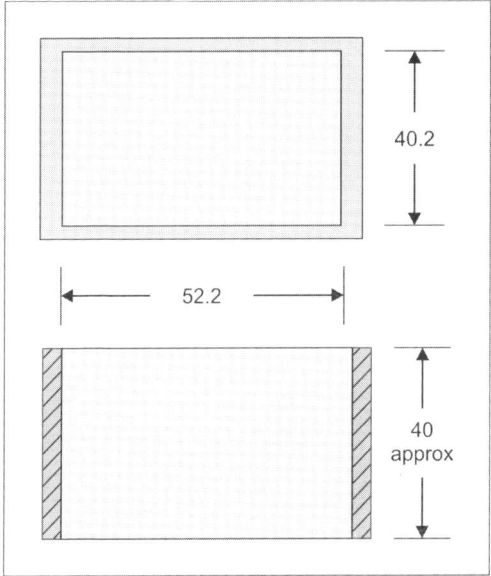

Fig. 12.8: Punch and die fitting jig

- Press punch lightly (see Fig. 12.7b)
- Remove block (4) and see if forming impression is made on brass sheet blank. Watch if there is any squeezing on impression.
- Remove jig and press combination again. Jig is removed to allow slight shifting of block as die and punch align themselves during pressing.
- Press more so that some depth in impression is formed.
- Note that sheet piece with partially formed shape acts as source of aligning die and punch, during fitting process in die set.
- Fit block (3) with punch on upper plate of die set on face F2.
- Put block (4) on upper face F4 of lower die set plate.
- Place formed piece of sheet over die opening so that form sits properly in die profile.
- Close the die set and clamp tightly by means of suitable clamping fixtures.
- Place the die set upside down. Now on top is surface F3.
- Face F3 is to be marked for centers of two screws and two dowel pins. Also mark outer periphery of block (4) in such a way that it just coincides block (4) which is under plate (2) face F4.

- Drill two tapping holes (5.6 mm) so deep that depth of hole in block (4) is about 10 mm.
- Increase holes in die set plate (2) only to 6.1 mm diameter by drilling.
- Remove block (4) and do M6 tapping in blind holes.
- Refix block (4) and close die set with formed sheet piece in between two block faces.
- Clamp the whole arrangement very tight.
- Tighten two M6 socket cap screws.
- Open die set. Put a fresh piece of 0.26 mm thick brass sheet.
- Press die set by means of a screw hand press.
- Open, examine the forming for any burnishing or squeezing. Note down the side on which squeezing is taking place.
- Tap the die block by a plastic hammer so that clearance between punch and die increases a little towards squeezing side.
- Retighten the die block and do a second trial forming.
- If found okay then drill dowel pin holes. Holes in block (4) should not cross, otherwise a mark would come in face of block (3), which is undesirable.
- Take out upper plate of die set together with block (3).
- Now pass the drill through dowel pin holes to cross block (4).
- Ream dowel pin holes while block (4) is still tightly fitted.
- Assemble dowel pins on both sides of die set tool.
- Tool is now almost ready for carrying out forming/stamping.
- Harden, temper and polish punch and blocks and assemble.

Tool is now ready for carrying out forming/stamping.

Brass sheet strip of 0.26 mm thickness and 19.5 mm wide is to be used to carry out first operation, which is forming cum stamping. Feeding of strip would be manual from left to right. Operator will operate the press by right hand and would feed the strip by left hand.

Figure 12.9 shows a small length of strip on which forming operation is performed. After each stroke, operator has to feed strip length so much that next forming could be done by next stroke. Feeding of strip should be done in a way that there is a gap between two impressions so much that after shearing of strip length of one impression, it has enough margin for trimming.

Figure 12.10 shows a sheared piece of strip having one impression with a distance of 3.5 mm between impression (form) end and strip edge. Hence, the total length of one piece of strip

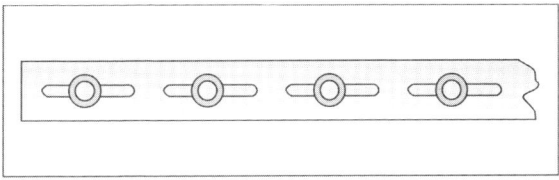

Fig. 12.9: Small length of formed strip

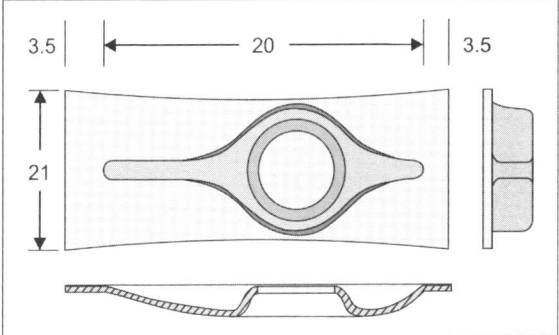

Fig. 12.10: Sheared piece of formed strip

with forming would be 27 mm. This is the pitch 'P'. Every time operator has to feed more strip by 27 mm before taking a forming stroke.

Since the length of block (4) is 52 mm, formed impression will be on plain surface of the block. If a stroke is given in this condition, impression will get crushed, so to avoid this, recess would have to be made as shown in Fig. 12.11. This recess would have to be machined on block (4) face before hardening.

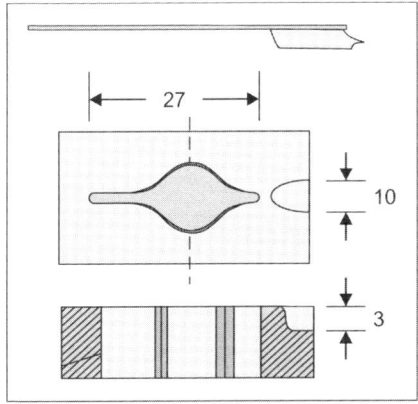

Fig. 12.11: Recess is shown

Now comes the question as to how operator would feed 27 mm every time. This difficulty is overcome by providing a limit stopper as shown in Fig. 12.12.

The operator would push the strip so much that formed impression end touches stopper 'S'. In this way, complete strip can be formed.

The above explained forming cum stamping system is highly undesirable as it is slow and extra man hours are required to cut formed pieces out of strip.

An improved system is briefly described below:

Die block (4) has to be of modified design, which is shown in Fig. 12.13.

• In modified system, cavity (die cavity) is to be machined only 2.8 mm deep, about 0.3 mm deeper than the depth of form which is 2.5 mm. Cavity can be machined by EDM.

• A shearing blade (6) of high carbon high chromium steel is to be fitted as shown. Thickness 'S' of blade may be 4.5 mm and distance T appropriate. The object of fitting a shearing blade is to replace when cutting edge becomes blunt.

• In this arrangement, pitch 'P' would be more than 27 mm. Hence, percentage of scrap in trimming operation would

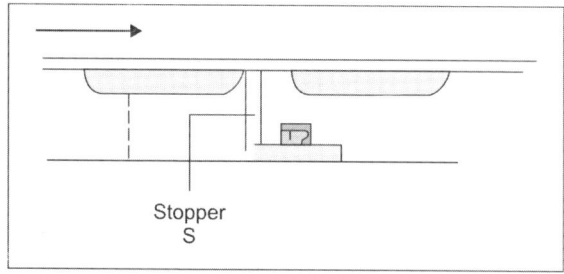

Stopper
S

Fig. 12.12: Limit stopper

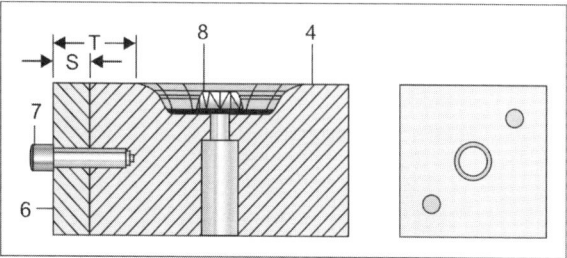

Fig. 12.13: Modified design of die block

increase. Although undesirable, but in the interest of combining shearing with forming, this arrangement may be done.

- Alternatively, instead of adding a shearing blade, edge of vertical face of block (4) may be used as cutting edge.

  The great disadvantage with this arrangement is that every time, face of block (4) would have to be ground to achieve sharpness in cutting edge. Consequently, depth of die cavity will have to be maintained once again.

- Working surface of block (4) becomes uneven due to stamping action. It may be necessary to maintain surface say, after fifty thousand to one lakh strokes while re-sharpening of shearing blade may become necessary say, after only thirty thousand strokes.

- Keeping all the above factors in mind, design may be adopted.

- A suitably designed place for top shearing blade would have to be provided to match lower shearing blade with shearing clearance of 0.02 mm for 0.26 mm, half hard brass strip.

Trimming

A piece of strip with formed profile is shown in Fig. 12.14.

It has to be trimmed (which is second operation) to get component as shown in Fig. 12.15.

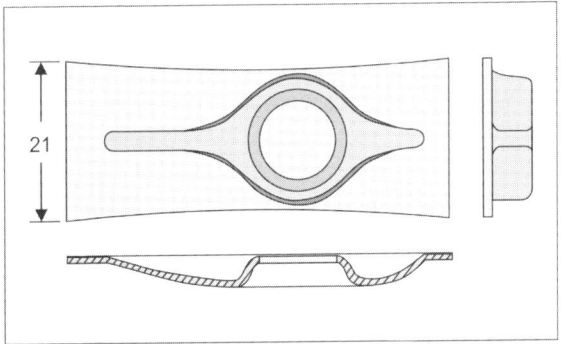

Fig. 12.14: Piece of strip with formed profile

The basic design of trimming tool is briefly explained with the help of Fig. 12.16.

List of parts is as follows:

1. Upper plate of die set   6. Nest cum ejector
2. Lower plate of die set   7. Ejector spring

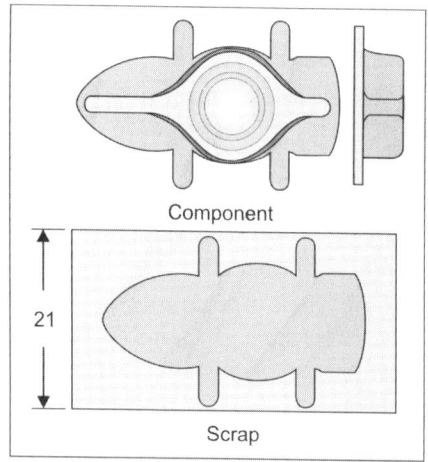

Fig. 12.15: Trimmed component and scrap

3. Punch holder plate
4. Trimming die disc
5. Trimming punch

8. Dowel pin
9. Socket cap screws
10. **Stem**

The basic idea of assembly procedure is already given while describing forming and stamping tool, so assembly procedure for this trimming tool is not described. It is hoped that designer and

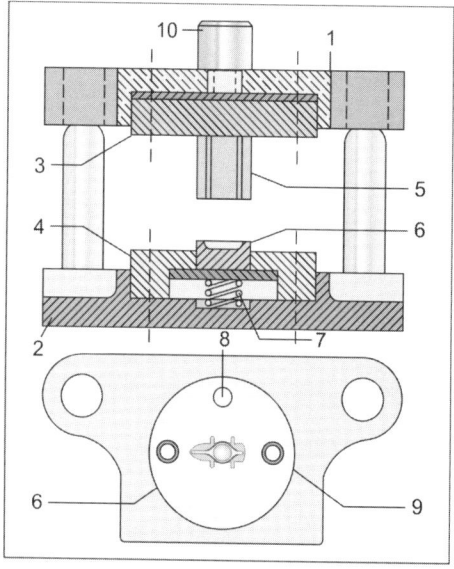

Fig. 12.16: Basic design of trimming tool

die maker may decide machining and assembly procedure for themselves. Hence, making of die ring (4) and nest (6) is described.

Trimming punch (5) may be made by one of the following methods:

- Conventional as was used for making form cum stamping punch
- Machining by punch shaper and final finish manually
- EDM
- Wire cut EDM

The final shape of trimming punch required after hardening, tempering and polishing is shown in Fig. 12.17.

For carrying out electric discharge machining, punch must be semi-finished by a punch shaper as per marking already done on the face of punch. A finishing allowance of 0.5 mm all around may be left to be finished by spark erosion. On one face of punch, a M4 tapping should be done for holding on a jig for spark erosion machining. For EDM, a female electrode of copper is to be made with 0.2 mm machining allowance. High carbon high chromium (HcHcr) punch may now be hardened and tempered, ready for electric discharge machining.

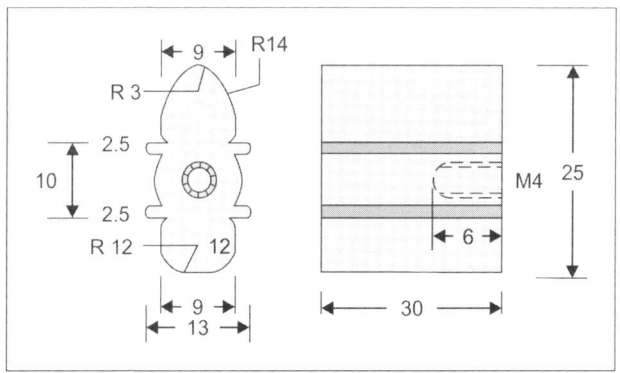

Fig. 12.17: Final shape of trimming punch

The basic set up for EDM is shown in Fig. 12.18, where part (2) is a electrode plate and (1) is a holding jig. On upper side, there is a stem to hold the jig in machine head. Part (3) is a hardened punch mounted on a cylindrical column, diameter about 7.5 mm.

Electrode plate (2) will keep on spark eroding the punch till full length is completed. If finish cut is given then there would be hardly any need to polish the surface of punch.

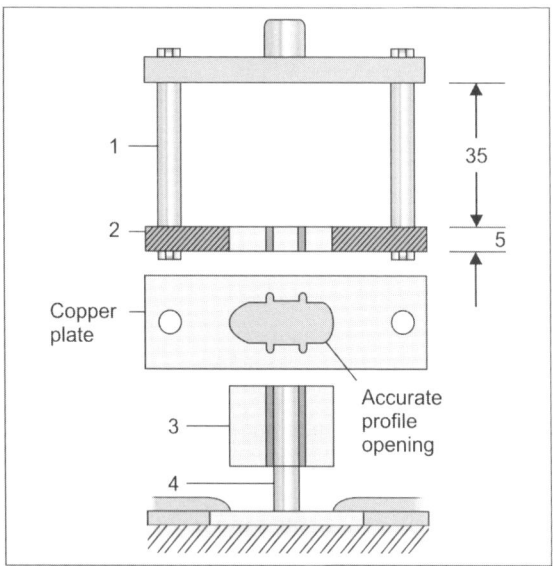

Fig. 12.18

Die Making Procedure

First, complete the fitting of punch in punch holding plate. Punch holding plate (round disc) may now be fitted in upper plate (1) of die set. Now fit trimming die round disc in plate (2) of die set. Fitting means screw and dowel fitting. Close the die set so that punch face sits on plain surface of disc (4). Do marking of contour of punch. Take out the disc and remove as much material as possible taking care that about 0.5 mm material still remains to reach marking. Fit the die disc again and load the tool on heavy hand press. Give few hard strokes till a little deep (say, 0.2) impression of punch is formed in die disc. Take out the disc and file out shaved steel which is pressed inside. By repeating the process few times, a skilled die maker may finish the die opening nicely so that punch travels inside the die without excess or less clearance. Trimming of two or three 0.26 mm thick brass sheet may be tried out. Once satisfied, die disc may be hardened. It is very important that trimming should take place in such a way that position relationship of form profile and trimmed profile is correctly achieved during regular production.

Position accuracy can be achieved by using a nest (6) which moves up and down in die opening accurately. The shape and size of nest is shown in Fig. 12.19.

Form profile of component may be machined on top surface of nest cum ejector in such a way that component sits completely

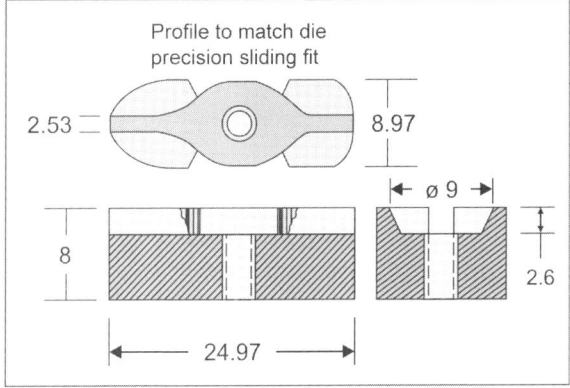

Fig. 12.19: Shape and size of nest

inside profile hollow without shaking and plain surface of component sits properly on surface of nest.

After final assembly of tool, nest must be protruding over the die face by 3–3.50 mm. Another point worth noting is that straight trimming portion in die should not be more than 4 mm. Rest of the thickness of die should be tapered out to make clearance. This point is illustrated with the help of Fig. 12.20.

Once the tool is finally assembled, formed pieces of strip may be placed over the nest and then trimming stroke of punch is given.

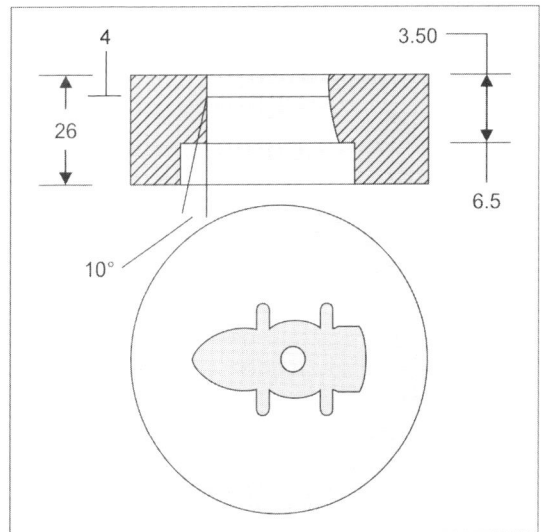

Fig. 12.20: Straight cutting length in die

Down position of press RAM should be so set that punch does not enter die more than 3–3.5 mm.

When punch is taken up, spring loaded nest pushes the component up free from trimming die set for manual or air jet removal of trimmed component.

Scrap portion of strip goes up with punch. After moving some distance up, an ejector blade stops movement of scrap while punch still moves upwards. Operator then puts another piece and process is repeated. As production process is manual, practically seven to eight thousand components could be produced per eight hours shift.

COMPLICATED PIERCING TOOL

A simple piercing tool is shown in Fig. 12.21. It may be an open tool with total height to suit a light hand press such as an eccentric operated with vertical wheel. (see Fig. 12.4).

Gap 'G' between RAM face and tool placing surface would depend on the size of press. Let minimum value of 'G' is 60 mm and maximum be 85 mm, so the shut height (A + B) of piercing tool may be 65 mm.

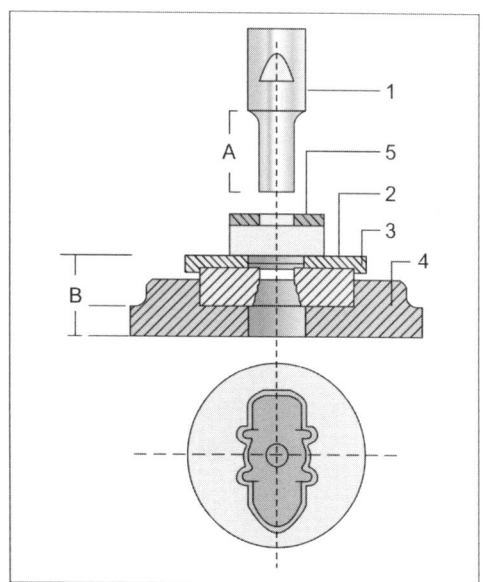

Fig. 12.21: Simple piercing tool

Referring to Fig. 12.21, part (1) is hardened and ground piercing punch of $3.99 + 0.0, - 0.01$ diameter. Part (2) is a nest cap which has an exact trim profile with minimum clearance so that trimmed component sits conveniently on face of piercing die (3). Nest can be suitably fixed to die. After piercing stroke, slug is pushed down the die and component comes up with punch. A suitable stripper (5) may be fitted on die block (4). Once again, feeding and removal of component is manual, hence time consuming.

By now, it must have been realized that the whole process with individual operation is not viable for economical production of **large quantities**.

For this reason, a tool has to be designed and made so that all three operations viz. forming, trimming and piercing may be combined. Tool has to be designed to run on a low tonnage power press, say a 25 tons power press. Moreover, there should be arrangement for automatic feeding of strip in roll form. There may be an arrangement in tool as a last operation to shear the piece of scrap on each stroke so that removal of scrap is convenient. This type of tool is called a 'Follow on' tool. While designing it, the following design aspects may be kept in mind:

- What should be the sequence of operation?
- Should the raised form impression be towards down or upside of tool?
- How strip may be indexed for operations one after the other?
- In what way indexed operation would be at accurate place on the next station of operation?
- What would be the guiding system for strip roll while passing through tool?
- What would be the methodology to eject component at last operation?
- How would die and punch combination for each operation station is assembled in die set?
- Ease of maintenance of tool may also be taken into consideration while designing the tool.

After doing deep thinking, rough sketches may be drawn on paper for analyzing the ideas once again.

It is almost obvious that forming cum stamping should be the first operation because then 'form' would be the guiding features for piercing and trimming operations. Piercing cannot be the first operation because pierced hole would get distorted if forming is the next operation. Similarly, trimming cannot be the first operation because trimmed features would badly get distorted on forming operation. Now it is decided that the first operation would have to be formed feature. Further comments are presented here with the help of sketches shown in Fig. 12.22.

Unavoidable necessity is that in third operation of trimming, sunken form side of strip should be towards face of trimming punch. How can it be the other way as shown in option 1? Form feature of component would crush, hence no question of trimming.

Hence, the first station has forming features. Forming punch would be on lower side of tool and forming die on top. Second station would be piercing station where piercing punch would be on top and die on lower side of tool so that slug (small disc) of sheet cut by piercing punch may fell down. Here it is necessary that round depression form is located concentric to piercing punch before actual piercing take place. This arrangement is shown in Fig. 12.23.

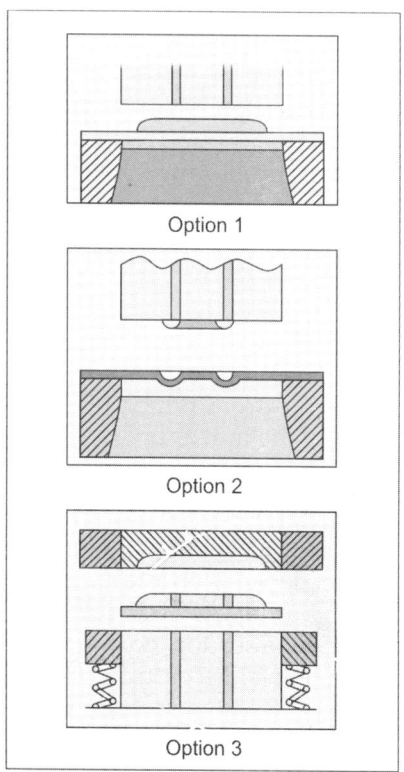

Option 1

Option 2

Option 3

Fig. 12.22: Trimming method options

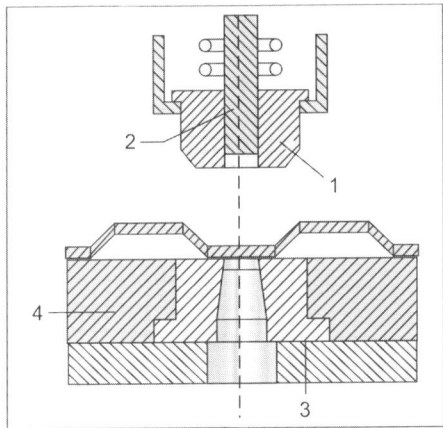

Fig. 12.23: Guide sleeve arrangement

Referring to Fig. 12.23, a guide sleeve (1) accurately slides around piercing punch (2). End of sleeve is matched with the round formation (depression) on strip which has taken place at first station of forming cum stamping. When upper portion of tool moves down, centering sleeve (1) first aligns the form on proper location and then piercing punch advances and after piercing returns back to up position. Hence, the strip is now free to be indexed forward for the next operation which is trimming.

It is already mentioned that trimming punch has to be on the lower portion of tool and die on upper portion. See option 3 in Fig. 12.22. Nest in the die slides accurately and is advanced by about 4 mm. This means that formed features would be aligned first before trimming takes place. Stripper plate around trimming punch would strip scrap from punch and component would be ejected when trimming die moves up for some distance. An air jet would throw the component in a collection chute.

Figure 12.24 shows typical construction of completed tool. It is also briefly explained.

Selection of a die set depends on total length and width of station blocks. In Fig. 12.24, four blocks of 27 mm long and 40 mm wide are shown in plan view after top portion of die set is removed. 21 mm long space on both sides of 'train' of blocks is provided for any possible additions of any part. Hence, total length comes to 150 mm and the length of die set plates (upper and lower) should be around 150 mm. A standard die set may be chosen which has a value of L around 150 mm.

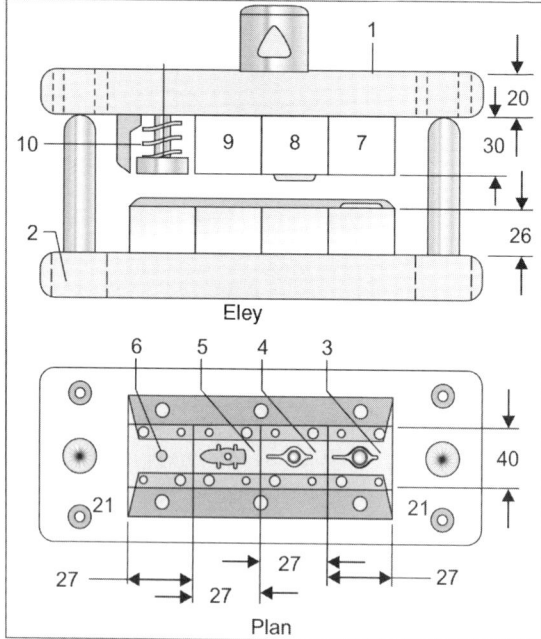

Fig. 12.24: Arrangement of blocks

There are following four blocks (stations):
- Part (3) is a block for forming station having punch fitted to it.
- Part (4) is a block for piercing of hole. Piercing die is fitted to this block, having passage for dropping of slug and to be ejected out of block by means of an air jet.
- Part (5) is a block for trimming. Trimming punch is fitted to this block.
- Block (6) is meant for chopping off scrap strip.

All the four blocks are precisely housed in a shallow slot in lower plate (2) of die set. It is for better accuracy in assembly of blocks on die set plate (2). Since the center of formation (circular depression) of piercing and trimming are accurately positioned in each block, therefore, pitch of indexing of strip would have to be 27 mm. All the four blocks (3), (4), (5) and (6) are 40 mm wide and 26 mm in height. Hence, top faces of all the blocks would be in one plane.

Blocks (7), (8), (9) and (10) are fitted on face of plate (1) of die set. Here a slot in die set plate is not needed. All blocks would have to be fitted with proper matching with lower blocks. Matching method is already discussed.

Upper blocks (9) and (10) are little complicated and need a detailed discussion. Block (9) is being discussed here with the help of Fig. 12.25.

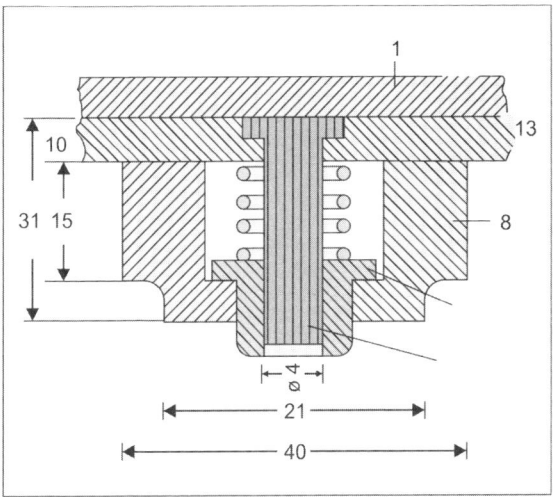

Fig. 12.25: Upper block arrangement

Block (8) would be 21 mm thick with an additional plate (13). Combination of block (8) and plate (13) would be assembled together to plate 1 of die set.

Figure 12.24 shows strip guide radius with a distance of 21.3 mm between them. Since the strip width would be 21 mm, therefore, it can move freely in a gap of 21.3 mm.

Since the faces of upper and lower blocks are to meet, width of blocks is reduced from 40 to 21 mm up to a depth of 6 mm so that it does not foul with guide rails.

Station 5/9 of trimming operation is somewhat complicated. Pushing component from trimming die by means of spring is not viable as component will remain stuck in strip. There should be some mechanical arrangement by which nest cum ejector is pushed down when upper portion of tool moves up till TDC of RAM. This will push the component out and removed by air jet.

A possible arrangement may be as shown in Fig. 12.26. The upper block (9) has a rectangular opening through and through. A cross bar (10) passes through rectangular opening. The position of opening is such that cross bar rests on the back of nest cum ejector. Height of rectangular opening is such that cross bar can move up at least 8 mm. When trimming strokes take place, pulling chains (14) loosen and trimming punch from below pushes nest cum ejector up, say by about 5 mm. On return stroke, the component remains stuck in die opening and cross bar raised by 5 mm. While reaching near the top end, chain (14) stops cross bar from moving

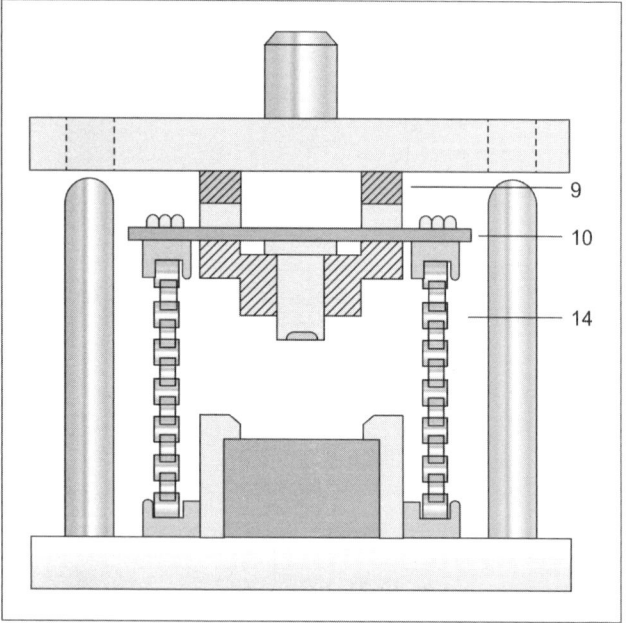

Fig. 12.26: Positive ejection system

a few millimeters. This causes ejection of component from die. At this very moment, an air jet blast throws the component towards collection chute.

COMPONENT 2

Operations

Component shown in Fig. 12.27 may be produced in three operations with tools for individual operations. The operations are:

• Blanking
• First bend
• Second bend

Blanking may be done on a number 2 screw hand press. Tool may be a two impression for the sake of reduction in percentage of scrap.

First of all, blank size has to be calculated as there is a curved end in the 'Tail' of component.

In triangle EOB (Refer Figs 12.28 A and B),

$$\frac{EO}{OB} = \cos (90° - \theta°) = \cos 65°$$

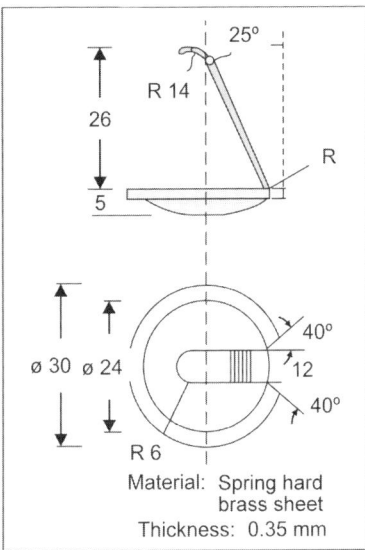

Fig. 12.27: Component 2

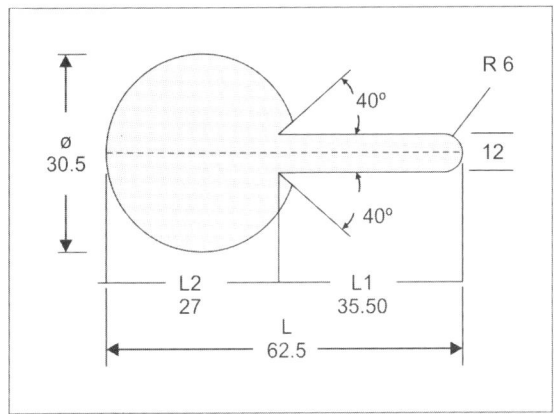

Fig. 12.28 A: Blank for component

$$EO = OB \cos 65°$$
$$= r. \, 0.4226$$
$$= 14 \times 0.4226$$
$$= 5.91 \text{ mm}$$

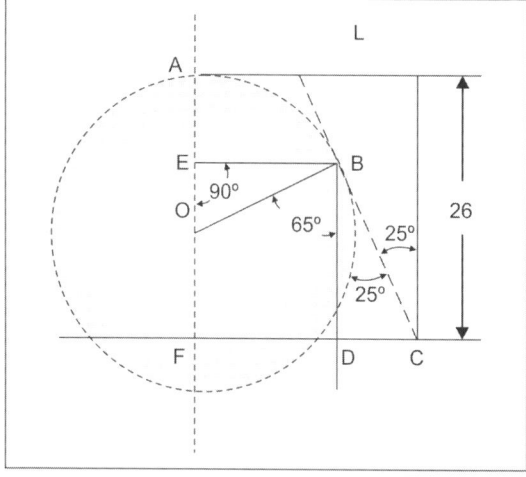

Fig. 12.28 B: Geometry for calculation

Therefore,   AE = AO – EO

= 14 – 5.91

= 8.04 mm

So,   BD = AF – AE

= 26 – 8.04

= 17.96 mm

Formula for length l or arc AB

= r. θ. 0.001745

= 14 × 65 × 0.1745

= 15.87

In triangle DBC,

$$\frac{DB}{CB} = \cos\theta$$

DB 17.96

Therefore, $CB = \dfrac{DB}{\cos\theta} = \dfrac{17.96}{\cos 25} = \dfrac{17.96}{0.9063}$

= 19.81 mm

Hence, total length of 'Tail'

$$= \text{Arc length} + \text{CB}$$
$$= 15.87 + 19.81$$
$$= 35.58 \text{ mm}$$

Shapes and Sizes of Left Over Scrap

Referring to Fig. 12.29, pitch 'p' should be kept so much that sheet tie 'T' is 3 mm. This much tie should also be there at the edges of strip. Since the diameter of round portion is 30.5 mm and tie 'T' as 3 mm, pitch would then be 33.5 mm. The width of strip may be kept as 68.58 (62.58 + 6).

Now question arises, how to fix two punches in a punch holding die? The fixing of punches with a pitch of 33.5 mm is not viable because then the thickness of steel between two dies would be 3 mm, which is too small a steel thickness from strength point of view. Hence, the solution may be to keep the two punches with two pitch distance between the center lines of two blanking die openings. It is shown in Figs 12.30 and 12.31.

Duel blanking tool may be designed to have the following features:

• It has to be made in a back pillar die set.
• It should have the facility of chopping off scrap strip.
• There should be suitable guide rails to guide strip through tool.
• There should be stripper plate to facilitate stripping of scrap strip from blanking punches.
• Calculate the force required to blank two pieces at a time.

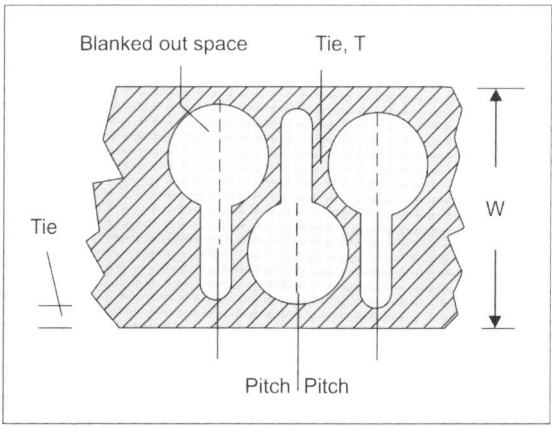

Fig. 12.29: Pitch of blanked portions in strip

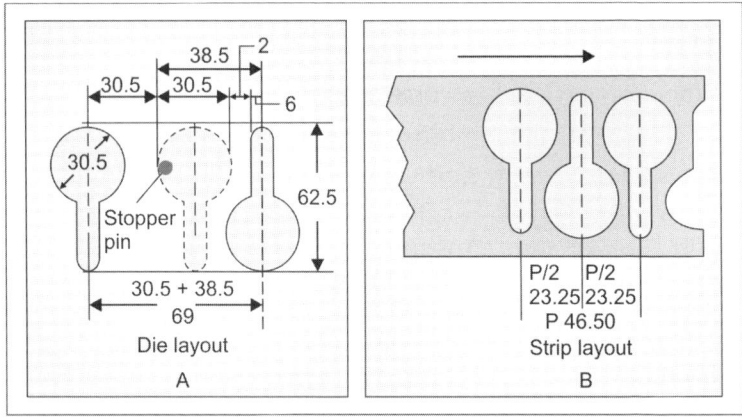

Fig. 12.30: Location of two punches and stopper pin

- Select a power press to suit tonnage requirement and space availability to load tool.
- Ejection arrangement for blanks ejection.
- Strip feed indexing system in synchronization with press RAM movement.

  Calculation of Blanking Force

  Referring to Fig. 12.28 a,

$$F = Area \times f_s$$

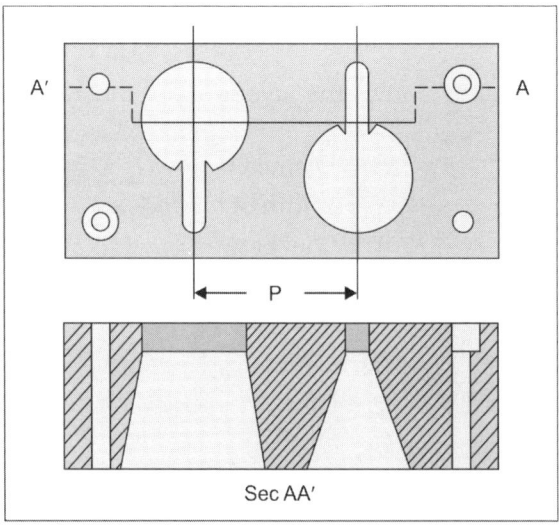

Fig. 12.31: Crs of punches

$$= \text{Length of cutting profile} \times \text{Thickness} \times f_s$$
$$= L_f \times t \times f_s$$

Where
$$F = \text{Force in tonn}$$
$$t = \text{Sheet thickness, cm}$$
$$f_s = \text{Shear strength of brass, kgf/cm}^2$$

Now
$$L_f = \pi D + 2L1 + \pi/2 \times 7$$
$$= \pi\, 3.05 + 2 \times 3.558 + \pi/2 \times 7$$
$$= 9.58 + 7.12 + 11$$
$$= 17.8 \text{ cm}$$

Therefore,
$$F = 17.8 \times 0.038 \times 2970$$
$$= 1887.1 \text{ kg}$$
$$= 1.887 \text{ ton}$$

Since there are two blanks to be cut simultaneously, therefore, blanking force required is 3.77 tons. Taking a factor of safety as 2, required tonnage may be assumed as 7.5 tons.

Now it has to be checked as to which tonnage power press is available nearest to 7.5 tons. What is the size of its bolster plate to make sure if die will fit in?

For example, a power press of 10 tons has a table of 500 mm × 400 mm. Stroke is 250 mm with open height of 350 mm.

So first of all, size of tool in die set has to be worked out. We have to take a start from Fig. 12.30. Let the layout of strip be as follows:

Figure 12.30 'B' shows the layout of strip where the following sizes are shown:

$$\emptyset = \text{Diameters of circular portion of blank}$$
$$W = \text{Width of 'tail'}$$
$$T = \text{'Tie' between two blanked out space}$$

Formula for calculating pitch, P
$$P = \emptyset + 2T + W \text{ ———————— (1)}$$

Referring to Fig. 12.32
$$2\emptyset + T + W/2 = Crs$$

Dimensions as per Fig. 12.28 (a) are,
$$\emptyset = 30.5 \text{ mm}$$
$$W = 12 \text{ mm}$$
$$T = 2 \text{ mm}$$

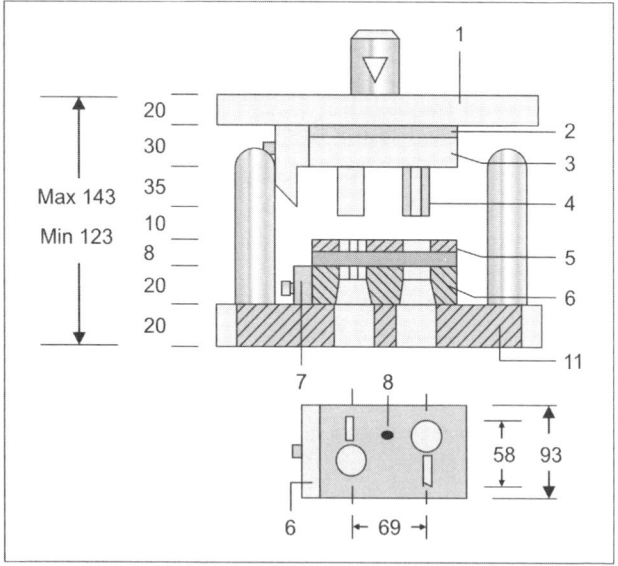

Fig. 12.32: Two impression blanking tool with height determining dimensions

As per equation (1)

$$P = 30.5 + 2 \times 2 + 12$$
$$= 46.5 \text{ mm}$$
$$Crs = 2 \times 30.5 + 2 + 6$$
$$= 69 \text{ mm}$$
$$S = 30.5 \text{ mm}$$

Following are the parts of two impressions blanking tool (Fig. 12.32) for components shown in Fig. 12.27:

- Upper plate, die set
- Punch holding block
- Stripper plate/block
- Chopping plate stopper pins
- Dowel pins
- Back plate for punches
- Punches
- Die block
- M8 socket cap screw

Distance between the center lines (Crs) of two punches and the location of stopper pin is already calculated which is shown in Fig. 12.32. First of all, make punches holder block (3) and die block (6). Make all holes, tapered hole as per drawing. Die holes profile and same profile in punches holder block have to be machined by wire cut electric discharge machining. For this, a starting hole has to be provided for both the profile holes. Before hardening die block, make sure that dowel pin holes are reamed for accurate size. Now harden and temper die block (6) and carry out wire cut operation in both the die profiles (openings). All dimensional details and Crs must be saved in PLC of wire cut EDM machine. This wire cut operation has to be repeated for punches holder block (3). Make sure that faces of blocks are precisely parallel and set on machine with zero deflection, but in this block, dimensions have to be less by 0.01–0.015 mm so that punches are fitted with light interference fit.

Pair of punches is also to be machined by wire cut from a block of suitable size for holding purpose. Now start making holes or any other holes for holding the blocks on machine. Once all the above preparations are done, block has to be hardened and tempered and then a pair of punches has to be cut out of block. Fine cutting operation should be done so that there is hardly any need for polishing the punches and a cutting clearance is achieved between already machined die block and punch. Cutting clearance should be 0.015 mm (4–5% of brass sheet thickness). Once the punches are machined, check their matching with die profile. If it appears that there is slight interference then about one fourth of punch length may be polished with fine diamond paste so that punch moves in and out smoothly in the die.

Figure 12.33 shows the set up for fixing punches into punches holder plate. Punches are to be fixed one by one.

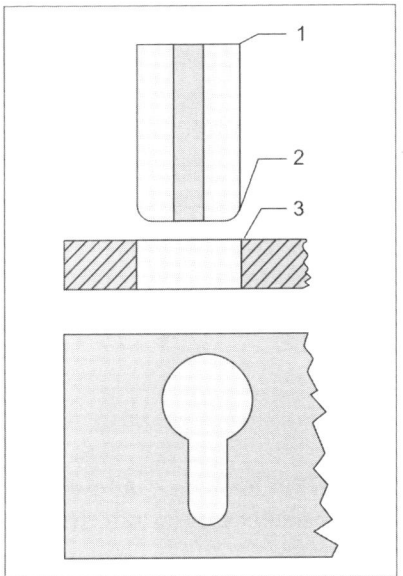

Fig. 12.33: Set up for fitting punches

End (1) of punch is cutting end and (2) is the end which is to be pressed inside the opening in punch holder block (3). Note that edges of end (2) of punch are rounded off so that it is pressed in block (3) without any shaving action. Generally, hand press is used to press the punch. Before pressing, silicon oil layer may be applied on surface of punch and hole. It is to facilitate fittings. Great care has to be taken to ensure that punch is in right angle position when pressing is started. While insertion is in progress, keep checking with the help of a small precision trisquare if punch is still in right angle position. It is very important for successful assembly of tool.

After having fitted two punches in punch hold plate, check if both the punches enter into die conveniently. Actually, it should happen if fitting of two punches is accurate. Once matching of fixed punches with dies is ensured, fit punch holder block (3) to die set plate (1) with hardened and tempered back plate (2) (having suitable holes for passing of screws and dowel pins). The location of holder block (3) on upper plate (1) of die set should be according to drawing.

Now it is the turn of fitting die block (6) to die set plate (11). It has to be done in such a way that after complete fittings, punches should enter the dies when upper plate of die set is slided towards lower plate (11). This is to be done manually in first place.

This condition may be achieved by following the procedure below:

- Place die set upright on table with punches block (3) fitted on die set plate (1).
- Raise top die set plate (1) up to make room for die block to be placed under punch faces.
- Slowly bring the die set plate (1) down till punches enter die by about 3 mm.
- Put some wooden stay in between two plates of die set so that punches do not go down further.
- Take marking of M8 socket cap screw holes in die block (6) on die set lower plate (11).
- It can be done by using a 6 mm diameter center punch about 35 mm long.
- Take out upper portion of die set, allowing free access to drill taping holes for M8 socket cap screws through marked center.
- Tap the holes.
- Bring the punches again inside the die with wooden stay.

- Tighten two socket cap screws lightly. Check the movements of punches inside the dies.
- Once satisfied, tighten screws fully and recheck matching of punches and dies.
- Now extend dowel pin holes from die block (6) to lower plate (11) of die set. Holes should be reamed without removing block (6).
- Fit dowel pins.
- Make two number of stripper plate as per drawing (Refer Fig. 12.34).
- Fit one stripper over each blanking die opening. Profile on stripper plate is made loose over the punch so that punch does not rub against stripper.
- Fit stopper pin in the hole (8) already made in die block (6).
- Fit scrap chopper plates (7) and (12) in such a way that there is a shearing clearance of 0.012 mm.
- Stripper plates have to rest on die block (6) and tighten on die set plate (11). Spacer of suitable thickness has to be placed between stripper leg and die set plate for proper tightening. Spacer is specially provided so that it could also be ground in case die face is ground for maintaining sharpness.
- A very accurate indexing system is needed so that on every stroke, strip is moved forward by 46.5 mm when strip is just released from punches by stripper while punches are moving up.

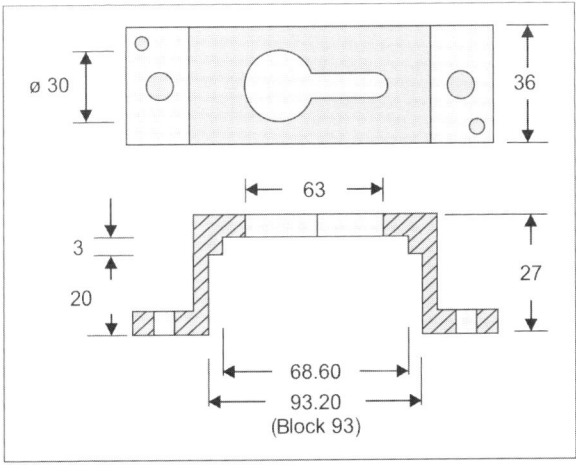

Fig. 12.34: Stripper plate

- On return stroke, punch faces would first push the strip down and already blanked space edge will slide down the stopper pin having round end.
- For first stroke, leading edge of strip roll has to be manually inserted to reach stopper pin.
- Then the blanking process may go automatic.
- It is worth noting that opening in die set plate (11) is already made for blanks to fall down in a container through opening in bolster plate and press bed.

Bending Cum Forming Operation

Blanks produced, have to pass through first bend cum forming operation which may be done on a No. 3 screw hand press or 10 tons power press. Shape and size of component required after this operation are shown in Fig. 12.35.

First bend tool of single impression may be made for a hand press or a power press. Feeding of blank and taking out of component may be manual with a forcep or doing a little automation. The basic tool design idea is given by means of Fig. 12.36.

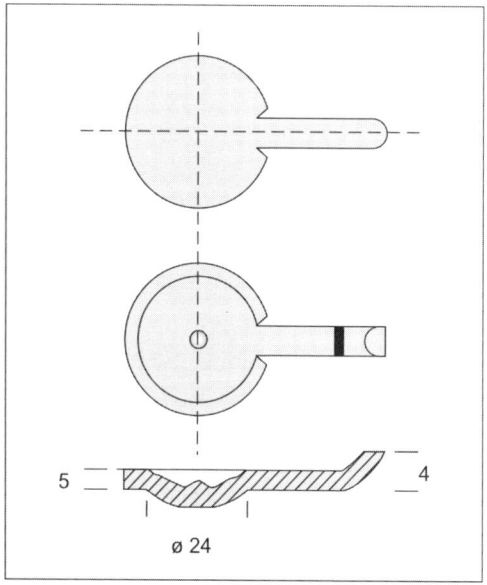

Fig. 12.35: First bend operation

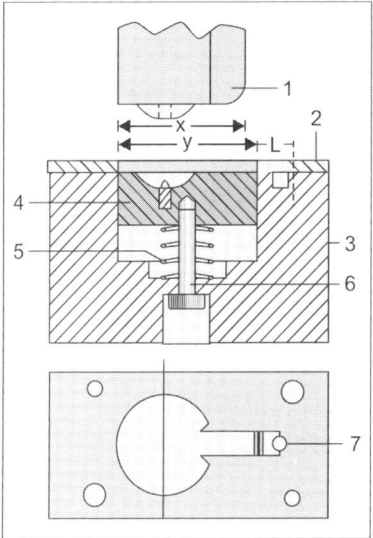

Fig. 12.36: Bending tool design

Detailed dimensioning may be done according to choice of press. In this case, an open tool (without die set) may also be made.

Referring to Fig. 12.36, part (1) is a forming punch which forms dome as well as tail bent with radius, as shown in component drawing. Part (2) is a guide plate in which blank is placed. Clearance between blank and guide plate should not be more than 0.1 mm. There is also a small blind hole to facilitate removal of already placed blank if needed. A needle may be used to lift blank. Part (3) is a die block in which forming die insert cum ejector (4) is housed. When stroke is given, firstly tail end is bent to form radius as in drawing. Dimension Y is bigger than X by the thickness of strip which is 0.38 mm in component Fig. 12.27. Further, down movement of punch pushes down the forming cum ejector insert (4) till it sits on block seat. Now again the forced movement of punch creates forming of dome.

When punch goes up, component formed tail projects over guide plate surface. Now component can be removed by tweezers.

Dimensions of blocks may be worked out according to fitting arrangement, 'daylight' and stroke of press.

Second bend Operation

The basic construction of tool for second bend operation is shown in Fig. 12.37. The objective is to provide concept of tool design. Details may now be worked out by tool designer.

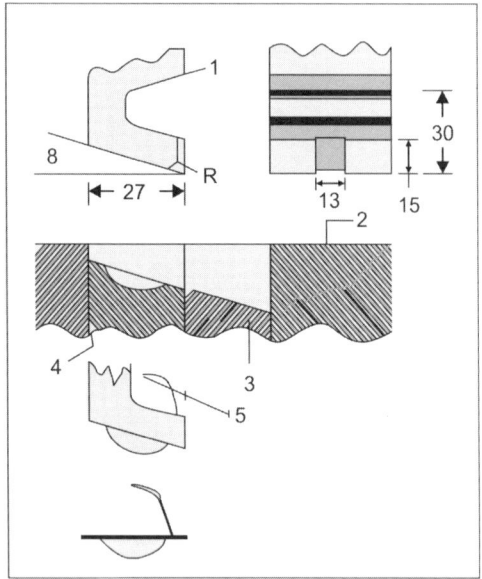

Fig. 12.37: Second bend tool design

The main features of this tool are as follows:

The Lower portion of tool consists of die block (2) suitably sized for a screw hand press or a low tonnage power press. Part (4) is a spring loaded sliding bend forming insert of shape and dimensions shown in Fig. 12.37. Part (3) is a stationary strip to affect 'tail' bending when component is moved down by the punch. Dimension and shape of punch is also shown in Fig. 12.37.

One of the important points of design is the value of angle θ. An angle of 25° has to be achieved in the component, so the question is what value should be given to angle θ so that 25° is achieved in component. In most of the cases, there is a 'spring back', due to which angle gets increased in the components. Therefore, smaller angle of bent has to be provided to compensate for 'spring back'.

Spring back allowance in angle may be calculated by the formulae available on Internet. It needs a detailed study. Write 'sheet metal bending spring back formula' on Google search engine, click search. There are a number of sites which may be opened and studied.

It is not always necessary that calculated value of θ would give correct bend in component as specified. This is due to the fact that value of hardness/tensile strength and grain direction assumed may not be exactly that in actual sheet.

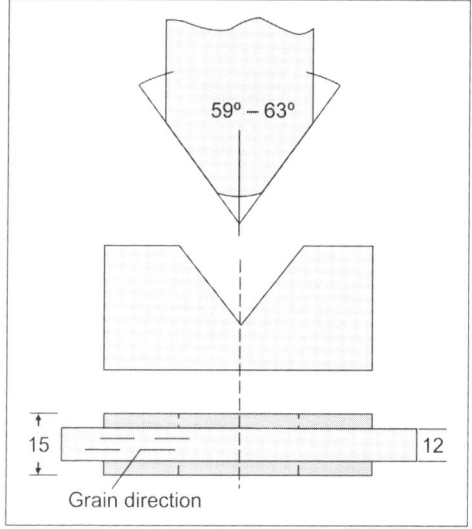

Fig. 12.38: Bending trial for spring back

It is very important to make sure that the value of θ in bending punch is so much that desirable results are achieved in first attempt of bending operation. It is because those modifications in tools may be costly and time consuming, so it is suggested to determine value of θ by practical trial by means of simple jig, something like the one shown in Fig. 12.38.

Correct angle may be determined after two or three trials.

COMPONENT 3

Component shown in Fig. 12.39 is a hinge used for a panel of a big size cover of a diesel engine.

Component may be manufactured by individual tools for each operation or a 'follow on' tool to be run on a power press with suitable automation.

Let us first discuss individual operation by individual tools. **Sequence of operation** may be as follows:

- Cut outs
- Pierce
- First bend
- Second bend
- Shear off

Before designing tool, it is necessary to calculate the

- Uncurled length of arm
- Force required to blank 'cut out'

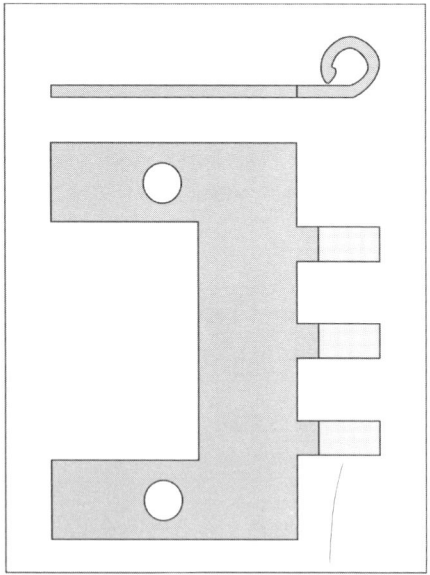

Fig. 12.39: Component 3

Calculation for Length of Uncurled Arm (Refer Fig. 12.40)
Straight Length = 3/4 of Circumference (Internal diameter of round)

$$H = 3/4 \times 3.14 \times 9$$
$$H = 21.2 \text{ mm}$$

Calculation for Force, F

$$F = \text{Total cutting length} \times \text{Thickness} \times f_s$$
$$= 38.52 \times 0.19 \times 3515$$
$$= 25725 \text{ kgf}$$
$$= 25.7 \text{ tons}$$

Taking a factor of safety as 1.5 then force required is equal to 38.5 tons. Hence, a standard power press of 40 tons force may be used. A standard 40 tons power press may have the following specifications:

- Maximum daylight ———————————— 340 mm
- Stroke ———————————— 60 mm
- Shut height ——————— - ————— 280 mm
- Bolster plate length ———————————— 800 mm
- Bolster plate width ——————————— 450 mm

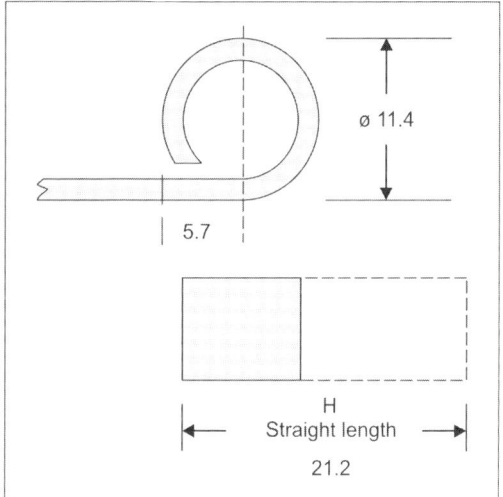

Fig. 12.40: Straight length for curled portion

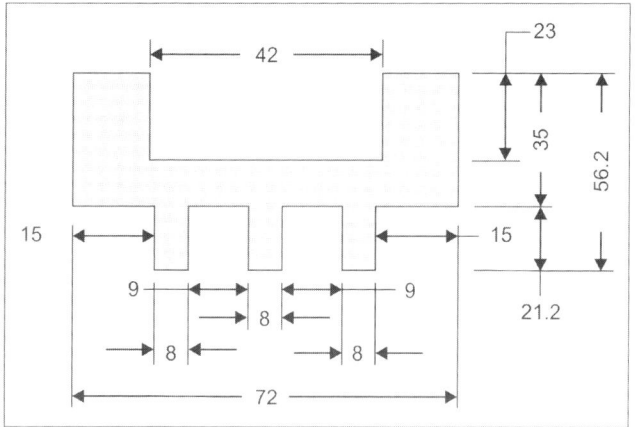

Fig. 12.41: Blank dimensions

Hence, tool's close height has to be kept as 290 plus by an adjustment allowance of approximately 30 mm. Depth of punch entry into the die may be adjusted during final setting. This adjustment is normally provided in RAM connecting rod assembly. This point is already discussed in power press chapter.

Blank dimensions would be as per Fig. 12.41.

With single impression tool, blanking may be performed by two methods. In first method, width of strip is bigger than punch width, say with a tie of 2.5 mm on both sides of the punch. In second method, all the blanks will be of same size, i.e. 56.2 ± 0.02 mm, but there would be loss of material in shape of scrap.

If dimensions 21.2 mm and 35 mm are not critical, that means a variation of plus or minus 0.2 mm is permissible, then width of strip may be kept as 56.2 mm. There would be a play of about 0.15 mm between strip width and guide plates, so this play would be the reason for dimensional variation as explained above. The advantage of this method is that there would be no loss of material in terms of ties.

In both the above methods, one punch and die would be duly matched and fitted in die set. The basic construction of such a die is shown in Fig. 12.42.

Tool consists of the following parts:

- Blanking punch (1)
- Blanking die (4)
- Punch holder plate (2)
- Strip guide cum stripper plate (5)
- Punch back plate (3)
- Stopper pin (6)

Drawings of all the above parts are provided vide Fig. 12.42. The object of providing drawing is that a better concept of construction of tool is attained.

Refering to Fig. 12.43, the length of punch is chosen as 62 mm. Reason for this selection is that a length of 24 mm will remain in punch holder plate and remaining 38 mm will be enough for its

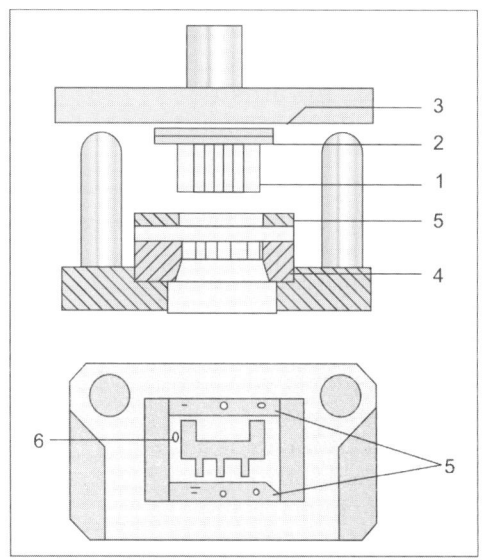

Fig. 12.42: Basic blanking tool construction

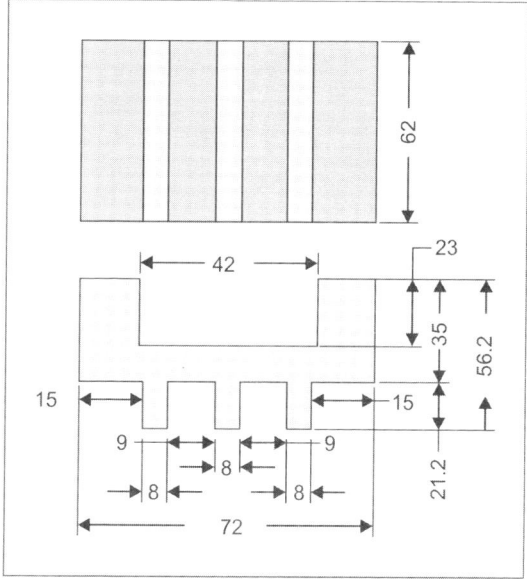

Fig. 12.43: Blanking punch

face to reach blanking die face after passing through guide cum stripper plate (5) in Fig. 12.42.

Punch may be machined by wire cut EDM as already described earlier. It may also be machined by milling. A question might be put, why punch holding plate (2) is to be 24 mm thick, and why not 8 mm thick. Well, theoretically it may be feasible, but experience in tool making points towards thick punch holding plate for sturdiness. A thumb rule is that punch holder plate thickness may be 1/4 to 1/3. of length of punch.

Blanking die profile may be machined by wire cut only if punch is also machined by wire cut (see Fig. 12.44). It is necessary because coordinates of points along periphery may be utilized in setting coordinates for die. In fact only cutting clearance is to be provided. This can easily be done by the use of CAD and CAM systems.

If punch is machined by milling then die has to be finished by taking impression of punch and filing out excess material. This method is already explained.

Stripper Plate cum Guide (5) is a part which has two functions to perform. One, to guide strip through the tool and the other to keep strip down while punch is moving up after blanking.

Guide plate and stripper plate may be separate parts, but for the ease of assembly, both the functions are combined. Figure 12.44 shows the drawing of stripper cum guide part (5).

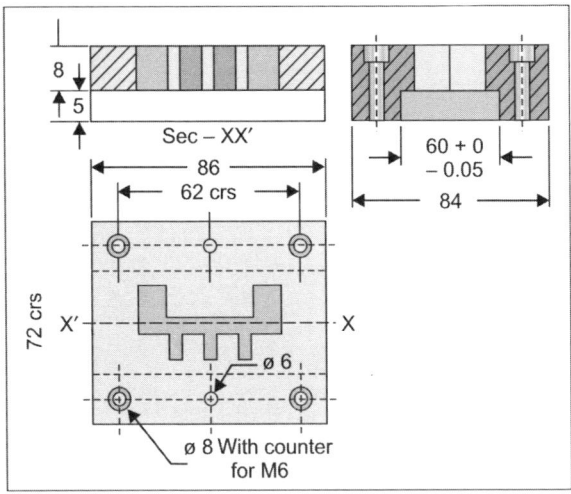

Fig. 12.44: Blanking die

Provision for M8 threaded holes and dowel pin holes should be made on die block (before hardening) for tightening stripper cum guide plate (5) on it.

Stopper Pin

The function of stopper pin is to place leading edge of strip for the start of operation, blanking. Afterwards edge of blanked strip is placed, touching stopper pin. This concept is illustrated in Fig. 12.45.

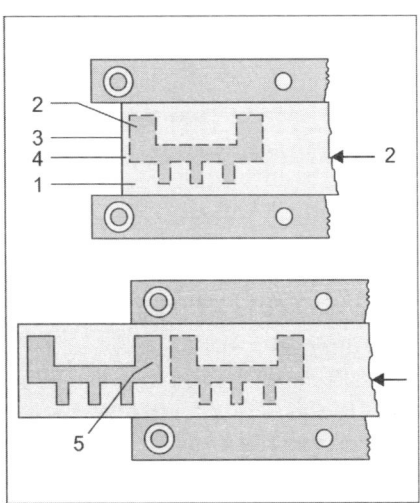

Fig. 12.45: Blanking tool with stopper

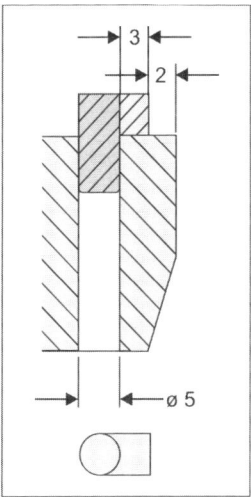

Fig. 12.46: Stopper pin design

Part (1) is a strip, leading edge of which is touching stopper pin (3). Profile (2) of blanking die is shown by dotted lines as there is strip over blanking die. Distance (4) between edge of die profile and stopper pin is 2 mm. Attention is drawn to the fact that if pin hole is made so near to cutting edge of die then it may weaken the material (steel) near cutting edge and there is a great likelihood of cracking of edge. Hence, stopper pin may be made and fitted as shown in Fig. 12.46.

From Fig. 12.46, it is clear that distance between cutting edge and pin surface is 5 mm instead of 2 mm, so now the chances of cracking of steel is remote.

Operation 1st Bend

The ultimate objective is to curl three 'legs' of component to a round shape suitable for putting a pin during assembly of hinge. This objective is achieved in three operations, viz. 1st and 2nd bends and finally curling.

The first bend is explained with the help of Fig. 12.47.

It is necessary to hold the component blank under pressure before end of 'legs' are bent with a radius. This requirement is achieved by placing blank (2) over pressure pad (3). It has three rectangular holes (7) which are about 3 mm wider than the width of legs which is about 8 mm. There are three radius forming punches (4) which are slightly less wider than rectangular holes in pressure plate. Hence, pressure plate can move up and down freely.

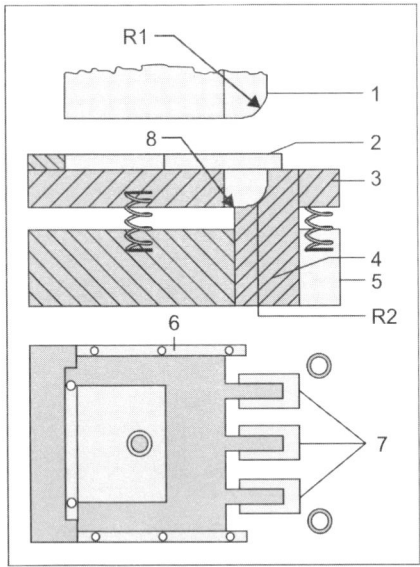

Fig. 12.47: First bend tool

Part (5) is the base plate in which forming punches (4) are fitted. Further, there are compression spring housed between pressure plate (3) and base plate (block) (5). Gap between (3) and (5) is maintained so much that surface (8) of forming punch (4) is in level with upper surface of pressure pad (3) when completely pressed down by forming punch (1). Forming radius R1 in punch (1) is to be made 2.6 mm and in forming lower punch (4) as 4.5 mm.

After 1st bending, component would look like the one shown in Fig. 12.48.

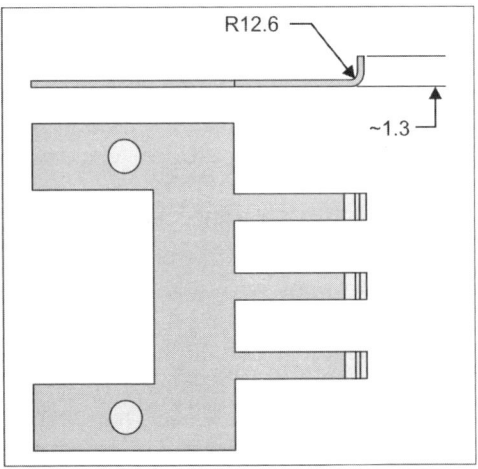

Fig. 12.48: Component after first bend

Operation 2nd Bend

The Objective of operation is to change the shape of component to the one shown in Fig. 12.49.

The Shape of component as shown in Fig. 12.49 may be achieved by a tool similar to one which is used for operation 1st bend.

The only difference would be in shape of the main punch (1) and lower punch (8) of Fig. 12.47. The shape of both the punches is shown in Fig. 12.50.

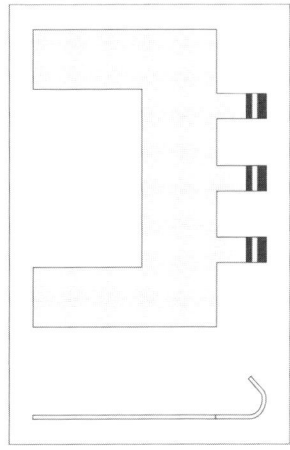

Fig. 12.49: Shape of component after second bend

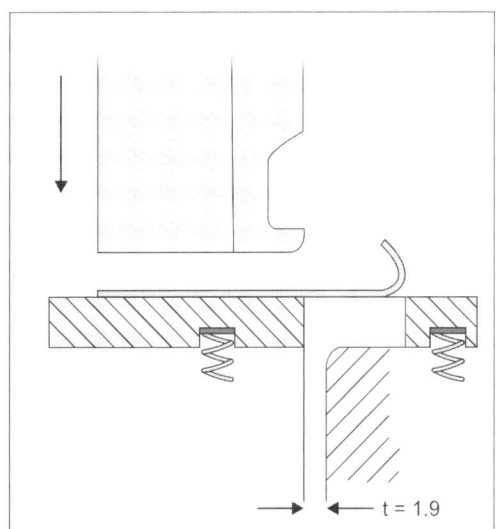

Fig. 12.50: Second bend punch shape

Operation Curling

In this operation, the shape of component as shown in Fig. 12.49 is changed to one shown in Fig. 12.51.

Arrangement required in the tool to get the shape of leg as shown in Fig. 12.51 is described below with the help of Fig. 12.52.

Curling punch (1) is fixed in punch holder plate block (4). This block also has four blind holes in which compression springs are housed. Part (5) is lower block of tool which has component

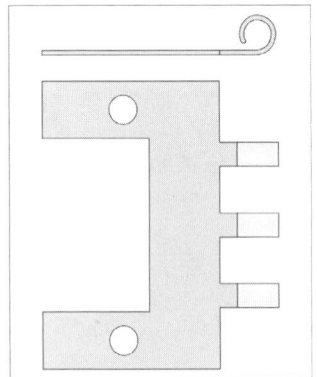

Fig. 12.51: Component after curling operation

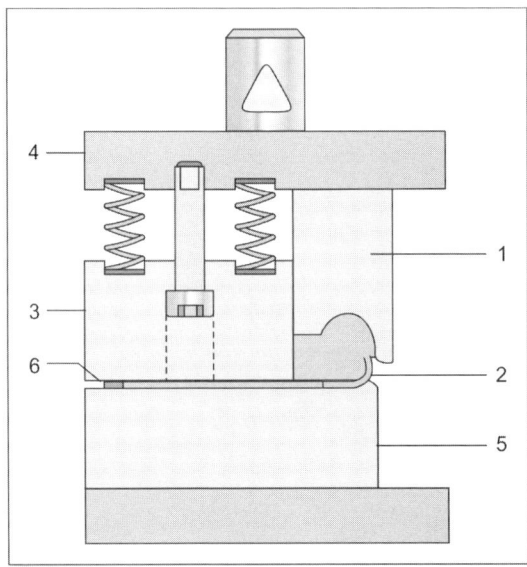

Fig. 12.52: Curling tool

positioning pin. Component (2) is shown placed over block (5) duly guided by guide pins (6). It is necessary that component should be pressed before curling starts taking place. For this reason, pressure pad (3) is attached with upper portion of tool.

Dimensions of various parts of the tool may be worked out.

The curling operation may even be carried out on screw hand press No. 4. In case tool is designed for a 10 tons power press, then placement and removal of component should be done by tweezers.

Progressive Tool for Component 3, Hinge

The design of this tool is somewhat complicated because progressive development needs component to be automatically

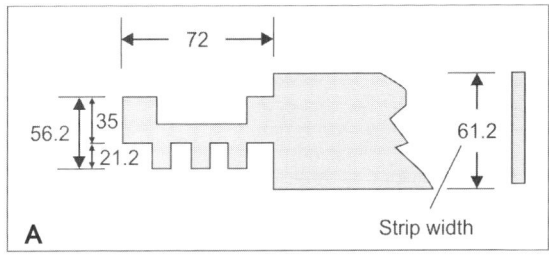

Fig.12.53 A: Component attached to strip

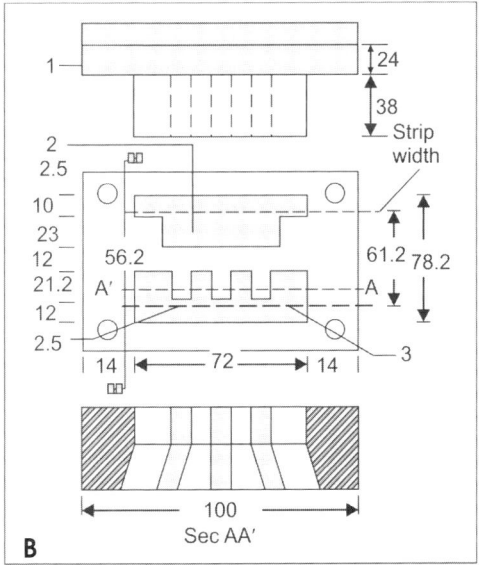

Fig. 12.53 B: Cut off nibbling punches and die

shifted from one station to other. To do so, component has to remain attached to the strip coil. This is shown in Fig. 12.53 (a). At last station, the component would be sheared off. For this purpose, there have to be two cut off punches. After cutting off by two punches, one on each side of strip forms shape of blank as shown in Fig. 12.53B.

The sequence of operation would be as follows:

- Cut off to form blank
- Piercing
- 1st bend
- 2nd bend
- Curling
- Shearing off

'Cut off' blank formation with the help of two cut offs/nibbling punches is shown in Fig. 12.53B.

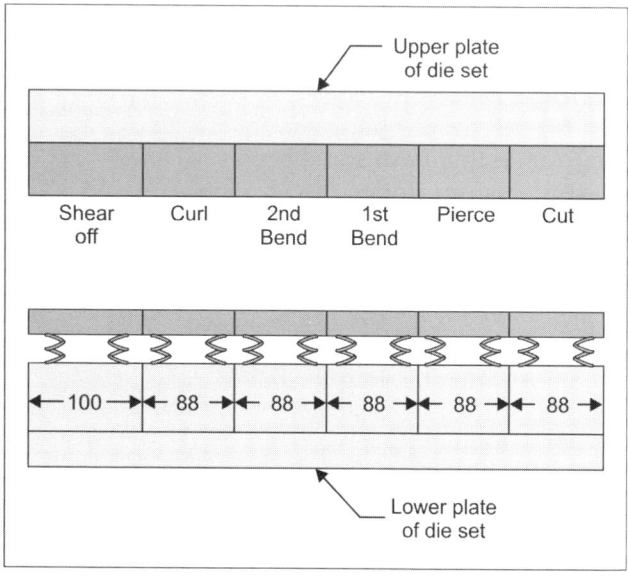

Fig. 12.54: Arrangement of stations for progressive tool

Selection of Die Set

Figure 12.54 is showing typical assembly of five stations for progressive tool. All the five stations are fitted in a rectangular die set of four pillars. Individual stations hardly need detailed description. However, the assembly of upper and lower blocks of

1st bend station is already shown in Fig. 12.48 to provide concept of functioning of station. Once the concept for this station is clear then other stations can also be built accordingly.

While designing and making this progressive tool, the following points may be taken care of:

- Precision machining and assembly should be done.
- Each station has upper and lower blocks.
- Lower blocks are fitted with spring loaded strip after cut offs guide plates. Level of all guide plates face strip slides should be precisely in one level.
- Strip entering edges of guide channels should be rounded off so that strip does not stick while indexed forward tool opens.
- Compression springs should be strong enough so that there are no chances of guide plate remain stuck with guide pins or forming inserts.
- Shearing off of component is combined with last curling station.
- Shearing inserts should be so made and assembled that it is possible to take them out for sharpening and re-fitting after placing spacer shim under the shearing cutter to maintain height.
- Upper and lower blocks should be fitted in die set in such a way that matching alignment of punch and die is achieved with precision.

To gain a better understanding of tool, it is recommended that videos on internet may be watched. Availability of video as on July 14, 2013 is as follows:

- Open 'You Tube' search engine on Internet.
- Write on search bar, 'sheet metal press tools'.
- Enter
- A page would open, having many websites with a small view box and very brief description.
- Website to be clicked is where hinge is visible in view box.
- A video box would open and a video of about 3 minutes shows the tool in action.

There are other websites also which may be viewed for general knowledge about functioning of various types of tools.

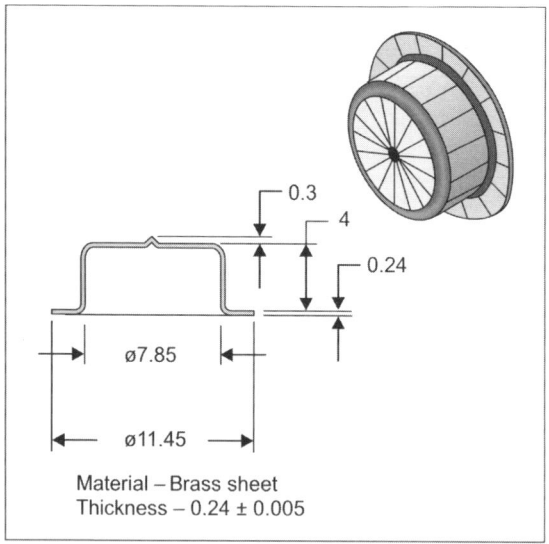

Material – Brass sheet
Thickness – 0.24 ± 0.005

Fig. 12.55: Fourth component

COMPONENT 4

This component is a sheet metal cup which may be used in production of dry cells. Figure 12.55 shows the cup.

This component may be produced either in small or **bulk quantities**. In either case, it has to be produced progressively in five stages which are as below:

• First dome draw
• Redraw of dome to increase height of dome and reduce diameter
• Straight wall draw to final height
• Sizing of bore
• Trimming (punching down)

In case, this component is produced by single impression tool then the width of strip will be enough to get a tie margin at trimming station. During draw, material would be easily drawn from the width of strip and partially stretched. For **bulk production**, a five impression progressive tool may be made. This means that five impressions component would be produced by each stroke of press. The speed of press may be as high as 120 per minute on a special high speed press.

The design of a typical tool of five impressions would be briefly explained with the help of Fig. 12.56.

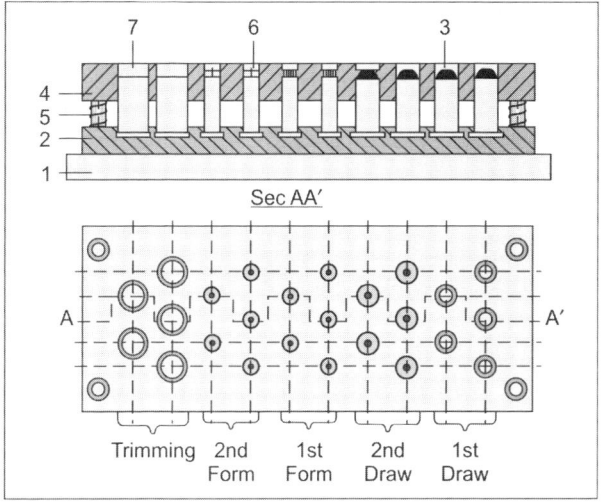

Fig. 12.56: Five impression tool layout

Figure 12.56 shows lower half of tool, i.e. lower portion. It consists of a base plate (1) of tough carbon steel of approximately 210 BHN hardness. Plate (2) holds all the punches and trimming dies over plate (1). Parts (3) are draw punches and (7) are trimming dies. Plate (2) is also a tough plate having hardness of about 210 BHN. Plate (4) is a **floating plate** which can move up and down freely. It is a tool steel plate of high carbon, high chromium steel. After all the machining and fitting operations are done, it is hardened and tempered to 59 HRC. Extreme care has to be taken that no bending of plate takes place during heat treatment process which obviously includes quenching. Finally this plate has to be given a fine surface grinding cut to make both faces of plate perfectly parallel and smooth. This floating plate (4) is held over a number of compression springs and held by suitable floating screws which hold the upper surface of plate about 3 mm above top of punches and trimming dies. Moreover, springs should be so housed that those are not crushed when floating plate is pressed down during draw operation. Down travel of plate would be about 9 mm. Hence, a gap of 15 mm may be maintained without pressing.

It is worth mentioning that total compression force of all the springs should be enough to create an under holding pressure on strip to avoid wrinkle formation during draw action when upper portion of tool comes down.

In Fig. 12.56, it is shown that there are five impressions group for each station, i.e. 1st dome, 2nd dome, forming, sizing and trimming. While designing the tool, pitches in longitudinal and transverse direction have to be worked out. Two important factors have to be taken into account, which are as follows:

- Pitches depending upon design of trimming die inserts, i.e. their diameter and placement.

- Space on sheet metal available between two close formations which is necessary for partial stretching while dome formation is taking place.

Figure 12.57 shows the scale drawing for determining the pitch 'P'. Pitch is the distance whose strip would index so that the first station five dome formations exactly lie over the second station set for second formation, and similarly for other stations. Please see the sectional view of trimming die inserts. Its wall thickness works out to be 3.7 mm.

Trimming die inserts would be located in lower half of tool. Trimming punch would be located in upper 'half' of tool. This arrangement is necessary for components (cups) to drop down by gravity.

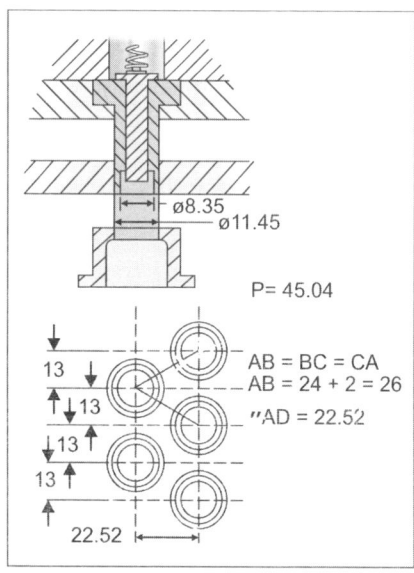

Fig. 12.57: Pitch determination

Figure 12.58 shows sectional view of trimming punch.

Punch is provided with a hollow space of 8.35 mm bore and 5 mm deep. Diameter of punch is to be kept as 11.45 mm which is trimmed diameter of component. It is to be noted that wall thickness of punch works out to be 1.5–5 mm in depth. There is no possibility to increase wall thickness, therefore, its heat treatment has to be done very carefully. High carbon high chromium steel may be used. A hardness of 60 HRC may be achieved in hardening. Part may then be tempered to bring the hardness down to 58 HRC. Tempering may be at 180°C for one and a half to two hours. The objective of this exercise is to avoid breakage of punch as well as have sharpness of cutting edge to last long.

Tool described above works very well when components are produced from brass. If components are to be produced from 0.24 steel strip then a modified design of tool is needed. Practically it was found that percentage of cracked component was high as compared to 0.3% in brass component. It was due to the fact that stretching in steel sheet was not good, so the modification needed is that there should be free material for forming of domes and further formation. This is achieved by introducing another starting operation of lancing.

Figure 12.59 shows **blank diameter** duly **lanced** in two circles, each circle having three unlanced small portion, say about 1.2 mm

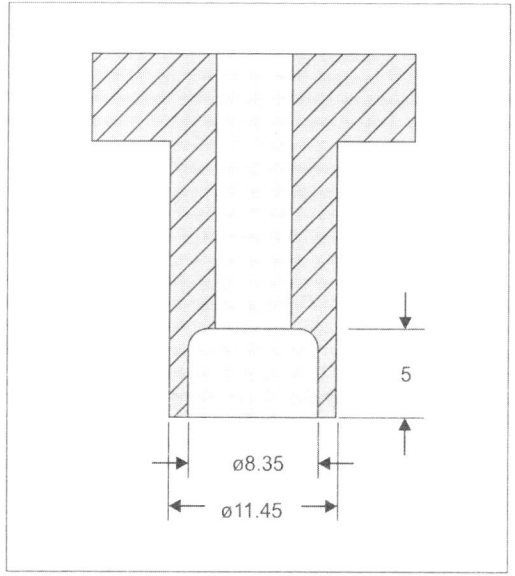

Fig. 12.58: Sectional view of trimming punch

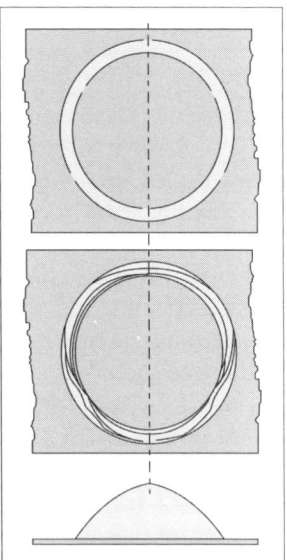

Fig. 12.59: Blank diameter duly lanced

wide. These portions are 120° apart. Another circle which is about 2.4 mm smaller than first circle would also have three small unlanced portions, 120° apart, but there should be an annular shift of 60° between unlanced points of two circles. At the time of drawing, blank sized disc would be drawn in with attaching ribs of 1.2 mm width.

COMPONENT 5

This component is a hexagonal calot of zinc for producing cans. The shape and size of a typical calot is shown in Fig. 12.60.

The designer of tool may decide to get 5 components in one stroke. Hence, there has to be 5 punches and dies. Layout of punches should be such that minimum possible left over (scrap) is generated. Please note that this left over or scrap is not rejection. It is inherent to blanking as per layout. Figure 12.61 shows typical layout of punches and dies which may be adopted by the designer.

Let 2R be the size of calot from corner to corner and 2r be the size of calot from side to side. For the calot shown in Fig. 12.60, then

$$R = 16 \text{ mm}$$
$$r = 0.866 \text{ R}$$
$$= 13.85 \text{ mm}$$

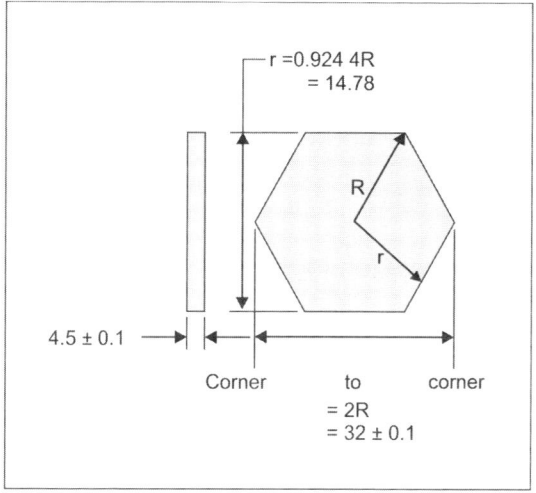

Fig. 12.60: Hexagonal calot

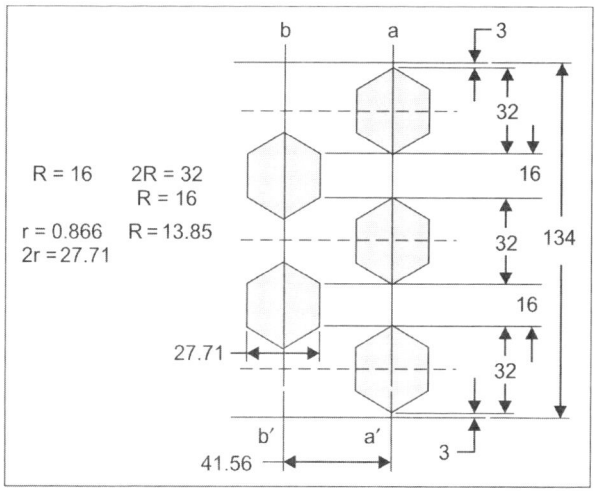

Fig. 12.61: Typical layout of punches

Formula for calculating length S of side of hexagon when corner to corner dimension 2R is given

$$S = R$$
$$= 16 \text{ mm}$$

Distance from flat to flat of calot is equal to 2r, which comes to 27.71 mm.

Punches are arranged in two rows. First row has three punches and second has two punches. aa' is central line of three punches and bb′ is the central line of two punches. Distance between the two central lines is kept as one and a half times flat to flat size of calot, which is 27.71 mm, so the distance between two central lines comes to 41.56 mm (27.71 + 13.85). Total distance between hexagon punch corner to corner of third punch works out to be 128 mm. There should be tie of 2–3 mm in the width of strip, hence, width of strip may be 134 (128 + 6) mm.

Determination of Press Tonnage

Formula

$$F = 1 \times t \times f_s$$
$$= N \times 6 \times S \times t \times f_s$$
$$= N \times 6 \times S \times t \times f_s$$

Where

F = Force required to blank, tons

S = Length of one side of hexagon, calot

T = Thickness of calot in cm

f = Shear strength

N = Number of punches

Let us consider two options

- 5 up tool
- 9 up tool

In 5 up tool, there would be two rows of punches, one of 3 punches and the other of 2 punches.

So, force required for 5 up tool

$$= 5 \times 6 \times 1.6 \times 4.5 \times f_s$$
$$= 30 \times 1.6 \times 0.45 \times 4218$$
$$= 91108 \text{ kgf}$$
$$= 91.108 \text{ tons}$$
$$= 91 \text{ tons}$$

Taking a safety margin, a 150 tons power may be selected.

Again force required for blanking 9 calots in one stroke

$$F = N \times 6 \times S \times t \times fs$$
$$= 9 \times 6 \times 1.6 \times 0.45 \times 4218$$
$$= 164 \text{ tons}$$

Taking a safety margin, a 200 tons capacity press may be selected. Bed sizes for power press may be as per Table 12.2:

Table 12.2: Bed sizes for power press

Power press capacity (tons)	L to R mm	F to B mm	Remark
150	1100	600	
200	1250	800	

For a 9 up calot blanking tool, bed size of 200 tons press is quite sufficient to accommodate tool. Approximate dimensions of tool would be 170 mm long x 250 mm wide and close height of tool would have to be about 20 mm more than shut height of press, which would be about 450 mm.

Typical shut height of tool will be as shown in Fig. 12.62.

1. Top plate to be fixed with RAM face
2. Punches holder plate
6. A space between two plates
3. **Stripper plate**
4. Blanking die
5. Base plate

Close height of tool is short by 210 mm (450 – 240). A bolster plate of 210 mm thickness may be used to compensate for shut height of power press. Press has to be fitted with roller indexing system which has to be fitted with actuator by an eccentric on press crankshaft. Eccentric has to be stepless adjustable so that accurate indexing could be achieved.

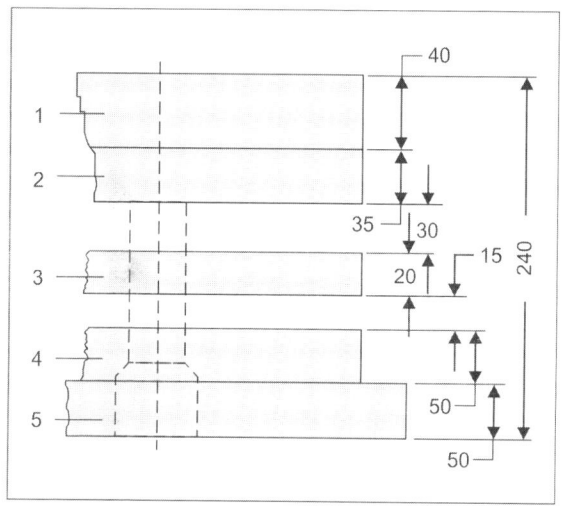

Fig. 12.62: Tool height dimensions

Generally, 4.5 mm thick sheets of zinc are in the form of strips and not roll. Consequently, operator has to handle the last portion of strip when it comes out of roll feed.

COMPONENT 6

This is a component similar to one which is fitted in front wheel of many models of motorbikes. It may be produced by either stainless steel or carbon steel. Let us consider it to be produced by stainless steel (Refer Fig. 12.63).

In case, strips of stainless steel are used then percentage of left over (scrap) will be high, so there may be a possibility of using round blanks which may be supplied by stainless steel sheet producers. If so, then it may be an economical proposition.

This component may be produced in the following operations:

- **Fine blanking cum pierce**
- Bending/forming
- Holes piercing

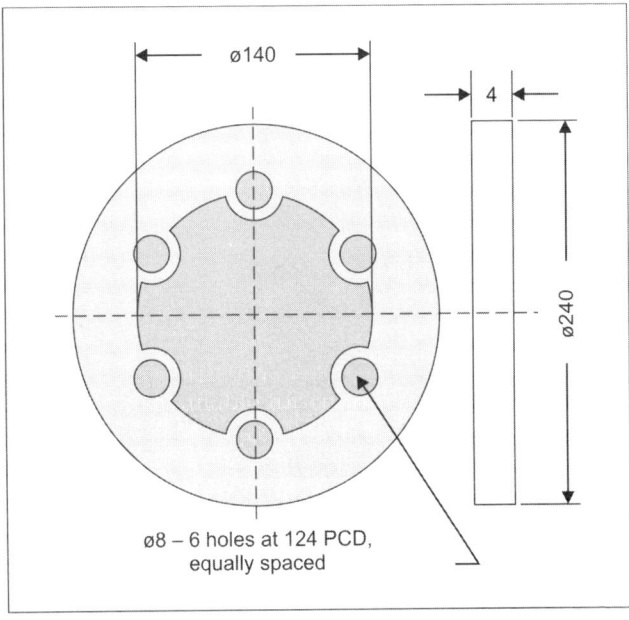

ø8 – 6 holes at 124 PCD, equally spaced

Fig. 12.63: Sixth component

Blanking cum Piercing

Tool design for blanking cum piercing may be as per Fig. 12.64.

Blank cum pierce is a compound tool in which outer circle of component is blanked and inner feature is pierced. Part (1) is a blanking die in which a counter is machined which is concentric to blanking bore. Piercing punch (2) has a collar which is machined while diameter of punch is machined. This precaution ensures concentricity of piercing punch and blanking die. Obviously faces are also machined while diameters are machined. It is also a point to ensure concentricity. Part (3) is a blanking punch cum piercing die. Feature of blanking die has to match with piercing punch and at the same time, diameters should also match precisely. Since sheet thickness is 4.5 mm, therefore, a cutting clearance of 0.5% may be provided, which comes to 0.225 on diameter.

Part (4) is a pressure pad which keeps the sheet pressed while blanking is taking place. Just after completion of blanking, piercing of an internal feature of component takes place.

The operation of this tool can be done on a hydraulic press with auxiliary jack to provide sufficient force through pressure pins (5) on pressure pad (4) which also works as ejector to remove component from piercing punch and blanking die.

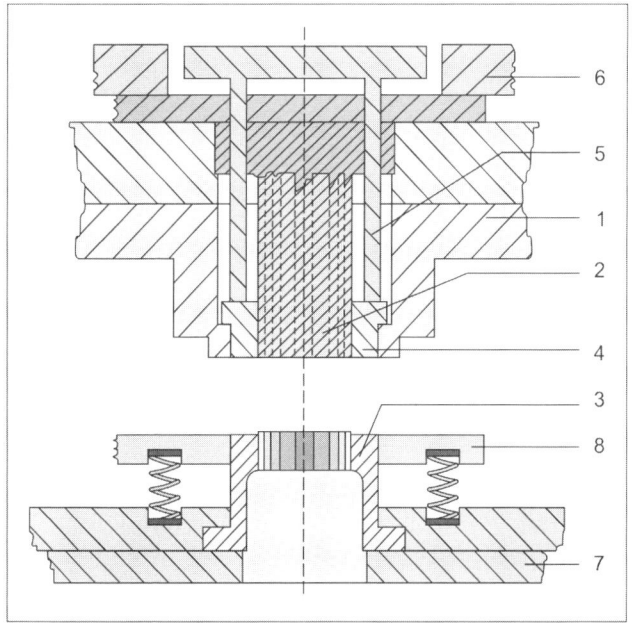

Fig. 12.64: Blanking cum piercing tool

Since the operation is heavy, tool must be constructed strong. For example, top plate (6) and lower plate (7) may be of steel having a Brinell hardness of 220 BHN. Thickness of plates may be 60 mm or more.

The function of part (8) is to provide holding pressure on sheet by face of blanking die (1) and to push out strip from blanking punch (3). Other parts may be suitably dimensioned. A 4 pillars die set may be used to assemble the tool.

Forming and piercing tool may be a compound tool having two stations. One station for stamping/forming and the other for piercing of six holes.

The operator has to feed the blank in first station of tool and then stroke will be given for pressing six internal features of component. Tooling is such that only 2 mm pressing would take place. When tool opens, operator transfers the component from 1st station to 2nd station and also places next component on 1st station. In this way, operations keep on taking place. This tool is also run on hydraulic press.

Required tonnage of power press may be calculated by means of usual formula.

F = Total length of periphery × Thickness × Shear strength of material

$$F = L \times t \times f_s$$

In this case, total length of periphery 'L' works out to be

$$= \pi \times 24 \times \pi \times 14 + \text{Half of internal}$$
$$\text{periphery allowance for features}$$
$$= 22/7 \times 24 \times 22/7 \times 14 + 22$$
$$= 75.42 + 44 + 22$$
$$= 141.4 \text{ cm}$$
$$= 141.4 \times 0.4 \times 4921$$
$$= 278 \text{ tons}$$

Giving safety margin, a hydraulic press of 300 tons may be used.

COMPONENT 7

This component is a kitchen washbasin made from stainless steel sheet of 1.0 mm thickness. The shape and size of a typical component is shown in Fig. 12.65.

This component may be produced in the following operations:
• Cutting of blank on a guillotine (shearing machine)
• Drawing   • Stamping
• Trimming on guillotine   • Piercing   • Bending

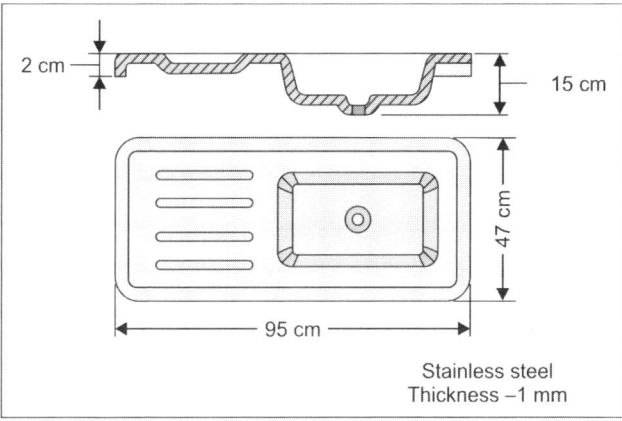

Fig. 12.65: Seventh component

Component has 1.5 cm wall on all the four sides. Corners are not drawn, but cut. Size of blank to be cut on shearing machine would be 110 × 55 cm. For drawing operation, tool size would be around 160 × 105 cm. The shut height of tool would be about 50 cm. To maintain shut height to suit press, a special support block (bolster block) may be used.

A hydraulic press of 200 tons is needed. Under holding pressure for drawing of sheet would be provided by auxiliary hydraulic jack of press. Bed size of a 400 tons hydraulic power press may be used.

Right to left ————— 1000 mm

Front to back ————— 1000 mm

COMPONENT 8

Figure 12.66 shows a container/shell drawn from aluminum sheet.

First of all, **blank diameter** would have to be **calculated**. Formula for a straight shell may be as follows:

$$\emptyset D^2 = \emptyset d^2 + 4 \emptyset dh$$

Calculation (See Figs 12.66 and 12.67)

$$D^2 = 22^2 + 4 \times 22 \times 17.5$$
$$D^2 = 2024$$
$$D = 44.98$$

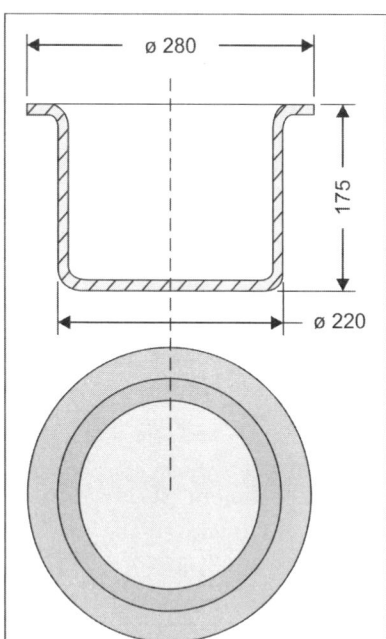

Fig. 12.66: Eighth component

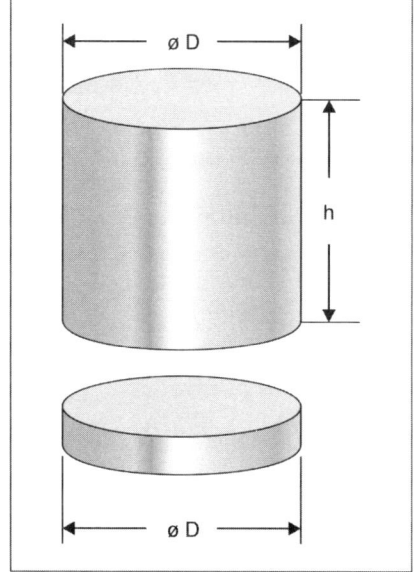

Fig. 12.67: Calculation for tonnage

Say 45 cm, add a margin of 6 cm for flange.

Hence, final blank size comes to 51 cm.

Force required to blank a disc of 51 cm diameter from 0.38 cm thick aluminum sheet may be calculated in the following manner:

$$F = \text{Circumference} \times \text{Thickness} \times f_s$$

where
$$F = \text{Force required to blank in tons}$$
$$C = \text{Circumference in cm}$$
$$F_s = \text{Shearing force, kgf/sq cm}$$
$$= 3.14 \times 51 \times 0.38 \times 2104$$
$$= 128035 \text{ kgs}$$
$$= 128 \text{ tons}$$

Force required to draw cup
$$F = n \times \pi \times d \times t \times f_{uts}$$

Where n = Drawing coefficient (0.7 to 0.95)
$$d = \text{Cup/shell diameter}$$
$$t = \text{Material/sheet thickness}$$
$$f_{uts} = \text{Ultimate tensile strength}$$
$$= 0.8 \times 3.14 \times 22 \times 0.38 \times 2249$$
$$= 47.472 \text{ tons}$$

A hydraulic press of 500 tons is recommended.

COMPONENT 9

This component may be produced by the following operations (Refer Fig. 12.68):
- Blanking
- Piercing
- Both bending

Blanking and piercing tool may be designed and made on the same principles as discussed earlier for other components. The main feature in this component is two bending. Both the bending may be done in one operation. 19 mm high wall is to be bent with 20° inclineation with vertical.

It is a unique situation due to two reasons:
- How an angle of 20° would be achieved?
- How component would be ejected from punch.

Straight withdrawal of bending punch is not possible due to tilted wall. The component has to be ejected from the side of the punch.

The arrangement in tool for achieving 20° bent is shown in Fig. 12.69.

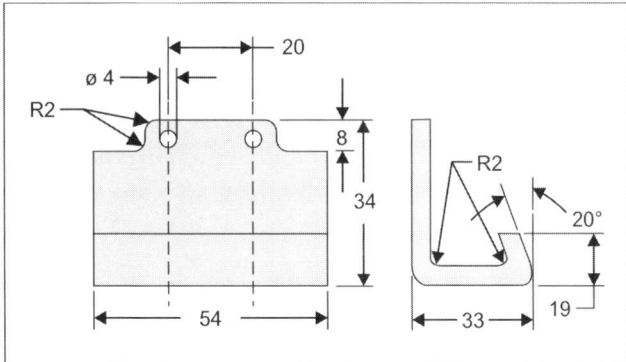

Fig. 12.68: Ninth component

The description of bending tool is as follows with reference to Figs 12.69 and 12.70.

Part (1) is the bending punch with a face width of 31 mm, length 54.2 mm, corner radii of 2 mm. One side of punch has a taper of 20°. Lower portion of tool consists of parts (2) and (3). These die inserts have radii at the edge to facilitate movements of blank inside the die. On the top of die, there is a guide plate (7) for keeping the

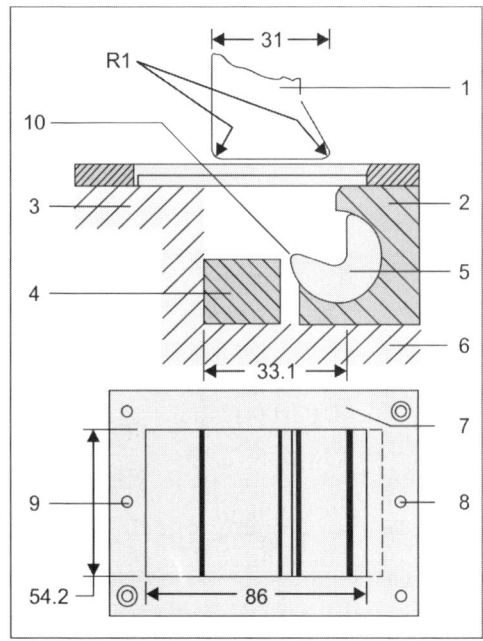

Fig. 12.69: Bending tool

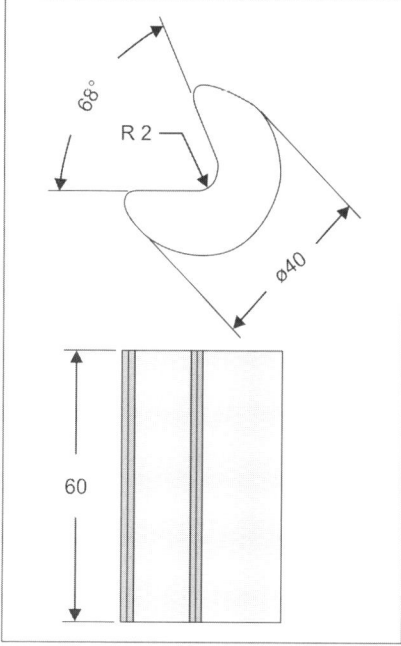

Fig. 12.70: Rotary former

blank in place. Dimensions of guide plate is such that blank is placed easily. There is a clearance of about 0.2 mm in width and length.

If blank is pushed down in a normal die by punch, bent component will have bends approximately at right angle, but it is desired that 19 mm high portion is bent inside by 20°. To achieve this, a unique arrangement of rotary former (bending device) (5) (Fig. 12.70) is provided. It has a 68° V shape cut up to its center. Rotary former (5) is housed in die block (2). Finish and fitting is such that it rotates precisely. There is an arrangement of a spring and stopper pin so that position of (5) remains as shown in Figure. When a punch goes down with blank, it presses point (10) of rotary former. The other point of former presses the 19 mm long portion to attain 20° from vertical. Part (4) is a limiting block, beyond which punch cannot move down.

It is worth noting that, center of former lies in line with top surface of stopper (4). The angle of former is made 2° less than an angle on component which is 70°. It is done to compensate for 'spring back after bend'.

When punch is withdrawn, it will come out of die with component. Component would have to be slided out of punch manually or some sort of automation.

COMPONENT 10

Brass Cup

The principle of circular cup drawing is already discussed. Formulae for determining blank diameter and force required for blanking and drawing is already given in connection with description of component 8.

The construction of a cut and cup tool is shown in Fig. 12.71. This tool consists of upper and lower portions. Both the portions are inter related. There are many dimensions related to cup diameter and height, such as Ø3, Ø4, Ø5 and Ø6. Values of some dimensions such as Ø1, T and d depend on the experience of tool designer.

The importance of various **dimensions** and how they are to be decided is briefly explained in the following paragraph.

X_1 and X_4 ———— should be approximately equal to 1.25 cup height.

X_2 ———— 3–8 mm

$H_u + H_l$ ———— It has to be maintained so much that shut height of tool should suit shut height of press and its stroke.

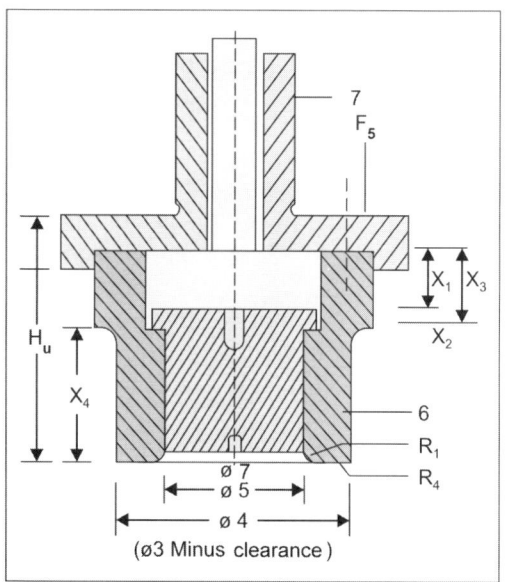

Fig. 12.71: Cut and cup tool

There is a small hole Ø7 in ejector (8). Hole is provided to put a small rubber bit to push the component from the face of ejector. In practice, it is found that sometimes component remain hanging on the face of ejector if surface of strip is lubricated.

Cutting punch cum draw die may be made of high carbon high chromium steel, hardened and tempered to 59 HRC. It should be so precisely assembled to stem (7) that its diameter and face F_5 may be checked on a lathe.

Drawing radius of draw die R_1 may be four times the sheet thickness. Cutting edge of hardened punch cum draw die should be sharp.

Coming to lower portion of tool, the following points may please be noted:

Ø3 ——— It should be equal to Ø4 plus cutting clearance, which is normally kept 10% of sheet thickness.

Ø1 and Ø2 ——— Machining of die block (1), draw punch (4) and cutting ring (2) should be so precisely done that both the diameters are concentric. Care should be taken that faces F_1 and F_2 are straight.

D ——— This size may be 1.5 times the length of cup.

d, j, T, ——— These dimensions may be decided by designer, depending on his experience.

Pressure pins (5) ——— The length of these pins may be so much that lower faces remain inside the die block (1) by about 1 mm when pressure pad (3) is on maximum top position, touching cutting ring (2).

Pressure pad (3) ——— The face of pressure pad should be about 1–3 mm over the cutting ring face F_3.

R2 ——— It is the draw radius of punch 3. It may be 3 to 4 times the thickness of sheet.

Cutting ring (2) ——— The fitting of cutting ring to die block may be from three side set screws or by a ring to be tightened from top. For this, the design of cutting ring would have to be modified.

Cut and cup tool may be an open tool or in a die set. In both the cases, cutting punch and cutting ring has to be very accurately matched otherwise cutting edge may get damaged.

COMPONENT 11

Trimmed Cup

Drawn cups normally have unequal length at various places. Figure 12.72 shows cup in two conditions–one untrimmed and the other is trimmed.

Trimming of round cup is generally done by a rotatory trimming machine, which is an additional operation, so to avoid cost increase due to additional operation, an innovative method of trimming cup in the cut and cup operation may be introduced. It is called pinch trimming.

Pinch trimming is explained with the help of Fig. 12.73 which has three views 'A', 'B' and 'C'.

Referring to Fig. 12.73 'A' part (1) is a draw die of bore Ø and (2) is the draw punch. Its height is exactly equal to length of finished cup. Draw punch (2) is mounted on a trimming punch (3) of Ø2. Bore of draw die is Ø1 which is 0.015 mm bigger than Ø2. Object is that Ø1 should have a precision sliding fit over Ø2. Figure 12.73 'B' shows the position when draw is just completed and the left over flange of component is under pressure between the faces of draw die and pressure pad. At this stage, distance between faces F_1 and F_2 is equal to sheet thickness. Further movement of draw die will create a shearing force on flange of component by sharp edge of trimming punch (3). Here the movement of draw die would completely shear a ring as shown in Fig. 12.73 'C'. Ring of material cut out is scrap and it is pushed out of punch by means of pressure pad. Component goes up with draw die to be ejected by an ejector.

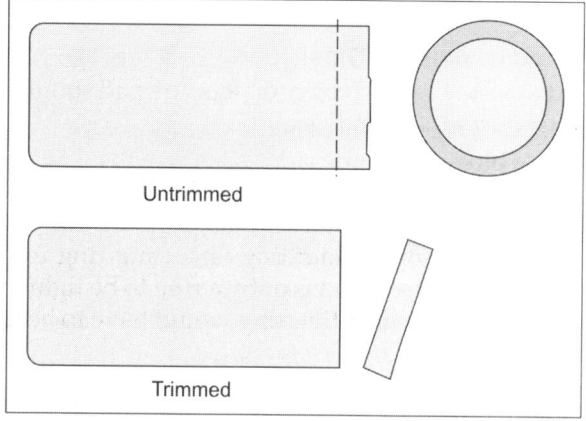

Untrimmed

Trimmed

Fig. 12.72: Untrimmed and trimmed cup

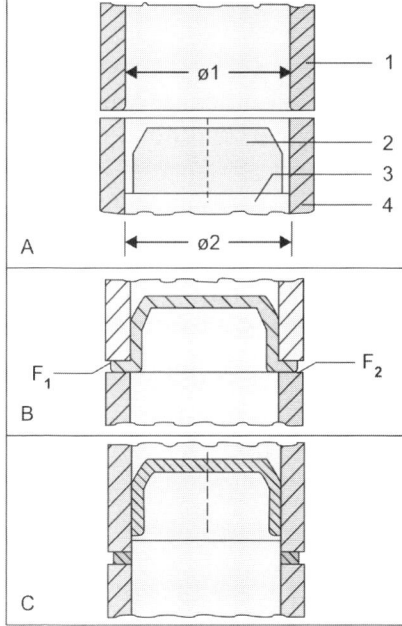

Fig. 12.73: Punch trimming

COMPONENT 12

Redrawn Cups

Finished redrawn cup is shown in Fig. 12.74. This component may be produced in two stages. First step may be cut and cup and the next step would be redraw. The shape and size of cup would be as shown in Fig. 12.74.

Figure 12.75 shows component and redrawn tool. Lower portion consists of a die block (1), guide ring (2) and reducing die ring (3). Bore of guide ring is 0.05 mm bigger than cup diameter which is 34 mm. Object is that cup is easily fed inside the guide ring. Bore of reducing ring is 29 mm which is the outer diameter of cup's lower portion. Bore of reducing die ring (3) has a chamfer which matches the component shape. There is a radius of 0.6 mm where chamfer meets the bore. It is to facilitate draw. Draw punch is carefully set in such a way that complete shape of redrawn cup is achieved. Component goes up with redraw punch. It is ejected by means of **ejector rod** (5).

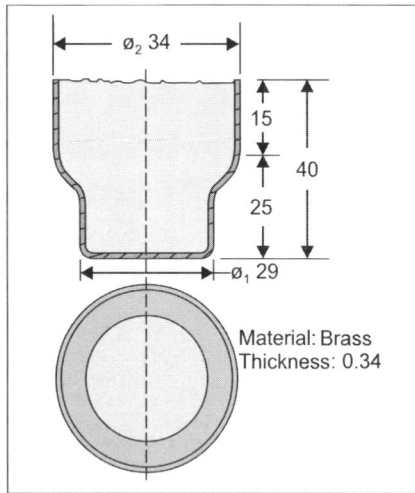

Fig. 12.74: Redrawn cup

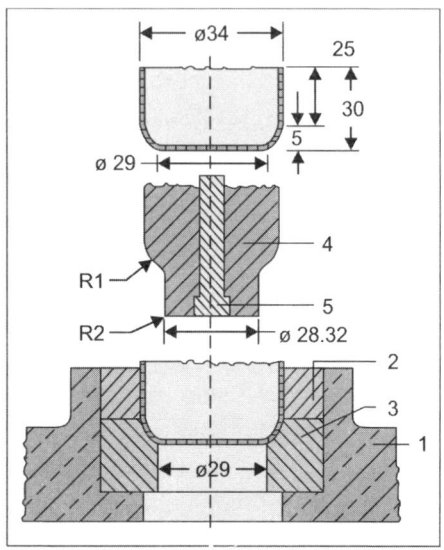

Fig. 12.75: Redrawn tool

COMPONENT 13

Flashlight Reflector

A typical reflector is shown in Fig. 12.76. Reflector may be produced by the following operations:

- Dome drawing
- Stamping
- Trimming cum piercing

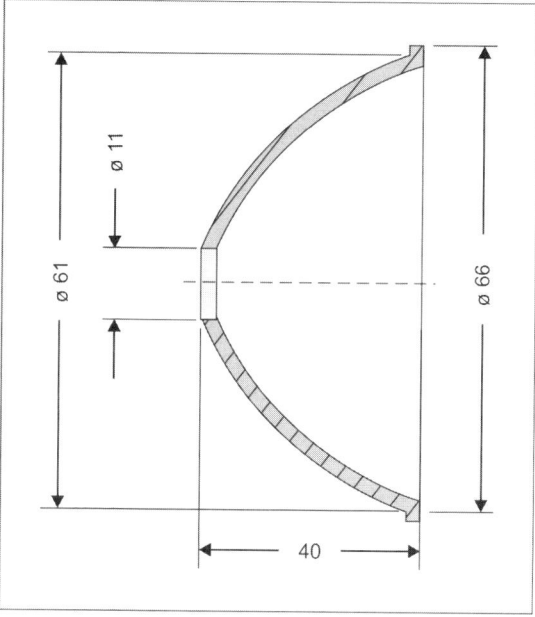

Fig. 12.76: Flashlight reflector

Dome Draw Tool

It is a conventional cut and cup tool with the only difference that there has to be a circular matching bead on the face of draw die and pressure pad as shown in Fig. 12.77.

When the tool was first time made and tried out, it was found that enough under holding pressure could not be provided by plane

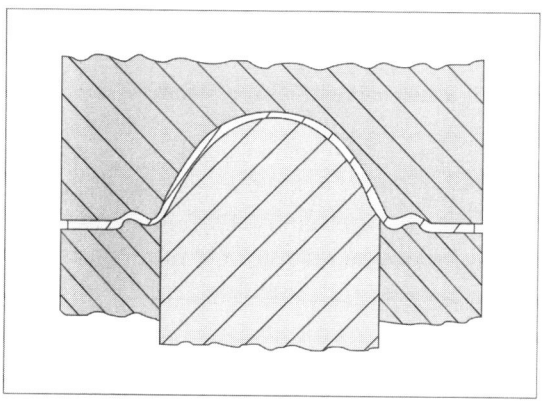

Fig. 12.77: Dome draw technique

surface pressure pad. Wrinkles could only be eliminated by providing a shallow bead. Too sharp a bead would not allow material to flow, consequently, rupturing of dome would occur.

Draw punch has a parabolic shape which is defined by equation

$$Y^2 = 4\,ax$$

Where x and y are coordinates and 'a' is a constant (focal point of parabola). It depends on design of reflector. Reflectors may be shallow or deep. Higher value of 'a' will make a reflector deep. Machining of draw punch may be done by CNC profiles generating lathe. In practice, it is found that good quality focus is not achieved if reflector is made from drawn dome. This is due to the fact that parabolic surface is not very smooth. There are some undulations. To smooth out undulations, a stamping tool is used. This tool is very accurate profile of punch and die.

Figure 12.78 highlights correct and incorrect matching of punch and die. 'A' shows that gap between parabolic surface of punch and die is not uniform, hence, uniform stamping would not take place. 'B' shows uniform gap between punch and die surfaces. This is necessary for uniform stamping. 'C' again shows non-uniform gap which is highly undesirable like 'A'.

Other dimensional detail of tools may be worked out as per principles already discussed in connection with other components.

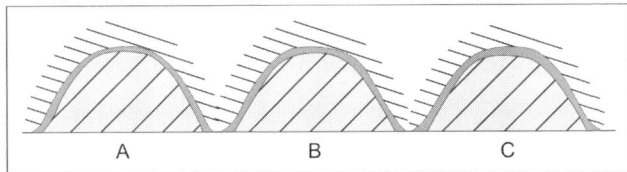

Fig. 12.78: Match and mismatch of parabolic punch and die

COMPONENT 14

Figure 12.79 shows a long tube or barrel of brass. There may be three methods of manufacturing long tube/barrel which are as below:

- Rolling and brazing
- Deep drawing
- Long drawn and cutting

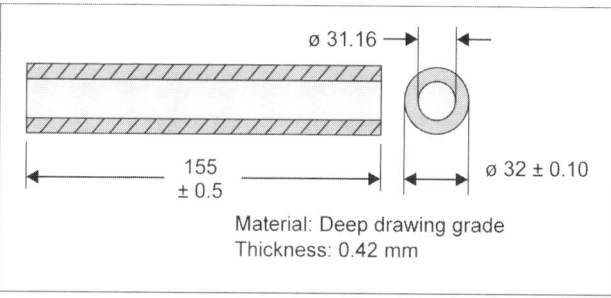

Fig. 12.79: Long sheet metal tube

Rolling and Brazing

First of all, blanks are produced by means of blanking die. Typical design of a blanking tool is shown in Fig. 12.80.

Sizes of die are so maintained that blank of 155 mm long and 100.57 width is obtained. Dimensions of die block may be 350 mm long, 300 mm wide and 90 mm thick.

A 60 tons eccentric power press with about 50 mm stroke adjustment may serve the purpose.

Blanks are then rolled to form tube in such a way that edges meet straight and evenly. Tubes are then tied with wire. Tubes are then placed over a jig of refractory clay and spot brazing is done at two or three places. Wires are then removed and full length brazing is done in one 'go'. Good quality brazing can be done by highly

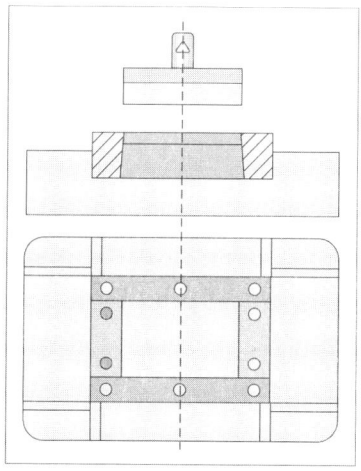

Fig. 12.80: Typical sheet blanking tool

skilled brazer. Brazed tubes are then inspected for any burn holes or over brazed material. In both the cases, defective barrels are repaired by rebrazing defective spots or filing away excess brazed material. In the end, surface of tubes/barrel are then polished on rotary polishing machines.

Deep Drawing

Production of barrels by redrawing method is done by cup making and three or four draws. A typical sequence of operation is shown in Fig. 12.81. These types of draws are generally done by long stroke horizontal presses, sometimes known as 'body expander presses'. Generally, there is a 'group' of four or five presses. Drawn component from first draw press is immediately fed to second draw press. Component is then fed to next press and so on till final draw. Care has to be taken to see that after certain draw, component is 'work hardened'. In such a case, stress relieving operation has to be carried out. Stress relieving operation is already explained which is heating of component lot in a furnace. Stress relieving may need heating of lot around 380°–400°C for about two hours. These stress relieving parameters are just a guide. Actual temperature and time which give desired result is determined by trials.

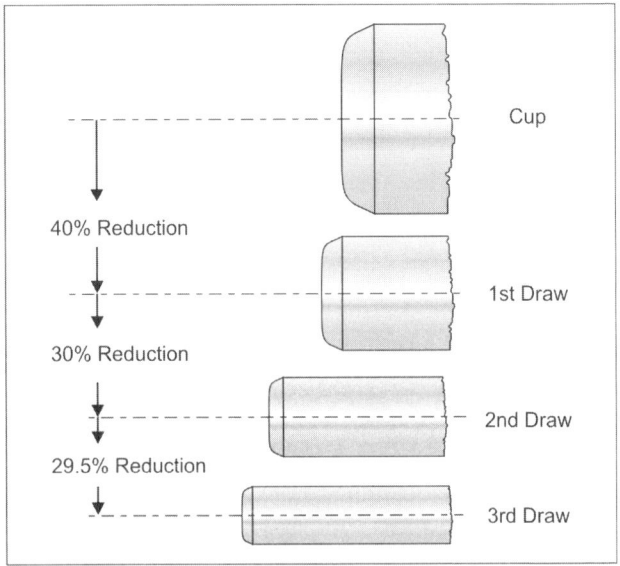

Fig. 12.81: Typical redraw sequence of operation

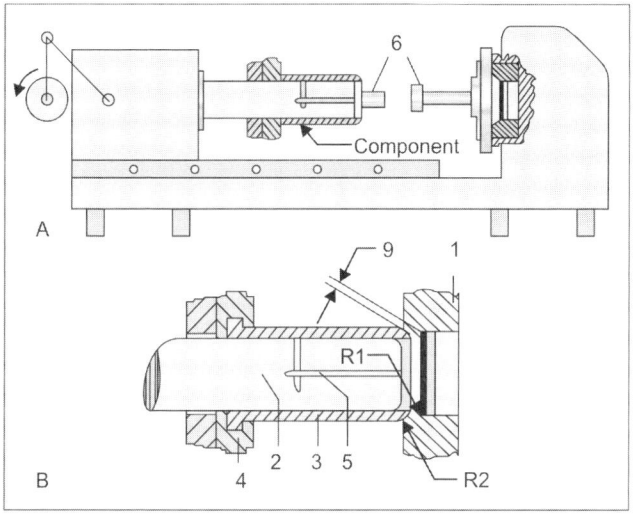

Fig. 12.82: Setting of body expander press

Setting of a body expander press is briefly explained with the help of Fig. 12.82.

Figure 12.82 'A' shows general construction of long stroke horizontal press. Figure 12.82 'B' gives details of tool setting. Part (1) is a redraw die fitted to vertical portion of press. It is casted very strong with thick walls and ribs. Part (2) is a redraw punch fitted to press RAM. Part (3) is a sleeve which is mounted on a cross strong plate. Diameter of sleeve is about 0.1–0.15 mm less than receiving component which may be a cup or a redraw component. When stroke is given punch, cross plate moves towards die. There are two stoppers (6), which are adjusted to achieve gap 'g' equal to wall thickness of component which is fed for redraw operation. If gap 'g' is greater than material thickness then wrinkles would form and base of component would rupture and punch would continue to move forward without drawing. If gap is too small then drawing would not take place and bottom of component would rupture.

Long Drawn Tubes

In this system, first of all, thick wall tube is produced by hot extrusion of brass. Typically a tube of 38 mm diameter with a wall thickness of 2.8 mm and about 1 meter long is produced. Tubes so produced are drawn three to four times to achieve tube of wall thickness as 0.4 mm.

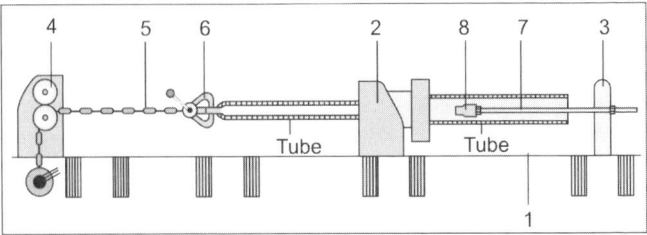

Fig. 12.83: Tube draw bench

A special draw bench is used for drawing operation. These benches may be about 9 meters part long. The basic construction and working of draw bench is explained with the help of Fig. 12.83.

Part (1) is a long bed of draw bench. Part (2) is a draw die holding block, an integral part of bench. Part (3) is a vertical column which holds bar (7). There is a provision to set the length of bar so that draw guide punch (8) reaches to die, just leaving a space (one side) equal to wall thickness of tube to be redrawn.

To start the process, one end of 38 mm diameter tube is manually manipulated to make it pass through die up to so much length that it can be gripped by strong gripper (6).

End of strong sprocket chain is fitted to gripper block (6). The other portion passes over a sprocket wheel. Now draw guide punch is pushed forward towards die inside the tub.

When machine is started, sprocket wheel rotates slowly, thus, pulling the gripper. Barrel is pulled through the die. Diameter gets reduced and length increased. This process may be carried out two to three times with reduced bore of draw die till the final die to give and outer diameter of tube as 38 mm and the length of tube becomes three to four meters. These **long tubes** are then **cut off** by high speed circular saw. Cut lengths of 155 mm are achieved.

In practice, a variation of 0.03 mm is found in wall thickness and 0.1 mm in diameter of tube.

This process is quite expensive and needs a lot of setup and space. There may be very few companies in India which exclusively produce various types of tubes for their clients.

COMPONENT 15

Round Components with Hexagonal Panels

Such a component is shown in Fig. 12.84.

Figure 12.84 'A' shows round component of 32 mm diameter, 12.5 mm high and having a pierced hole of 28 mm. Hexagonal panel as shown in Fig. 12.84 'B' is to be produced.

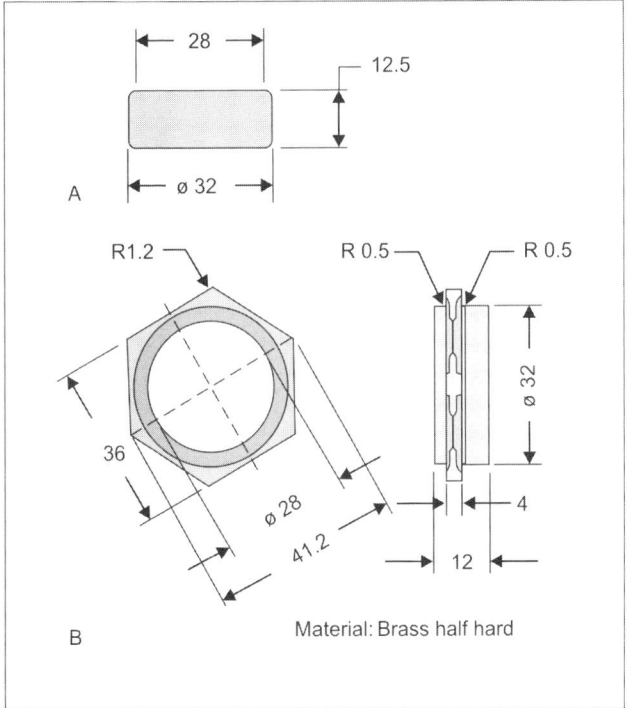

Fig. 12.84: Component with hexagonal panels

Hexagonal panels may be produced by means of a special machine, known as paneling press. The operation of this machine is quite typical. The working of machine is briefly explained with the help of Figs 12.85 and 12.86.

Figure 12.85 shows vital parts of tooling which gives shape of hexagonal panel to the component. Part (1) is a plate of 4 mm thickness and having octagonal hole with corner size as 36.59 mm and side to side size as 33.8 mm. Part (2) and (3) are side plates. Part (4) is a panel forming punch. Its corner to corner size should be about 4 mm less than minimum bore of components which is 31.32 mm, hence, corner to corner size of punch would be 27.32 mm, so the component may easily be placed and taken out of panel forming punch. Axis of punch and die ring (1) normally coincides when punch is not moved down. There is a provision in the machine to move the punch radially up and down by means of handles. Flat to flat dimensions of octagonal hole ring (1) should be such that component is easily fed on punch without any interference by octagonal rings (see Figs 12.85 and 12.86).

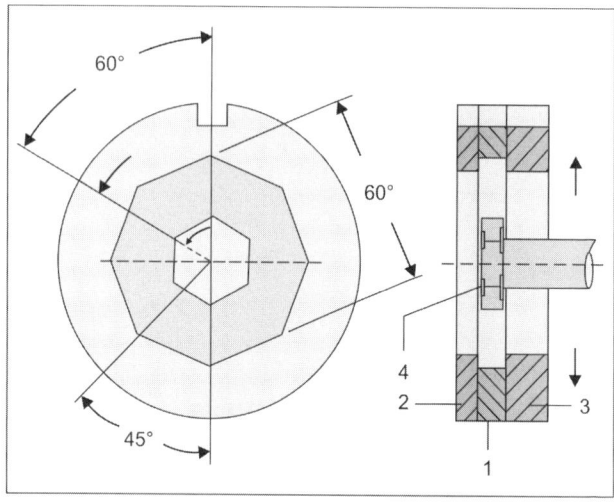

Fig. 12.85: Panel rolling arrangement

Formula for Hexagon

> R = Radius of circumscribed circle
> r = Radius of inscribed circle
> R = S = 1.155 r
> r = 0.866 S = 0.886 R
> S = R = 1.155 r

Formula for Octagon

> R = Radius of circumscribed circle
> r = Radius of inscribed circle
> R = 1.307 S = 1.082 r
> r = 1.207 S = 0.924 R
> S = 0.765 R = 0.828 r

Calculation for side of hexagon of punch

> S = R = 27.32 /2
>
> = 13.66 mm

Let size of side of octagonal hole in ring is equal to 13.66 mm plus three times the thickness of component wall 0.34 mm, so it comes to 14.00 mm.

Corner to corner distance = 2 (1.307 S)

> = 2 (1.307 × 14)
>
> = 36.59 mm

And side to side distance = 2 (1.155 r)

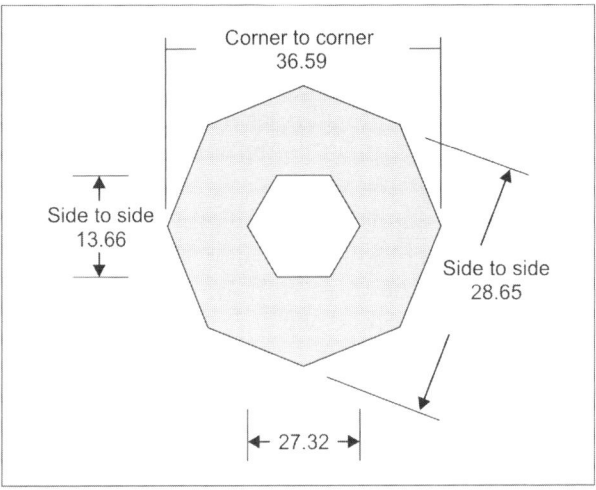

Fig. 12.86: Hexagonal punch and octagonal die ring

$$= 2 \ (1.155 \times 0.886 \ S)$$
$$= 2 \ (1.155 \times 0.886 \times 14)$$
$$= 28.65 \ \text{mm}$$

Difference between hexagon and octagon

In sides ——————— $28.65 - 2 \ (0.765 \ R)$
$$= 28.65 - 2 \ (0.765 \times 27.32 + 2)$$
$$= 28.65 - 2 \ (0.765 \times 13.66)$$
$$= 28.65 - 20.89$$
$$= 7.76 \ \text{mm}$$

Difference between hexagon and octagon

In corner to corner
$$= 36.59 - 27.32$$
$$= 9.27 \ \text{mm}$$

Holes in discs 2 and 3 should be 1.5 mm less than side to side dimensions of octagon, which is 28.65 mm.

Operation

- Component is fed on the hexagonal punch.
- The backrest in a free rotating disc form is moved towards punch so much that component is just free in between the faces to shift up and down, left or right.
- Punch and die ring are rotating in same direction, say clockwise.

- Downwards movement of punch is limited by means of a stopper bolt below the upper swinging body of machine, and it is adjustable.
- Bring the punch down to stopper by operating a handle. This action will form hexagonal panel while component is rolling together with punch and die ring.
- It takes hardly 3 seconds to complete the forming operation.
- On raising the punch up by means of handle, component with formed panels become free and normally moves out of punch by itself, otherwise a probe of a wire may be used to extract the component.

It is important to mention that rotation of punch and die ring is synchronized by a gear train in the machine. Ratio of gear train is so maintained that when punches rotate by 60° then die ring rotates by 45°.

Sharply formed panels look good, but it is necessary to provide radii in punch hexagon edges so that material does not crack or grains opened, so the practice is to start with just by removing the sharpness. By two and three attempts, minimum radii are achieved where no cracking or grain opening takes place.

Hope, examples of components given would prove to be source of guidance in designing and making of sheet metal press tools.

Appendix

Useful Websites

S.No.	Topic	Website address	Remarks
1.	Atomic structure of metal elements	http://www.askphysics.com	
2.	Bench top chemical crystallo-graphy system	www.rigaku.com	
3.	Cup blank calculation conversion formula	Write on Google search bar, Cup blank calculation Google-write on search bar-conversion formula. There are many sites	Click search Click
	Cutting speed for various materials	Wikipedia-write on search bar the topic	Click
	CNC milling machines	Google-write on search bar-American machine tool co	There are many sites for various machines
4.	Die sets	www.dmeindia.com; http://www.royalclamps.com /ecatalog	
	Deep drawing formability	Google-write on search bar the topic	Click
5.	Engineering related questions	http://iamechatronics.com	

Contd.

Contd.

S.No.	Topic	Website address	Remarks
	Engineering dictionary	Engineering Directory.com	Find high strength steel stamping, Design manual
	Engineering information	www.engineeringedge.com	
	Engineering unit conversion factor	Google-write the topic. There are many sites	Click the topic
6.	Fine blanking	Write on Google search bar, sheet metal fine blanking.	Click search
		Youtube-write on search bar, Fine blanking processes	There are many sites, see Hartford metal tech co, Ltd
	Fasteners	www.fortress.co.nz	
7.	Global Spec, Inc (the engineering search engine)	http://www.globalspec.com	
	Grinding wheel	Google-write on search bar-Grindwell Norton Ltd	Click
8.	Lecture on Titanium	www.periodicvideos.com	Youtube
9.	Material chemistry and physics	www.journal.elsevier.com /materials.chemistry-and-physics	
	Measuring instruments	www.aber.ac.uk/en/imaps /research/materials http://www.mitutoyo.co.jp	
	Metallography	En,Wikipedia.org/wiki/ Metallography;	Click, page opens with
		www.metallography.com/ types.htm;	many images
	Metallography	youtube	Write on search bar, metallography
	Machine tool	www.renishaw.com	

Contd.

S.No.	Topic	Website address	Remarks
	Milling machine	www.glacern.com www.Machine Shop Secrets.com	
	Metal forming milling machine diagram, svg	www.matelformingmagazine.com Email: info@wikimedia.org	
10.	Nelson files	www.commons.wikimedia.org	
11.	Optical equipment	Praka Engineering prakavision@gmail.com	Through Google Email address
12.	Press tool dies	youtube	Write on search bar, Type of press tool dies
	Power presses	www.mankoopresses.com www.milappresses.com www.keshavmachine.in www.rajeshpowerpressesindia.com www.ravipowerpress.com	
	Press tools, sheet metal	Write on Google search bar, types of sheet metal press tools	Click search
13.	Rotary bending die	www.readytechnology.com	Taken from youtube
	Rolling of metal	Write on Google search bar	Click,
14.	Scientific reports, unraveling the atomic structure of ultra fine iron clusters	www.nature.com/srep/ 2012/121218	
	Surface finish	http://www.obsnap.com	Also see videos on youtube
	Sheet metal processing equipment	www.schulergroup.com	Useful information
15.	Tool design Typical selection of steels and useful notes	www.wisetool.com	
16.	Zinc alloy	www.markusfarkus.com /reference/Zinc.htm	

Note: Write on Google search-Internet search engine list-Click search. Many websites would appear-Click 'The Search Engine List (www.thesearchenginelist.com).

A list of hundreds of search engines would appear. It contains short description also.

In fact, almost any subject or topic may be searched.

General Properties of Steels

The following table lists the typical properties of steels at room temperature (25°C). The wide ranges of ultimate tensile strength, yield strength and hardness are largely due to different heat treatment conditions.

Table A.1: Typical properties of steels at room temprature

Properties	Carbon steels	Alloy steels	Stainless steels	Tool steels
Density (1000 kg/m³)	7.85	7.85	7.75–8.1	7.72–8.0
Elastic Modulus (GPa)	190–210	190–210	190–210	190–210
Poisson's Ratio	0.27–0.3	0.27–0.3	0.27–0.3	0.27–0.3
Thermal Expansion (10^{-6}/K)	11–16.6	9.0–15	9.0–20.7	9.4–15.1
Melting Point (°C)			1371–1454	
Thermal Conductivity (W/m-K)	24.3–65.2	26–48.6	11.2–36.7	19.9–48.3
Specific Heat (J/kg–K)	450–2081	452–1499	420–500	
Electrical Resistivity (10^{-9}W-m)	130–1250	210–1251	75.7–1020	
Tensile Strength (MPa)	276–1882	758–1882	515–827	640–2000
Yield Strength (MPa)	186–758	366–1793	207–552	380–440
Percent Elongation (%)	10–32	4–31	12–40	5–25
Hardness (Brinell 3000kg)	86–388	149–627	137–595	210–620

Source: www.efunda.com
As on November 10, 2013

Table A.2: Cutting speeds for various materials using a plain speed steel cutter

Material type	Meters per min (MPM)	Surface feet per min (SFM)
Steel (tough)	15–18	50–60
Mild steel	30–38	100–125
Cast iron (medium)	18–24	60–80

Contd.

Table A.2: Cutting speeds for various materials using a plain speed steel cutter (*Contd.*)

Material type	Meters per min (MPM)	Surface feel per min (SFM)
Alloy steels (C1008–C1095)	20–37	65–120[3]
Carbon steels (C1008–C1095)	21–40	70–130[4]
Free cutting steels (B1111-B1113 and C1108-C1213)	35–69	115–225[4]
Stainless steels (300 and 400 series)	23–40	75–130[5]
Bronze	24–45	80–150
Leaded steel (Leadloy 12L14)	91	300[6]
Aluminum	75–105	250–350
Brass	90–210	300–700 (Max. spindle speed)[7]

Source: Wikipedia as on November 12, 2013

Measurement of Thickness of Sheet Metals

The thickness of sheet metals is specified by gauge. The table below shows the actual thickness of sheet steel in millimetres and inches as compared to gauge size. These dimensions vary slightly between different materials such as stainless steel and aluminum but should be accurate enough for most everyday purposes.

Table A.3: Sheet steel gauge conversion chart

0	0.324"	8.2 mm
1.	0.300"	7.6 mm
2.	0.276"	7.0 mm
3.	0.252"	6.4 mm
4.	0.232"	5.9 mm
5.	0.212"	5.4 mm
6.	0.192"	4.9 mm
7.	0.176"	4.5 mm
8.	0.160"	4.1 mm
9.	0.144"	3.7 mm
10.	0.128"	3.2 mm
11.	0.116"	2.9 mm

12.	0.104"	2.6 mm
13.	0.092"	2.3 mm
14.	0.080"	2.0 mm
15.	0.072"	1.8 mm
16.	0.064"	1.6 mm
17.	0.056"	1.4 mm
18.	0.048"	1.2 mm
19.	0.040"	1.0 mm
20.	0.036"	0.9 mm
21.	0.032"	0.8 mm
22.	0.028"	0.7 mm
23.	0.024"	0.6 mm
25.	0.020"	0.5 mm
26.	0.018"	0.45 mm
27.	0.0164"	0.42 mm
28.	0.0148"	0.37 mm
29.	0.0136"	0.34 mm
30.	0.0124"	0.31 mm
31.	0.0116"	0.29 mm
32.	0.0108"	0.27 mm
33.	0.0100"	0.25 mm
34.	0.0092"	0.23 mm
35.	0.0084"	0.21 mm
36.	0.0076"	0.19 mm
37.	0.0068"	0.17 mm
38.	0.0060"	0.15 mm
39.	0.0052"	0.13 mm
40.	0.0048"	0.12 mm

Source: www.zygology.com
Click Engineering resource

Table A.4: Conversion Formulae

inch x 25.4 = mm
foot x 304.8 = mm
mile x 1.609 = km

mm x 0.03937 = inch
meter x 39.37 = inch
km x 0.6214 = mile

Fahrenheit to Celsius: $^\circ C = {}^\circ F - 32 \times 5 \div 9$
Celsius to Fahrenheit: $^\circ F = {}^\circ C \times 9 \div 5 + 32$

M i l l   a n d   L a t h e   C o n v e r s i o n s

To Find:			Formulae
Revolutions per Minute	RPM	=	(SFM x 3.8197) ÷ D
Surface Feet per Minute	SFM	=	RPM x D x 0.2618
Surface Meters per Minute (Metric)	SMPM	=	SFM x 0.3048
Inches per Minute Milling Feedrate	IPM	=	FPT x T x RPM
Feed per Tooth Mill	FPT	=	IPM ÷ (T x RPM)
Feed per Revolution	FPR	=	IPM ÷ RPM
Inches per Minute (Lathe)	IPM	=	IPR x RPM
Metal Removal Rate	MMR	=	W x d x F
Advance per Revolution (Inches)	ADV/R	=	F ÷ RPM

T h r e a d s

Mill Tapping Feedrate	IPM	=	1 ÷ TPI x RPM
Lathe Threading Feedrate (Thread Lead)	IPR	=	1 ÷ TPI

Tap Drill Size $=$ Major Dia. of Tap $-\dfrac{\% \text{ of Thread Height x .01299}}{\text{TPI}}$

Percent of Full Thread $=$ TPI $\times \dfrac{\text{Major Dia. of Tap - Drill Dia.}}{.01299}$

Mill Tapping Feedrate (Metric) $=$ 1 ÷ RPM x Metric Pitch

Tap Drill Size (Metric) $=$ Tap Major Dia. (mm) $-\dfrac{\% \text{ of Thread Height x Metric Pitch}}{76.980}$

Percent of Full Thread (Metric) $=\dfrac{\text{Basic Major Dia. (mm) x 76.980 - Drilled Hole (mm)}}{\text{Metric Pitch}}$

M i s c e l l a n e o u s

Radius of Circle	=	Circumference x 0.159155
Diameter of Circle	=	Circumference x 0.31831
Circumference of Circle	=	D x 3.1416
Area of Circle	=	R2 x 3.1416
Cutting Time in Minutes (Mill)	=	L ÷ IPM
Cutting Time in Seconds (Lathe)	=	$\dfrac{\text{Distance to go x 60 sec}}{\text{IPR x RPM}}$

A b b r e v i a t i o n s   a n d   M e a s u r e m e n t   U n i t s

D	=	Diameter of Milling Cutter or Lathe Part	SFM	=	Surface Feet per Minute
d	=	Depth of Cut	SMPM	=	Surface Meters per Minute
FPR	=	Feed per Revolution (in Inches)	T	=	Number of Teeth in the Cutter
FPT	=	Feed per Tooth (in Inches)	TPI	=	Threads per Inch
IPM	=	Inches per Minute (Table Travel Feedrate)	W	=	Width of Cut
IPR	=	Inches per Revolution	°C	=	Degrees Celsius
L	=	Length of Cut (Inches)	°F	=	Degrees Fahrenheit
RPM	=	Revolutions per Minute (Spindle Speed)			

08-01-02

Source: Google>Hass conversion
formula>SFM
Formula chart 1
as on November 25, 2013

Table A.5: Standard Prefixes

Prefix symbol	Multiplier	Exponential	
yotta	Y	1,000,000,000,000,000,000,000,000	10^{24}
zetta	Z	1,000,000,000,000,000,000,000	10^{21}
exa	E	1,000,000,000,000,000,000	10^{18}
peta	P	1,000,000,000,000,000	10^{15}
tera	T	1,000,000,000,000	10^{12}
giga	G	1,000,000,000	10^{9}
mega	M	1,000,000	10^{6}
kilo	k	1,000	10^{3}
hecto	h	100	10^{2}
deca	h	10	10^{1}
		1	
deci	d	0.1	10^{-1}
centi	c	0.01	10^{-2}
milli	m	0.001	10^{-3}
micro	μ	0.000001	10^{-6}
nano	n	0.000000001	10^{-9}
pico	p	0.000000000001	10^{-12}
femto	f	0.000000000000001	10^{-15}
atto	a	0.000000000000000001	10^{-18}
zepti	z	0.000000000000000000001	10^{-21}
yocto	y	0.000000000000000000000001	10^{-24}

Google > Standard Prefixes >
Standard Prefix
as on December 5, 2013

Index